Human Factors in Flight

To Suzette

Human Factors in Flight

Frank H Hawkins

second edition

Edited by
Harry W Orlady

Ashgate

Aldershot • Brookfield USA • Singapore • Sydney

© Frank H Hawkins 1987
© Uitgeverij Uniepers Abcoude 1993

Published by
Ashgate Publishing Limited
Gower House
Croft Road
Aldershot
Hants GU11 3HR
England

Ashgate Publishing Company
Old Post Road
Brookfield
Vermont 05036
USA

Reprinted 1993, 1995, 1997, 1998

A CIP catalogue record for this book is available from the British Library
ISBN 1 85742 134 5 (Hardback)
ISBN 1 85742 135 3 (Paperback)

Design and production: Uniepers bv, Abcoude, The Netherlands
Typesetting: Uniepers DTP/Megaset, Aalsmeer, The Netherlands
Printed in The Netherlands by Koninklijke Wöhrmann bv, Zutphen

Also now available from the publishers:
Human Factors in Flight Student Workbook 0 291 39831 6
(and also an *Instructor's Guide*)
by Craig Funk, Embry-Riddle Aeronautical University

Contents

Acknowledgements

I wish to express my appreciation to those who have provided assistance in reviewing chapters in the book and suggesting revisions and amendments. In particular, I am greatly indebted to Professor Elwyn Edwards and Dr Mary Edwards for their support, encouragement and detailed analytical reviews of each chapter.

Others who have generously contributed their time and expertise to reviewing the book critically as a whole or in part, include:
Dr M Carruthers; Director, Positive Health Centre, Harley St, London, UK.
Mr K G Dougan; Operational Adviser, Smiths Industries Ltd, Cheltenham, UK.
Captain B A Draper; Training Manager, Flight Instructor, Stocksfield, UK.
Dr J A Horne; Department of Human Sciences, Loughborough University, UK.
Dr P Naitoh; Psychophysiologist, San Diego, USA.
Captain E R Parfitt; Airline Senior Route Instructor, Gata de Gorgos, Spain.

The tireless and skilled assistance of Suzette Burghard during the years of development of the material and its preparation for publication has been invaluable and has made a significant contribution to the project. Acknowledgement must be given, too, for the contribution of graphic artist Michael Hawkins in preparing the many illustrations for publication.

I am grateful to the administrators of the US Aviation Safety Reporting System (ASRS) and the UK Confidential Human Factors Incident Reporting Programme (CHIRP) for their kind permission to quote examples from their published reports to illustrate Human Factors principles presented in the text.

Illustrations

Acknowledgement with thanks is due to the following for their kind permission to use illustrations or tabulated data.
    Academic Press, Figs 3.5, 9.3, 9.4; American Institute of Physics, Fig 7.5; American Psychological Association, Fig 9.2; Aviation, Space and Environmental

Medicine, Figs 3.2, 3.3, 3.4, 4.3; Band J H, Figs 5.12, 6.4, 8.1, 9.6, 11.3, 11.7, 13.1, 13.4, 13.5; British Airways, Fig 3.6; Central Office of Information (UK), Fig 8.2; Chorley R A, Fig 12.4; Department of Health and Social Security (UK), Fig 8.2; Department of the Army (USA), Fig 13.2; Department of Transport (UK), Fig 8.2; Edwards E Prof, Figs 2.2, 6.2; Elseviers Scientific Publications, Figs 10.1, 10.2; Flight Safety Foundation, Fig 5.15; Health Education Council, Figs 4.1, 8.2; Her Majesty's Stationery Office, Fig 8.2; Human Factors Society, sample notices in Chapter 10; IPC Science & Technology Press, Figs 10.4, 11.5; Klein K E Dr, Tables 3.1, 3.2; Kryter K D Dr, Fig 7.1; McGraw-Hill Book, Figs 4.2, 11.1; NASA-Ames, Fig 11.2; Plain English Campaign, Fig 10.5; Prentice-Hall, Fig 7.4; Shell International Trading Company, Figs 5.11, 12.5, 12.7; University of Colorado, Fig 13.6.

Every effort has been made to trace copyright owners. However, if advised of any inadvertent omission, the publisher and I would be pleased to incorporate proper acknowledgement in future editions.

F H Hawkins

At the time of his untimely death Captain Hawkins was in the midst of preparing an updated edition of *Human Factors in Flight*. Virtually all of the material he had compiled has been included in this edition which is dedicated to his pioneering work. The contribution made by the late Captain William Price, who had been working with Captain Hawkins, in this updating of the first edition of *Human Factors in Flight* cannot be adequately acknowledged. The revision of this edition, beyond the work of Captain Hawkins, was done by Captain Harry W. Orlady.

The Publisher

Preface

Since 1940, data have been published periodically showing that three out of four aircraft accidents apparently result from inadequate performance of the human component in the aircraft man-machine system. This proportion has persisted in spite of years of exhortation to pilots to perform more consistently and with fewer errors. This exhortation has been notably, but predictably, ineffective in modifying the relative magnitude of this source of accidents.

Scientists, particularly in the USA and the UK, have been engaged in Human Factors research relevant to aviation during and since the Second World War, but only part of this research has found its way into flight operations and unfortunately, very little of it involved our major problem – the operational behaviour of pilots flying multicrew transport aircraft. The 1947 work in Dayton of Fitts, Jones and Grether on the misreading vulnerability of three-pointer altimeters is a good example. By 1980, the National Aeronautical and Space Administration (NASA) felt obliged to declare that despite these and other related studies, it was unlikely that this type of altimeter would be replaced in older operational aircraft. And in 1984 they were able to report yet another 10 000 ft misreading and flight deviation with this kind of altimeter.

On the other side of the Atlantic, at Cambridge in England, valuable Human Factors work was being done in the early 1940s on what became known as the Cambridge Cockpit. Yet more than a quarter of a century later, an eminent aeronautical society was able to make a policy statement declaring that it was too early for them yet to take ergonomics or Human Factors seriously.

When the first edition of this book was written, it was unusual to find a degree-qualified Human Factors specialist anywhere in the aircraft operational field, even in larger companies with more than 10 000 staff on the payroll. In the aircraft manufacturing industry, few of the world's civil aircraft constructors could claim properly qualified Human Factors staff, and even then, their role was normally only an advisory one and there were many reasons why their advice could be overruled. Most manufacturers were poorly endowed in this respect. A major European aircraft builder, for example, found it appropriate to seek help from a motor car manufacturer in the design of its new flight deck. This was in spite of the totally different environments in which the vehicles operate, the quite different levels of skill of the operators involved and the fundamental difference in the team-work task on a large transport aircraft compared with the single-operator task in a car.

In the 1990s, things look considerably better. Most manufacturers of large transport aircraft have qualified Human Factors experts on their staff and Human Factors considerations are given considerable weight at all stages of design and construction.

From the pilot side, a recently enacted International Civil Aviation Organization (ICAO) international pilot licensing specification requires that, in the future, all pilots be familiar with 'Human Performance and Limitations' as they relate to their flying activities and the formal privileges of their licence (ICAO, 1989). ICAO has also published a series of Human Factors Digests which are available on request and can be of considerable help to interested parties or to Contracting States. Human Factors also directly affect airline training. Many of the more progressive airlines have legitimate Human Factors experts on their training staffs.

Today there is no question that Human Factors is identified as a 'core technology' in air transport aviation. Particularly for new transports or other new components of the aviation system, Human Factors has become a major consideration in licensing, operation, and training. The overall problem, of course, is to make sure that all levels of the industry get and continue to get high-quality Human Factors in the design and operation of the transport aircraft they fly. It is essential that the emphasis presently given Human Factors remains more than just lip service, and that present concern is constructively recognised in all aspects of the industry.

It was not until the 1970s that some serious interest began to be shown in generating a greater awareness of Human Factors amongst those responsible for the design, certification and operation of civil aircraft. In fact, most of the research work done previously had been sponsored by the military. It had by this time become abundantly clear that a fundamental knowledge gap existed in civil aviation. At international symposia, few could even define what was meant by Human Factors; to one it meant post-mortems on accident victims and to another, simply aircraft handling procedures or other flight techniques. Medicine and Human Factors were widely confused and this confusion, for various reasons, dies hard.

By 1975, the International Air Transport Association (IATA) in its 20th Technical Conference in Istanbul concluded after a week of diverse but well intentioned discussions, that 'the wider nature of Human Factors and its application to aviation seem still to be relatively little appreciated. This neglect may cause inefficiency in operation or discomfort to the persons concerned; at worst, it may bring about a major disaster'. This ominous statement was followed just 17 months later by the double Boeing 747 disaster at Tenerife in which 583 people died and which cost about $150 million. It is significant that this accident – then the greatest in aviation history – resulted entirely from a series of Human Factors deficiencies.

Little more than a year after that IATA conference, the other side of industry, the pilots, also called for action to improve the application of Human Factors in their working environment. In their 1977 Washington symposium, the International Federation of Airline Pilots Associations (IFALPA) concluded that 'there is widespread concern at the frequent failure to apply Human Factors knowledge ... and a current lack of expertise in many areas of the commercial aviation industry'.

Although by this time there were short residential courses at the Universities of Aston in the UK and Southern California in the USA, these were only intended to meet the needs of senior or supervisory personnel and their cost prohibited a wider application. It was left to the national airline of the Netherlands, KLM, to initiate in 1977 a systematic educational programme to increase an awareness of Human Factors amongst its operational staff. This programme, now acquired by numerous other airlines and aviation bodies, consisted of a 15-unit audio-visual course, with each unit being designed to fill a basic educational gap in a different aspect of Human Factors. But the limited amount of time available in such formal courses to assimilate the very considerable information presented, stimulated a demand for more information in an easily readable documented form. This was needed for use as a textbook during the course and then for reading at leisure later. It is the purpose of this volume to meet that demand and also to provide the expanded information for those who are unable to do the basic classroom course but are nevertheless interested in acquiring more knowledge of the role of man in the aeronautical system.

To continue to move forward in achieving a better application of Human Factors in civil aviation, two broad principles must be accepted. Firstly, that questions about the role of man within complex systems are technical questions requiring professional expertise. And secondly, that adequate resources must be allocated to the design and management of man-hardware-software systems. While in many circles these principles are generally accepted today, neither of these principles has enjoyed widespread popularity in the past. The following chapters, while in no way constituting a comprehensive academic treatment of the subject, should provide a foundation of knowledge upon which a sound structure of Human Factors expertise and application in aviation can be built. For those not directly involved in civil air transport, this text should provide an insight into the role of man – passengers as well as crew – in the complex man-machine system which is today's transport aircraft.

Presentation format

Human Factors is an applied technology and so *Human Factors in Flight* is designed primarily for industry and aims to bridge the gap between academic sources of knowledge and the practical operation of aircraft. This orientation is apparent in the titles and sequence of chapters.

Chapter titles will be immediately recognised by those associated with aircraft operation as reflecting areas of Human Factors concern in their routine work. While the contents of each chapter and the appendices provide practical guidance, adequate references are included to enable the serious student to pursue his quest for knowledge beyond this introductory volume.

The chapters have also been sequenced with the interest of industry primarily in mind. In view of the widespread confusion which envelops the term Human Factors, it is essential to ensure from the beginning that we are all talking the same language – that we are all on the same frequency. Chapter 1 therefore provides this clarification and introduces a model which is utilised throughout the book as a tool for aiding understanding.

Safety is a theme never far beneath the surface in all activities connected with flying. For many, human error, which is inextricably linked with safety, has become almost synonymous with Human Factors and so very early attention is given to this subject, which forms the basis of Chapter 2. The principles discussed here will later be referred to in other chapters where the question of error is raised.

Reduced performance is increasingly being attributed by crew members to stress such as fatigue and the disturbance of sleep and biological rhythms. This forms, then, the subject matter of Chapter 3 and it is followed by two more chapters also with a physiological orientation.

Chapters 6 to 10 still retain the underlying concern with human performance but shift from physiological to more cognitive aspects of the technology. Throughout these five chapters the vitally important theme of communication is prominent.

Perhaps the most commonly recognised task of the ergonomist or Human Factors practitioner lies in the design of the workplace and this area forms the subject matter of the following three chapters. In transport flying, the workplace encompasses the cabin as well as the flight deck. Traditionally, the former has received less specialised Human Factors attention than the latter, so a complete chapter here is dedicated to the cabin. Not only does this approach the cabin as a workplace for the crew but also as a confined environment for large numbers of inactive passengers.

Finally, having provided an introduction to Human Factors in flight, we address the obvious and justifiable question which can be expected from readers in industry, '... now we have been introduced to the subject, where do we go from here?'

Referenced regulations, abbreviations and definitions

For an aircraft to have a world-wide sales market, it has been necessary that it be designed to meet the appropriate US regulations. This still did not guarantee that it would be accepted universally without modification, as other countries often had their own additional requirements. In order to establish a greater degree of standardisation and to facilitate the import and export of aviation products, the civil aviation authorities of certain European countries formed a Joint Aviation Authority (JAA) and agreed Joint Airworthiness Requirements (JAR).

In order to understand this standardisation process, some explanation of the classification system for US regulations is necessary. These regulations all form part of the Code of Federal Regulations (CFR) and although appearing under different Titles, for different national activities, they have a common numbering system. Title 14 is called Aeronautics and Space and it is this portion of the CFR which is of special interest to those engaged in aviation.

This Title contains parts promulgated by the Federal Aviation Administration (FAA) in Washington and also the former Civil Aeronautics Board (CAB), which is now incorporated in the Department of Transport. Those parts published by the FAA are known as Federal Aviation Regulations (FAR). These can be seen quoted variously as, for example, 14 CFR 25. 1309, Part 25. 1309, Section 25. 1309 or FAR 25. 1309. To avoid confusion, throughout this volume all such regulations will be quoted as, for example, FAR 25. 1309. The most common parts of these regulations encountered in civil aviation are 25 (Airworthiness), 61 (Certification: Pilots and Flight Instructors), 91 (General Operating and Flight Rules) and 121

(Transport Category Aircraft Operations). Part 121 is not applicable to non-US carriers.

The standardisation process between the FAA and JAA began with Part 25 of the FAR and this was selected to form the basis of JAR for aircraft over 5700 kg. It is referred to as the Basic Code. JAR 25. 1309 is therefore in principle the same as FAR 25. 1309. Amendments to this Basic Code are made by the FAA and considered for adoption for JAR by the JAA. In May 1989, the JAA established an Operations Committee which was given the task of producing a *European Book of Operational Rules*. These rules are known as JAR (Ops). Where reference is made in this volume to an FAR, it may or may not yet have a European equivalent.

While Part 25 FAR/JAR is applied to new aircraft, the regulations are not usually applied retrospectively to aircraft already in service. The operational requirements of Parts 91 and 121 are usualiy applied to all operations, although a time for compliance is normally given. In certain circumstances, a new aircraft may be delivered to a customer outside the regulatory jurisdiction of the country of manufacture, without fully meeting the airworthiness requirements of the manufacturing country, if this is requested by the customer. In such cases, a special fly-away permit is granted and the customer's national regulatory authority then accepts responsibility for the airworthiness of the aircraft.

In 1949, the Society of Automotive Engineers (SAE) in the USA formed a Flight Deck and Handling Qualities Standards for Transport Aircraft Committee, designated as S-7, with the aim of bringing some uniformity to these aspects of aircraft design. Since then the SAE has also formed Human Behavior Technology (HBT) and Aviation Behavioral Engineering Technology (G-10) committees. These committees are international in character, with their members being recruited, by invitation only, on the basis of their experience and dedication to aircraft development. The standards published by S-7 are not mandatory, though they are widely accepted. There is also some cooperation between the SAE and the International Organisation for Standardisation (ISO) in Switzerland. In addition to CFRs, therefore, reference will also be made in the text occasionally to an Aerospace Recommended Practice (ARP) as published by SAE and this will appear, for example, as (SAE ARP 4052).

In principle, international standard (SI) metric units have been used throughout. However, in civil aviation certain non-metric units are widely used internationally, such as feet for altitude, and so these have been quoted in the text. In some other cases, such as when the original data, as in CFRs, were specified in non-metric terms or where non-metric units for the particular item are in wide use internationally, then both metric and non-metric terms are given.

Many of the examples used to illustrate Human Factors aspects of flight have been taken from US accident reports. This in no way reflects accident distribution but is solely due to the generally greater accessibility of these reports and the cooperation of the National Transportation Safety Board (NTSB) which publishes them. Their openness and frankness make a valuable contribution to air safety for which all who work in aviation or travel by air should be grateful. A country which makes such accident studies and sees their accident investigation body as a servant of the public, benefits from a more solid foundation for air safety compared with those who practise a more secretive investigatory and public relations approach.

The first time an abbreviation is used in the text, its meaning is usually cited in full. In addition, for ease of later reference, a list of abbreviations used in the text is provided in Appendix 3. On the first occasion that a term is used which may not be familiar to aircraft operational staff, an explanation is given.

When it is found necessary to differentiate between flight deck and cabin crew and the inference is not clear from the context, then the distinction will be made specifically. Otherwise, crew or aircrew refers to all flying staff.

The word 'man' is used in its generic sense. Therefore it includes both sexes unless it is specifically indicated to the contrary. For example 'man' should be understood to mean people or humans including both males and females, 'he' should be understood to mean he/she, 'him' should be understood to mean him/her etc.

F H Hawkins
H W Orlady

1

The Meaning
of Human Factors

'Not only will men of science have to grapple with sciences that deal with man but – and this is a far more difficult matter – they will have to persuade the world to listen to what they have discovered.'

Bertrand Russell (1872–1970)

A hundred years of Human Factors

Man first began to make tools more than 5000 years ago. The fashioning of the handle of his axe to match the size and shape of his hand was an early application of elementary ergonomics, improving his efficiency at work. But the modern evolution of ergonomics or Human Factors as a technology can be said to go back just one hundred years.

It was in the 1880s and 1890s that Taylor and the Gilbreths started, separately, their work on time and motion studies in industry. And at about the same time academic work was being carried out by, amongst others, Galton on intellectual differences and by Cattell on sensory and motor capacities.

The First World War provided considerable stimulus to Human Factors activity as it became necessary to optimise factory production, much of which was being done by women totally new to this working environment. And in the USA from 1917 to 1918 two million recruits to the forces were given intelligence tests so as to assign them more effectively to military duties.

A psychological laboratory was established at Cambridge, England, towards the end of the 19th century and this later became the largest British centre for experimental psychology. Progress made during the war resulted, in 1921, in the foundation in the UK of the National Institute for Industrial Psychology which made available to industry and commerce the results of experimental studies.

An important milestone in this first century of Human Factors was the work done at the Hawthorne Works of Western Electric in the USA from 1924 to 1930. Here it was determined that work effectiveness could be favourably influenced by psychological factors not directly related to the work itself. This is still known as the 'Hawthorne Effect'. A new concept of the importance of motivation at work was born and this represented a fundamental departure from earlier ideas which concentrated on the more direct and physical relationship between man and machine.

The Second World War again provided a stimulus to Human Factors progress as it became apparent that more sophisticated equipment was outstripping man's capability to operate it with maximum effectiveness. Problems of selection and training of staff, too, began to be approached more scientifically.

At Oxford during the war, a Climatic and Working Efficiency Research Unit was established. At Cambridge, the Psychology Laboratory of the University was responsible for what might be seen as a second major milestone. They constructed a cockpit research simulator which has since become known as the 'Cambridge Cockpit'. From experiments in this simulator it was concluded that skilled behaviour was dependent to a considerable extent on the design, layout and interpretation of displays and controls. In other words, that for optimum effectiveness, the machine has to be matched to the characteristics of man rather than the reverse, as had been the conventional approach to system design. In 1945, the Applied Psychology Unit (APU) was established at Cambridge and this continues to be a source of much valuable Human Factors research work in the UK. At about this time in the USA, the much quoted research on three-pointer altimeter misreadings was beginning and this work was destined to be used as a standard illustration in discussions of design-induced error. And during this decade of the 1940s, aviation psychology centres were being initiated at Ohio State and Illinois Universities.

Perhaps a third milestone in this hundred years of Human Factors was the establishment of ergonomics or Human Factors as a technology in its own right. This was institutionalised by the founding of the Ergonomics Research Society in the UK in 1949, followed in 1959 by the formation of the International Ergonomics Association (IEA). In 1957 in the USA the Human Factors Society was founded and this was later affiliated to the IEA.

It was first calculated in 1940 that three out of four aircraft accidents are due to what has been called human failure of one kind or another. This figure was confirmed by IATA 35 years later. The next milestone in aviation Human Factors may be seen as the recognition that basic education in Human Factors was needed throughout the industry. In 1971 the 'Human Factors in Transport Aircraft Operation' two-week course was established at Loughborough University in England, later transferred to Aston University and also conducted elsewhere. In the USA a short course was established at the University of Southern California (USC), and at about this time, the US Air Line Pilots Association (ALPA) introduced, for selected representatives, an accident investigation course that stressed Human Factors and featured USC faculty. By 1978, KLM in the Netherlands provided the first 'Human Factors Awareness Course' for large-scale, low-cost, in-house indoctrination of staff in basic Human Factors principles. This course was later acquired by organisations in numerous other countries (see Chapter 14).

In the early 1970s, the International Air Transport Association (IATA), concerned with the basic problem of Human Factors in air transport operations, organised a Human Factors committee which later led to the 1975 Istanbul Conference. During this same period, United Airlines started a confidential non-punitive incident reporting system which led to the FAA/NASA Aviation Safety Reporting System (ASRS), (Reynard *et al.*, 1986). Both are covered in more detail in succeeding paragraphs.

As though to emphasise the urgent need for effective action, a fifth and tragic milestone was erected during this period when educational programmes were taking shape. In 1977 at Tenerife two aircraft collided at a cost of 583 lives and about $150 million, creating the greatest disaster in aviation history and resulting entirely from a series of Human Factors deficiencies (see Chapter 7).

An important aviation event in this first one hundred years of Human Factors was the establishment by FAA/NASA in 1976 of the confidential Aviation Safety Reporting System (ASRS), the principles and feasibility for which had been previously established by the United Airlines programme. The ASRS is operated by NASA for the FAA which finances virtually all of the costs associated with this breakthrough programme.

The ASRS recognised officially for the first time that it is unrealistic to expect to obtain meaningful and adequate information for analysis of human behaviour and lapses in human performance while at the same time holding the threat of punitive action against the person making the report. This basic change of attitude towards both pilots and administrators has been justified by the accumulation during the first 15 years of operation of the scheme of a data bank of more than 180 000 reports from which analyses were made and published periodically. During this period more than 2450 Search Requests were made and more than 1000 Alert Bulletins (ABs) were issued (ASRS Callback No. 143). After 15 years, ASRS was able to report that although statistical totals were increasing annually, event types, consequences and percentages remained generally consistent. The single exception being an increase in altitude deviations following the FAA inauguration of its Quality Assurance Program (QAP). Under this programme, any loss of separation in US airspace is automatically recorded by radar and requires examination of the incident by an FAA supervisor. An altitude deviation of more than 300 feet can cause a loss of separation if there is traffic at a conflicting altitude. Submission of an ASRS report provides immunity from FAA prosecution to the reporter in these instances. This increase in altitude deviations does not necessarily mean that there has been an increase in actual altitude deviations, but may only mean that there has been an increase in the number of altitude deviations reported.

The ASRS data bank now (mid-1992) includes over 210 000 reports. The programme receives an average of slightly under 3000 reports per month, the majority from airline pilots. This vast store of operational reports may have revealed, in principle, nothing new to those who have been for years concerned with the study of air transport Human Factors. However, it remains an extremely valuable exercise and has provided sufficient data for meaningful computer analysis – data that was previously unobtainable. The sheer weight of the evidence, backed by the authority and analytical facilities of NASA, has focused more regulatory and industry attention on the problem area than previously could be elicited.

Six years after the ASRS programme was set up in the USA, a similar scheme, the Confidential Human Factors Incident Reporting Programme (CHIRP), was initiated in the UK. In 1986 a similar scheme called the Confidential Aviation Safety Reporting Programme (CASRP) was initiated in Canada. At about the same time, New Zealand instituted an Independent Safety Assurance Team programme (ISAT) and in 1988 Australia instituted a comparable programme, the Confidential Aviation Incident Reporting (CAIR). Germany has recently established a similar programme which will serve as an anchor for a new European Community (EC) Directive which is now being formulated.

The monthly bulletin of ASRS is called 'Callback'. This safety bulletin has a wide circulation in about 60 countries. By 1991 some 60 000 copies were being circulated monthly. The periodic bulletin of the British CHIRP system is called 'Feedback'. Canada produces a bi-lingual bulletin 'In-Flight' or 'Aprç', while New Zealand's bulletin is called 'Flashback'. In Australia data and reports are included in its Basic Aviation Safety Information (BASI) publications.

In 1991, ASRS began a new publication called 'Directline' to meet the needs of operators and flight crews of complex aircraft. Distribution is directed to managers and management personnel, safety officers, and training and publications departments. Occasional reference will be made in the following chapters to examples taken from these bulletins to illustrate Human Factors problems.

A very comprehensive mandatory occurrence reporting system has been conducted in Australia for many years and this also has a provision for anonymous submission of reports. Although a very large number of incidents are reported annually, providing an impressive data base, and using computer processing since 1969, only limited analyses have been made compared with the later ASRS programme. Together, these reporting systems provide a wealth of statistics for the study of the role of the human component in aviation safety.

We now move into the second century of Human Factors. As we do so, it is fair to ask whether this technology has been adequately applied in aviation in the past and what can usefully be done to ensure adequate progress in the future. The chapters which follow will provide a basis upon which reasoned assessment of these issues can be made.

Defining Human Factors

Industry recognition

The 20th Technical Conference of IATA which was held in Istanbul during November 1975, and was entirely devoted to Human Factors, is seen by many as a turning point in official recognition of the importance of Human Factors in air transportation. Amongst members of its steering group were names of international repute in aviation Human Factors. Yet in spite of this input of expertise, attention of participants was stretched to cover a wide range of topics from medication and pilot psychiatric screening to windshear on approach and flight data recording.

The 600 delegates freighting home their 20 cm stack of conference papers might have been justified in feeling confused as to what Human Factors was all about. Few could have delineated the scope of the subject. Fewer still could have hazarded a definition of Human Factors as an applied technology.

Nevertheless, Istanbul generated two significant messages. Firstly, that something was amiss related to the role and performance of man in civil aviation. And secondly, that a basic Human Factors educational gap existed in air transport. Both messages called for urgent action. The great Tenerife disaster just 17 months later seemed to have been staged specifically to underline these messages and to call upon civil aviation to put its Human Factors house in order. These chapters are concerned to contribute towards filling the basic educational gap which Istanbul and other conferences have revealed.

Human Factors and ergonomics

In order to ensure that we are all talking the same language and can thus focus thought more constructively on the special problems associated with the role of man in the aeronautical system, we must start by defining clearly what we mean by Human Factors and what is its scope.

Human Factors is about people. It is about people in their working and living environments. It is about their relationship with machines and equipment, with procedures and with the environment about them. And it is also about their relationship with other people. While all definitions are man-made and rarely carry the force of law, they are useful in guiding enlightened discussion and crystallising professional activity. They should not be too restrictive, however, and should allow for development and new knowledge. The most appropriate definition of the applied technology of Human Factors is that it is concerned to optimise the relationship between people and their activities by the systematic application of the human sciences, integrated within the framework of systems engineering (Edwards, 1985). Its twin objectives can be seen as effectiveness of the system, which includes safety and efficiency, and well-being of the individual.

The term ergonomics was derived in 1949 from the Greek words *ergon* (work) and *nomos* (natural law) by the late Professor Murrell and used as the title of his textbook on the subject published in 1965. The word, or local adaptations of it, are now used in many parts of the world, such as *ergonomie* in Holland and in France, and *ergonomia* in Hungary and Brazil. He then defined it as 'the study of man in his working environment' and this comes very close to the later Edwards' definition of Human Factors. There is, however, a small difference in emphasis. Human Factors has come to acquire a somewhat wider meaning, encompassing some aspects of human performance and system interfaces which would generally not be considered in the mainstream of ergonomics. Nevertheless, both are primarily concerned with human performance and human behaviour and for practical purposes both can be taken as referring to the same technology.

In the USA, the word ergonomics is only gradually coming into use. The term Human Factors is more common there, though its use is not precise and there are still varying interpretations as to its meaning. A clarification may be found in the activities associated with NASA's Human Factors Research Division and the membership of the Human Factors Society (Appendix 1.1 and 1.2). Human Factors as a term has its problems because the words used in their vernacular sense can be applied to any factors related to humans. The use of capital initial letters when referring to the technology helps to minimise any confusion which may exist. In vernacular usage the terms human aspects or human elements are helpful alternatives to avoid ambiguity and aid comprehension.

Disciplines utilised in Human Factors

Before the availability of dedicated Human Factors courses, those wishing to enter the field originated from different disciplinary backgrounds. Most commonly, these were engineering, psychology and medicine. The motion studies mentioned earlier in this chapter were a good example, with Frank Gilbreth being an engineer and his co-worker wife, Lilian, a psychologist.

Because of these origins and the fact that Human Factors is a multi-disciplinary technology, certain myths and misconceptions have arisen. Perhaps the most persistent of these is that Human Factors is a branch of, or somehow related to, medicine. This confusion may have emerged because some of the earliest problems recognised in flying were physiological in nature. The physician was probably the person nearest at hand with a knowledge of physiology and it was natural that he should become concerned with these aspects of flight.

But since the pioneering days of flying, optimising the role of man and integrating him in this complex working environment has come to involve more than simply physiology. In particular, it has come to be concerned with human behaviour and performance; with decision-making and cognitive processes; with effective use of the equipment in all operating environments; with the design of controls and displays, and with flight deck and cabin layout; with communication and with the software aspects of computers, maps, charts and documentation. It is also concerned with the refinement of staff selection, and training and checking, all of which require skilled Human Factors input. Properly controlled experimental studies are now sometimes required to seek solutions to the more difficult operational problems, often involving very large sums of money. A shift of emphasis has clearly taken place from physiology towards psychology and this has introduced qualifications well outside the field of medicine.

Nevertheless, there are times, notably where physiology and health are concerned, when there should be effective communication between the physician and the ergonomist or Human Factors specialist. A typical example in aviation is in the use of medication to counter the sleep and biological rhythm disturbance induced by long-range flying, as performance on board as well as personal well-being are involved. The lumbar region back pain associated with sitting for long periods in uncomfortable seats may come to the notice of the aviation physician before the Human Factors specialist where no proper operational feedback system has been established. This also applies to accidents such as burns or abrasions which occur while working on board using badly designed equipment.

The predominantly psychological basis of Human Factors is illustrated not only in the professional disciplinary background of its modern practitioners, but also in the literature which supports it. Almost all of the Human Factors reference books and textbooks are authored or edited by psychologists from one or other branch of the discipline.

Nevertheless, this multi-disciplinary technology uses, as the Edwards definition makes clear, many sources of information from the human sciences. In spite of the shift in emphasis towards psychology, physiology is still an important source of Human Factors knowledge and is necessary, for instance, for an understanding of vision and hearing. Anthropometry and biomechanics, involving measures and movements of the human body, are relevant to the design of the workplace – in this case the flight deck and the cabin – and to the equipment in it. Biology and its increasingly important sub-discipline, chronobiology, are necessary for an understanding, for example, of the body's rhythms, which have an influence on performance. Genetics may be a factor in understanding the variations in behaviour and performance of people of different ethnic backgrounds. No proper analysis of data and no presentation of the results of surveys or experiments are possible without some basic understanding of statistics, an essential discipline in the education of every Human Factors practitioner.

In spite of the utilisation of academic sources of information from various disciplines, Human Factors is primarily oriented towards solving practical problems in the real world. As a technology, its relationship with the human sciences might be likened to that between engineering and the physical sciences.

A conceptual model of Human Factors

In order to clarify further the scope of Human Factors it may be useful to construct a simple model. From time to time in subsequent chapters references will be made to the interfaces shown in this model.

The original *SHEL* concept, named after the initial letters of its components, *Software, Hardware, Environment, Liveware*, was first developed (Edwards, 1972) using a different model from the one used here. The 'building block' model used below was first published some years later in a European Community paper (Hawkins, 1984). Both models, however, are based on exactly the same concepts.

Liveware

In the centre of the model is man, or the *Liveware*. This is the most valuable as well as the most flexible component in the system. Yet man is subject to many variations in his performance and suffers many limitations, most of which are now predictable in general terms. It might be said that the edges of this block are not simple and straight and so the other components of the system must be carefully matched to them if stress in the system and eventual breakdown are to be avoided.

In order to achieve such matching, a sound understanding of the characteristics of this central component is essential. As many of those concerned with Human Factors have an orientation towards engineering, it might be fruitful to describe the *Liveware* in this context in engineering terms.

Physical size and shape. In the design of any workplace and most equipment, body measurements and movement play a vital role. These will vary not only between ethnic, age and sex groups, but extensive differences can be expected to occur within any particular group. Fundamental decisions must be taken at an early stage in the design process as to the human dimensions, and consequently the population percentage, which the design is going to satisfy. Data to make such decisions are available from anthropometry and biomechanics.

Fuel requirements. In order to function properly, man needs fuelling with food, water and oxygen. Deficiencies in this fuel supply can affect his performance and well-being. This type of information is available from physiology and biology.

Input characteristics. Man has been provided with a vast system for collecting information from the world about him. He has means for sensing light, sound, smell, taste, movement, touch, heat and cold. This information is needed to enable him to respond to external events and to carry out his required tasks. Some senses involve more directional information; some are more sensitive than others. And all are subject to degradation for one reason or another. Physiology and biology are the main sources of knowledge here.

Information processing. While the sensing apparatus is vast, the information processing capabilities of man have severe limitations. It is fruitless to provide an operator with information from displays, for example, without an understanding of how effectively the information can be processed. Poor instrument and warning system design has frequently resulted from a failure properly to take into account the capabilities and limitations of the human information processing system. Short- and long-term memory are involved here and we also encounter other factors which can influence the effectiveness of the system, such as motivation and stress. Many human errors find their origin in this area of information processing. The source of background knowledge here is the discipline of psychology.

Output characteristics. Once information is sensed and processed, messages are sent to the muscles and a feedback system helps to control their actions. We need to know the kind of forces which can be applied and the acceptable direction of movement of controls. Speech characteristics are vital components in the design of efficient voice communication procedures. Here we look to biomechanics and physiology for support.

Environmental tolerances. People, like equipment, are constructed to function effectively only within a rather narrow range of environmental conditions. Temperature, pressure, humidity, noise, time-of-day, light and darkness, can all be reflected in performance and sometimes also well-being. In less tolerant individuals, performance can also be affected by heights (acrophobia), enclosed spaces (claustrophobia), and even by flying itself (flight phobia), to name just a few. A boring or a stressful working environment can also be expected to influence performance. Physiology, biology and psychology all provide relevant information on these environmental effects.

An important characteristic of this central component, which will become increasingly apparent as the reader progresses through these chapters, is that people are different. This will become particularly evident when we come to discuss motivation and attitudes. While it is possible to design and produce hardware to a precise specification and expect consistency in its performance, this is not the case with the human component in the system, where some variability around the normal, standard product must be anticipated. Some of the effects of these differences, once they have been properly identified, can be controlled in practice through selection, training and the application of standardised procedures. Others may be beyond practical control and our overall system must then be designed to accommodate them safely.

The *Liveware* is the hub of the *SHEL* model of Human Factors. The remaining components must be adapted and matched to this central component.

Liveware–Hardware

The first of the components which requires matching with the characteristics of man is the *Hardware*. This *L–H* interface is the one most commonly considered when speaking of man-machine systems. In one application, it concerns tasks such as designing seats to fit the sitting characteristics of the human body. The passenger who complains of uncomfortable seats, inadequate leg space, a non-adjustable backrest and too narrow armrest is concerned with this *L–H* interface, as is the crew member reporting lower back pain to his medical officer.

More complex problems may face the Human Factors specialist when designing displays to match the information processing characteristics of man. Controls, too, may create problems and there are many examples of incorrect actuation as a result of inappropriate movement, lack of proper coding, or poor location, all matters of Human Factors concern. The user may never be aware of the *L–H* deficiency, even when it finally leads to disaster. None of those who died misreading the notoriously vulnerable three-pointer altimeter would be able, if they could be consulted, to recognise what had caused their demise. The natural human characteristic of adapting to *L–H* mismatches masks but does not remove their existence and to that extent constitutes a potential hazard to which designers should be alerted.

Liveware–Software

The second interface with which Human Factors is concerned is that between the *Liveware* and the *Software*. This encompasses the non-physical aspects of the system such as procedures, manual and checklist layout, symbology and, increasingly, computer programs (Fig. 1.1). The problems are often less tangible than those associated with the *L–H* interface and more difficult to resolve. For example, symbology and conceptual aspects of the Head-Up Display (HUD) have attracted much attention, yet even after a quarter of a century are still resistant to universally accepted resolution.

Fig. 1.1 In the *SHEL* model, Software includes symbology which is reflected in this collection of signs which can be seen in and around airports.

Liveware–Environment

One of the earliest interfaces recognised in flying was between the *Liveware* and the *Environment*. Flyers were fitted with helmets against the noise, flying suits against the cold, goggles against the airstream and oxygen masks against the effects of altitude. Later, anti-*G* suits were added to protect against acceleration loads. These measures all aimed at adapting man to match the environment. As aviation progressed, a trend developed to reverse this process, to adapt the environment to match human requirements. Pressurisation systems were installed to eliminate the need for routine use of supplementary oxygen. Air conditioning systems were designed to control temperature and sometimes also humidity. Soundproofing brought noise levels down low enough to permit normal conversation in flight and to make helmets and earplugs redundant. But new environmental challenges have arisen to replace the old. Ozone concentrations at flight levels used even by subsonic jets can be hazardous. Beyond that, we are still concerned about the possible effects of high-altitude ultraviolet radiation and of the effects of the ionising radiation to which crews may be exposed. Each may require better measuring, the establishment of exposure limits, and additional training for the flight and cabin crews involved. And the problems associated with disturbed biological rhythms and related sleep disturbance and deprivation, have notably increased with the speed of transmeridian travel and the economic need to keep aircraft, and their human payload, flying 24 hours a day.

The effects of aviation's phenomenal growth is another part of the environment that is becoming increasingly critical. In the 1990s air transport must function safely and efficiently while it utilises teeming passenger terminals and congested gates, taxiways and runways. It must then operate in increasingly crowded skies.

Liveware–Liveware

The last interface with which Human Factors is concerned, is that between people, the *Liveware–Liveware* interface. Traditionally, questions of performance in flight have been focused on the characteristics of the individual pilot or crew member. Increasingly, however, attention is being turned to the breakdown of team-work or the system of assuring safety through redundancy. Flight crews function as groups and so group influences can be expected to play a role in determining behaviour and performance. In the cabin, interaction between passengers can also be expected to influence their behaviour and this can be important in emergency situations. We are concerned in the *L–L* interface with leadership, crew cooperation, team-work, and personality interactions. Such questions as instructor/student relationships also come within the scope of this interface, as do those of a staff/management nature. Current cockpit/crew resource management (CRM) programmes are beginning to deal seriously with these issues.

Other interfaces

There are within the aeronautical system other interfaces outside the field of Human

Factors. Interfaces between different *Hardware* can be seen in plugs and connectors. *Hardware–Environment* interfaces are apparent in equipment packaging and insulation. *Hardware–Software* interfaces are involved in certain aspects of equipment instruction manuals. These are not within the scope of the technology of Human Factors and so will not be discussed further in the chapters of this book.

A foundation for progress

The remark of the representative of a large independent airline at an international Human Factors workshop in Holland in 1984 was probably not one which in retrospect and with a little more understanding he would care to remember. In a declaration of his company policy he said that 'Human Factors is just an excuse for incompetence'. With equal validity he might have remarked that 'medicine is just an excuse for malingering'. There is, of course, incompetence and there are malingerers. But this provides no justification for neglecting either Human Factors or medicine. Human Factors attempts to research and explain the nature of human behaviour and human performance, using the human sciences. Armed with this knowledge it tries to predict how a person will react and respond in a given set of circumstances. It is most important to understand that it is not concerned to remove or replace the notion of individual responsibility or to condone incompetence or negligence.

Understanding the meaning of Human Factors is only the foundation, though an essential one, for the acquisition of Human Factors expertise. On reaching the end of this chapter this foundation will have been laid and it is now possible to progress in greater depth with acquiring a knowledge of this technology as it is applied to air transport operations.

Complete books have been written on subjects such as vision and visual illusions, yet we can allocate just a few pages. The same applies to motivation, communication, displays and workplace design. The following chapters must be seen, then, as only an introduction to the subject. In Appendix 2, books are recommended for further reading and the contents of a modest Human Factors library is proposed. The list of references provides an extensive store of research information for those who wish to apply more attention to one particular aspect of the subject. But already, we should all now be on the same frequency.

2

Human Error

The nature of error

Air safety in perspective

In aviation human error is closely associated in the eyes of the public – and in reality – with incidents and accidents. It might therefore be useful to set the stage for a discussion of human error by first putting the current air safety scene into some perspective.

The following data do not include accidents or fatalities in the former USSR, the Republic of China or the former East European bloc. Reliable statistics have traditionally not been available from these sources.

Civil jet transport total losses amount on average to about one aircraft a month. This has been rather constant in spite of the steady increase in air traffic, suggesting that some progress is being achieved.

There are various ways to represent fatalities statistically and there is far from an industry consensus on the optimum way to measure air safety. One improvement has been the Flight Safety Foundation's change from the rate per flying hour to the rate per flight sector (Lederer *et al.*, 1987). Flight sector is a better measure than simply flying hours.

Actually, one finds that with the exception of those measures that are directly related to growth (e.g., the actual number of accidents or fatalities), safety in air transport has increased in virtually any way that safety can be measured. However despite this, most observers believe that previous safety standards are now no longer acceptable. In the 1990s, we seem to be coming fairly close to achieving the maximum number of accidents or fatalities, each month or each year, that the public will tolerate without inducing political, and often emotional and unproductive action. The industry cannot simply stand still. Human Factors seems truly to be 'the Last Frontier of the airline safety problem.' (Hammarskjold, 1975). (See also 'The Public Perspective of Risk', p. 2,3.)

Accident penalties

In addition to the cost in loss of life – some 300–700 in an average year – air accidents are also expensive in financial terms. By 1987 it was estimated that aircraft hull losses had already cost the industry about $7 billion in less than 30 years since the start of jet transportation (Eastburn, 1987).

The settlement in US litigation for loss of life in air accidents had risen by 1990 to about $1 million per passenger billed (Hefti, 1986 and personal correspondence). Elsewhere, the value of a single human life has been estimated as high as $2 million (Klema *et al.*, 1987). It may be sobering for the captain of a large jet transport to know that the aircraft and its human payload for which he is legally responsible (FAR 91.3), may be valued at something approaching $1000 million.

Legal liability trials in the USA, following aircraft accidents, have resulted in huge damage claims, with money being sought from anyone even remotely connected with the accident flight – aircraft and equipment manufacturer, airline, pilots, FAA, airport authority, local government and so on. This kind of litigation has left its mark on the aviation industry. Liability insurance costs have risen dramatically and have reduced notably the number of new general aviation aircraft delivered.

Deregulation

In the USA, the Airline Deregulation Act of 1978 reduced airline protection from uncontrolled competition which had previously existed. The highly competitive environment which followed deregulation resulted in many new and marginally viable operators striving to take a share of the increased market. By 1984, the number of airlines in the USA had tripled to 150 and commuter airlines had nearly doubled to 269 (*Newsweek*, 30 January, 1989). Bankruptcies, struggles for survival and corporate take-overs proliferated. Cost-cutting became the name of the game. It was to be expected that ruthless cost-cutting would be applied wherever it would not violate a specific FAR. Minimum standards often became the norm, removing the padding which provided protection against the unexpected and the abnormal. This also occurred in the large UK summer vacation charter business where competition was intense. 'Captain's discretion' is allowed and the operator believes this should be applied for commercial reasons. The relevant document (CAP 371) unfortunately does not define 'Captain's discretion' (Feedback No. 21). The CAA was forced to intervene after publicity to abusers was revealed in the confidential CHIRP reporting programme during the 1980s.

It has always been difficult to quantify, in terms understood by an accountant, the benefits which accrue from a better application of Human Factors in operational areas such as emergency equipment, flight deck layout and standardisation within a fleet, non-technical crew training, and enlightened work/rest regulations.

Commercial deregulation effects, indeed, extends beyond cost-cutting and an economical approach to safety-related issues. Take-overs of one airline by another have meant that the predator airline struggles to digest – often simultaneously – several acquisitions, all of which have their own organisational culture (see page 134). For example, by the time the collapsing Republic Airlines in the USA was itself taken over by Northwest Airlines in 1986, it was suffering from the indigestion of consuming three other companies, in some cases following three different sets of procedures (*Newsweek*, 30 January, 1989). Mixtures of flight deck design, operational procedures and concepts, and distribution of crew duties create serious Human Factors problems for resolution. Seniority integration problems, professional downgrading to less advanced aircraft types, salary cuts, reduction in fringe benefits and job insecurity are all demotivating and do not create a working climate favourable to a high level of crew performance and behaviour, with an effective control of human error.

In spite of these potentially unfavourable effects on safety that commercial deregulation produces, this trend appears likely to continue. The abuse by airlines of monopoly power, with price fixing, capacity restriction and national protectionism has penalised the air traveller and has itself generated the public demand to deregulate.

In Europe, these restrictive practices have been particularly evident regardless of the fact that they are a violation of the Treaty of Rome and regardless of numerous attempts by smaller, independent carriers to break the monopolies throughout the 1980s and into the 1990s.

It must be remembered, however, that wherever commercial deregulation does arrive, the national regulatory authority will inevitably have to increase the scope of its operational regulations and the effectiveness of its supervisory function if safety is to be maintained.

The public perspective of risk

While the airline industry can justifiably boast of how much safer it is to travel by air than by road, the public does not react to safety in this way. Aircraft crashes make dramatic headlines, particularly those such as Tenerife (1977) when 583 occupants of the two colliding aircraft died and created the blackest landmark in aviation history. Or Lockerbie (1988) where wreckage and victims scattered across the Scottish countryside provided a field day for press and television cameramen. For the media, bad news sells better than good.

Public reaction to the disaster to the airship *Hindenberg* in 1937 with the loss of 36 lives, preceded by the crash of the British R101 in 1930 with 48 fatalities, was such that passenger transport by airship was halted for half a century. The dramatic radio running commentary of the *Hindenberg* conflagration and the spectacular press photos had an enormous impact on the public and ended the dreams of the 'flying hotel'. In the 4-day 1991 Easter holidays in Spain alone, 147 people were killed on the Spanish roads – normal for the vacation period. The news barely received a mention in the local newspaper.

Aircraft accidents deter some potential passengers from flying, while the daily carnage on the roads seems to have no equivalent effect. Many potential air passengers are already anxious about flying and a dramatic news presentation of an air crash simply confirms these fears (see also page 317). This public perception of the risk of flying is one with which the aviation industry must live.

The public is intolerant of aircraft accidents. It is also aware that in most cases, the loss of life could have been avoided if the human performance and behaviour involved had been better. In accident reporting, 'pilot error' is the aviation journalist's most overworked term.

It seems likely that with the rate of growth of air transport the total number of accidents each year will increase in the future even if we are able to achieve some modest improvement in the accident rate. Public intolerance of aircraft accidents may then generate dissatisfaction with the way the air transport industry is managing its affairs.

Errors and their consequences

In this chapter it is planned to examine the nature of human error and then apply this knowledge to meeting the challenge which error presents in a critical activity such as air transport. Once the nature of error is understood and its occurrence anticipated, it is then possible to devise protective measures.

Two thousand years ago the Roman orator Cicero declared that 'it is in the nature of man to err'. In this he was surely right. But he added that 'only the fool perseveres in error', and on this score he was on weaker ground. The remark suggests, in fact, that although he studied elocution, history and law, he had not done his Human Factors homework.

While a repeated error due to carelessness or negligence, and possibly even poor judgement, may be considered the act of a fool, these are not the only kinds of error made by man. Errors such as those which have been induced by poorly designed equipment or procedures may result from a person reacting in a perfectly natural and normal manner to the situation presented to him. Reacting in the way he has been created to react. Such errors are likely to be repeated and are largely predictable. When they occur, the strictures of Cicero should be directed to the designer and not the operator.

Human error has been a common and accepted element of behaviour throughout recorded history. Cicero himself may not have been immune, as his oratory and writing were brought to an abrupt end when his head and his hands were prematurely cut off. But the consequences of errors in today's technological society, where critical areas have been established in large-scale systems, are infinitely greater than ever before. The Three-Mile Island nuclear power plant incident in the USA in 1979 gave a timely illustration of the potentially catastrophic consequences of human error in a sensitive industry (Payne *et al.*, 1979). This potential was later demonstrated in the Chernobyl nuclear disaster in the former Soviet Union in 1986 where the full cost in human lives and human suffering may not be known for many decades. In this case, a catalogue of catastrophic human errors was gradually unveiled, though the full extent and nature of these may never be disclosed or even determined in their entirety.

In another environment, the Tenerife double Boeing 747 disaster in 1977, which cost 583 lives and about \$150 million, gave a preview of what can follow human error in aviation. In both industries, serious warnings of inadequate Human Factors and their possible devastating consequences were given shortly before the accidents. In the case of nuclear power stations, the alarm was given by the Rasmussen Report and the Electric Power Research Institute Survey. In the aircraft accident case, the warning was given in the conclusions of the IATA Istanbul Conference. In both cases, the accidents arose from human error occurring in a working environment deficient in Human Factors. In neither case had effective measures been taken in heed of the warning. Only 22 days before the Challenger Space Shuttle disaster in 1986, a potentially catastrophic human error was discovered and the launch of the shuttle Columbia abandoned with only 31 seconds to spare. More than 100 incidents had been reported in each of the previous two years, more than half of which were attributed to some type of human error. It cannot be said that NASA officials were unaware of the frequency or possible consequences of the human errors occurring (Presidential Commission, 1986).

Although the errors involved in these cases had very undesirable consequences, it is possible for quite similar errors to have insignificant results in other circumstances. Putting salt instead of sugar into the coffee at home is likely to raise only a grimace; possibly even a laugh if one has such a sense of humour. Essentially the same error transferred to the airport tarmac where a piston-engined aircraft has had kerosene instead of aviation gasoline put into its fuel tanks, and there are likely to be few of the aircraft occupants surviving to raise even a modest chuckle.

We are therefore already able to identify three basic tenets with respect to human error. Firstly, that the origins of errors can be fundamentally different; secondly, that anyone (even airline captains) can and will make errors; and thirdly that the consequences of similar errors can also be quite different. A recognition of these three tenets is an essential basis for making progress in meeting the challenge, the increasingly serious challenge, presented by human error to those concerned with the safe and efficient operation of aircraft.

Aviation is not immune from its share of mythology. Perhaps the most enduring of these myths is embodied in the term 'pilot error'. We do not read of 'surgeon error' following a death on the operating table. Or 'managing director error' for a business misjudgement. Neither is a faulty cut of meat classified as 'butcher error' or a wrongly installed carburettor called 'garage mechanic error'. All come under the heading of human error and, in fact, the death on the operating table may have resulted from essentially the same error as in the case of the faulty carburettor installation. The errors made by pilots are in principle no different from those made by everyone else. The use of the words 'pilot error' has suggested that somehow the nature of the errors made by this kind of operator is unique; that once an accident could be attributed to this 'cause', then the problem was solved and the case be filed. To many, this may not have been a wholly inconvenient procedure. This was illustrated in the case of the Mt Erebus DC10 crash in 1979 where the distorted initial findings of pilot error (AAR-79-139) were subsequently overturned by a New Zealand Royal Commission following a vigorous campaign of protest by pilots (Vette, 1983).

The pilot error concept also focuses rather more on what happened than why it happened and so for this reason, too, it has been unhelpful in accident prevention activity.

It is likely that persistence in the use of this term has impeded a more profound and rational examination of human performance on board and consequently obstructed progress towards greater flight safety and efficiency. The myth will not be given credence by further mention in this book.

Error rates

Having established that it is normal for man to err, it is reasonable to ask how often this is likely to occur; what error rates are considered normal. And a closely related though not identical question, how reliable is the human operator. In this respect, it should be remembered that man is a very flexible component, and if ergonomics has been properly applied to system design, he can serve to increase overall system reliability and not simply, through human error, decrease it.

On average, a person will make an error in dialling a telephone number on a round dial about once in 20 times. Performance is rather better with push-button selection. Many studies of human error rates during the performance of simple repetitive tasks have shown that errors can normally be expected to occur about once in 100 times or 10^{-2}. On the other hand, it has been demonstrated that under certain circumstances human reliability can improve by several orders of magnitude. An error rate of 1 in 1000 or 10^{-3} might be thought of as pretty good in most circumstances.

These are normal human error rates which are, so to speak, built into the human system. It is clear that they can vary widely depending on the task and many other factors such as fatigue, sleep loss and motivation. Nevertheless, this is the order of the problem we face in placing man into a complex man-machine system. It is clear that this kind of error rate, without some protective machinery, is quite unacceptable in a working environment in which the consequence of a single error can be disaster. While a direct comparison of the statistics would not be justifiable, it is worth recalling that the British Civil Aviation Authority (CAA) requirement for automatic landing equipment is that it shall not suffer a catastrophic failure more than once in 10 million landings or 10^{-7}.

As long ago as 1940, it was calculated that about 70% of aircraft accidents could be attributed to the performance of man (Meier Muller, 1940). A third of a century later, the causal role of man in accidents had shown no reduction (IATA, 1975) and human error remained the dominant theme (Fig. 2.1).

This situation, however, is not unique to aviation; it has been estimated elsewhere that 80%–90% of accidents are the result of human error (Drew, 1967).

Accident proneness

The concept of accident proneness originated early in the 20th century (Greenwood *et al.*, 1964). Since then the concept, and even its definition, has been the subject of controversy and it will not be possible in the short space available here to cover the area in detail. It will be sufficient for our purpose to accept that accident proneness

HUMAN FAILURE

Crew, ATC, Aircraft
Design, Maintenance

OTHER CAUSES

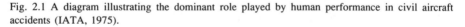

Fig. 2.1 A diagram illustrating the dominant role played by human performance in civil aircraft accidents (IATA, 1975).

is the tendency of some people to have more accidents than others with equivalent risk exposure, for reasons beyond chance alone.

Having said that, it is necessary to examine the kind of distribution which might be expected from chance alone. This can be deduced from what is called the Poisson Distribution (Reichmann, 1961). As an example, if 100 accidents were distributed at random over a number of years amongst 100 people, then it might be expected that the apportionment of accidents would look something like this:

0 accident	....	37 people
1 accident	....	37 people
2 accidents	....	18 people
3 accidents	....	6 people
4 accidents	....	2 people

If the example were extended to cover 1000 accidents distributed amongst 1000 people, then three might be expected to have as many as five accidents. There would then be a powerful temptation to label these three victims as being accident prone with discriminatory action being taken against them. On the grounds of frequency alone, such action would be unjustified. Similar caution must also be applied when assessing an airline's relative safety on the basis of the very small sample of accidents occurring.

It is clear, then, that one person can have more accidents than another purely by chance. However, there are also other reasons for differences in the number of accidents experienced by individuals, which may be more controllable. One such reason may simply be one of exposure to risk. A pilot flying routinely in bad weather with poorly equipped aircraft and into inadequate airfields is exposed to more risk than one flying in a good climate with modern avionics on board and into airports with proper facilities.

In addition to pure chance and exposure, a third reason for variation in the individual's accident record could be that he possesses some innate characteristics which make him more liable to have accidents. Many have seen this as a more appropriate definition of accident proneness. There has been some evidence to suggest that those who have accidents at work also have them at home (Newbold, 1964). And also that there was a significant correlation between accidents in a simulated kitchen and road traffic accidents (Guilford, 1973). Whether or not such an innate characteristic is subject to modification, if indeed it exists, is not established.

It is likely that in any given period one person makes more errors than another for reasons other than chance, exposure or innate characteristics. One housewife may consistently break more glasses while washing the dishes than another, and this may result from an inherent awkwardness, lack of muscle coordination or simply carelessness. But the glass-breaking frequency may change on a short-term basis with influences such as domestic or work stress, health variations, job satisfaction, or boredom.

It must be hoped that professional airline pilots with recognisable physiological, psychological, or personality deficiencies which would make them congenital 'glass-breakers' will have been mostly filtered out as a result of screening techniques applied during selection and training. However, some no doubt will slip through

33

this filtering process. The *ab initio*, or other low-time pilots can be a particular problem here. They are becoming a more important part of the aviation scene, and will not have been exposed to as much of the filtering process as have their predecessors. Improved testing and training have to take the place of this filtering process. For several years, European airlines have been particularly successful with their *ab initio* programmes.

We are then left mainly with chance, exposure and short-term factors, the last frequently resulting from changes in motivation.

When examining these short-term factors it is still important to recognise that response to a particular stressful influence varies from one person to another. A life-change event (discussed in more detail in Chapter 4) which causes distress and performance degradation in one pilot may be taken in his stride by another, who, in his turn, may be vulnerable to a different form of stress.

In view of the lack of consistency and agreement in the use of the term accident proneness, it might be wise to discard it altogether as a concept and refer simply to individual differences in involvement in accidents. This allows all of the influencing elements to be considered equally without assuming a basic pre-eminence of any one and avoids the conceptual confusion which has plagued accident proneness.

In particular, it allows full consideration to be given to shorter-term influences which may be a more fruitful approach in accident prevention. Such an approach will certainly involve an educational process, and this is outlined in Chapter 14.

The sources of error

SHEL interfaces

The *SHEL* model described in the previous chapter illustrated four interfaces and each of these can be seen as the location of error when there is a mismatch between components.

The interface between *Liveware* and *Hardware (L–H)* is often the source of errors. The old three-pointer altimeter was an early aviation example of equipment the design of which encouraged errors. Research as early as the 1940s analysed under laboratory conditions the frequency and magnitude of these errors (Grether, 1948). Knobs and levers which are badly located or lack proper coding also involve mismatch at this interface. Similarly with equipment which has been designed without proper regard to what we later refer to as Murphy's Law. Warning systems have long been an object of development and refinement in order that they should achieve more effectively their objective of providing alerting, information and guidance in abnormal situations (see Chapter 11). Yet they have often failed in one or more of these functions.

Deficiencies in conceptual aspects of warning systems come within the scope of the *Liveware–Software (L–S)* interface. So would an irrational indexing system in an operations manual which caused delays and errors in seeking vital information. Some early checklists were designed to encourage the pilot to do the check from memory, some had no written response on the list. These faulty concepts were often the source of errors. A published flight departure routing, where the Standard

Instrument Departure (SID) number could be mistaken for an initial departure heading which led into high terrain, is another case of an error generated by faulty matching at this interface.

Noise, heat and vibration are just three environmental factors which in certain circumstances can lead to an increase in errors and these are associated with the *Liveware–Environment (L–E)* interface. Also coming within the scope of this interface are errors resulting from reduced performance caused by the disturbance of biological rhythms in long-range flying, sometimes called jet lag. This is discussed in detail in the following chapter.

There is growing recognition that the organisational, regulatory and social aspects of the environment in which air transport operations are conducted are of considerable importance. These include the operating philosophy of the company, the general morale of other employees, and the general health of the organisation. A detailed discussion of these subjects beyond their partial treatment in Chapter 8, 'Attitudes and Persuasion' is beyond the scope of this book. However their importance should not be underestimated.

The last interface in the *SHEL* model is that between *Liveware* and *Liveware (L–L)*. Deficiencies in this area, in an activity where teamwork or crew cooperation is the foundation upon which flight safety today is built, can have disastrous results. As an example, the authority relationship between the captain and the first officer has been cited in many accidents and incidents. Tenerife (1977), the accident which can be quoted to illustrate so many Human Factors elements, is no exception. The 1982 Boeing 737 accident in Washington, in which the unassertive copilot's concern about the aircraft performance during take-off was ignored by the captain, is a further example (NTSB-AAR-82-8). In 1979, a commuter airliner crashed in the USA after the captain became incapacitated. He was a gruff company vice-president and the copilot a new recruit still on probation. The copilot failed to take over (NTSB-AAR-80-1).

There is an optimum 'trans-cockpit authority gradient' to allow an effective interface between pilots on the flight deck (Edwards, 1975). The gradient may be too flat, such as with two equally qualified captains occupying the two seats, or too

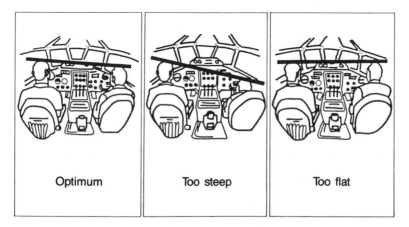

| Optimum | Too steep | Too flat |

Fig. 2.2 The trans-cockpit authority gradient (after Edwards, 1975).

steep, as with a dominating chief pilot and a junior and unassertive first officer. In such cases a reduced performance may result with a chance of errors going undetected and uncorrected (Fig. 2.2). A study in the UK of 249 airline pilots confirmed the importance of this aspect of flight deck communication. Nearly 40% of the first officers surveyed said they had on several occasions failed to communicate to the captain their proper doubts about the operation of the aircraft. Reasons appeared to be a desire to avoid conflict and a deference to the experience and authority of the captain (Wheale, 1983). Socio-psychological influences in practice can interfere with the proper exchange of information on the flight deck and thus with the safe operation of the aircraft (Foushee, 1982). Socio-cultural factors played a role in the 1977 Anchorage accident when young oriental flight deck crew members failed to intervene or express concern when the senior American captain was obviously under the influence of alcohol before take-off (NTSB-AAR-78-7).

A related problem with similar consequences is when a first officer clearly advises the captain of the error or discrepancy but the captain fails to take action. Examples of this phenomenon are cited in a later paragraph.

Problems in the *L–L* area are in a large part responsible for the importance given to CRM programmes by ICAO, by member States, and by individual airlines. In an effort to minimise 'trans-cockpit authority gradient' and other liveware problems, a great deal of effort is being given to redefining the role of the pilot-flying (PF) and the pilot-not-flying (PNF). This takes the allocation of duties away from the formerly utilised concept of 'captain's duties' and 'copilot duties'. Considerable effort is also spent in emphasising the authority of the captain and his role as the commander of the aircraft and the leader of its crews.

'Leadership' in its best sense requires both good interpersonal skills and good technical skills. Technical skills are reasonably evident and relatively easy to demonstrate. Interpersonal skills – including the ability to communicate well, to manage all resources effectively and to build and maintain a cohesive team, often under less than optimum conditions, are much more difficult to identify and to quantify.

An integral part of this concept is also recognition of the importance of 'followership' in a qualified and well-functioning crew, for crews should operate as teams. There is little doubt that the age of the 'autocratic captain', the too submissive copilot, and a 'one man band' operation have no place in a modern air transport operation. Both 'leadership' and 'followership' are helped by effective CRM training.

A remaining problem, which is not restricted to the old and experienced captain, is the intractable individual. If this person is a captain, he or she is usually a poor team leader and often does not recognise the importance of making their flight a team effort. If the individual is a subordinate crew member, he or she is not, for any number of reasons, an effective team player. Fortunately, these individuals are quite rare. Unfortunately, they are also a difficult problem for both their fellow crew members and for management because on a technical basis they can be highly-skilled individuals. There is some evidence that peer group pressure is an effective means for dealing with these individuals. (See also the discussion on conformity on page 181) Nearly all of these technically competent but interpersonally weak individuals want to be considered good pilots by their peers.

Information processing

Identifying the source of an error is an important step towards taking effective protective measures. At the centre of the *SHEL* model is the *Liveware* and a study of this component provides another means of identifying error source. A simplified model can be used to represent the human information processing system. Distinct stages can be recognised in this system and we can review them in sequence (Fig. 2.3).

Sensing

As already indicated in Chapter 1, the body has an extensive system for sensing information about the world around it. Different senses detect different types of signal and have different qualities as to the degree of directional information provided, sensitivity or the conditions of light or quietness needed for them to function properly. Individual sensory systems may be more sensitive in one person than another and are subject to deterioration as a result of physical and mental disorders, age and interference from lesions and other agents such as local anaesthetics.

Before a person reacts to a given situation, information about that situation must have been sensed. Here already is a potential source of error. The sensory systems

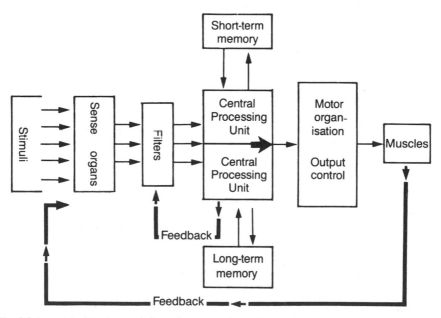

Fig. 2.3 A model of the human information processing system.

function only within a rather narrow range. Although an object may be photographed using an infra-red camera, it may not be visible to the eye due to the light frequency or brightness. The information is there but is not sensed. Similarly with sound. A dog can sense sound frequencies which are inaudible to the human ear. Someone with nerve deafness will never be able to detect certain sounds. So when a person concludes that there is no light or no sound, he may be in error. Similarly with other senses.

We should consider one other point. The sense organs are not designed to detect all information which may be important. A notable exception is related to velocity. While acceleration is sensed, primarily by mechanisms in the ear, consistent speed is a different phenonemon. An astronaut has no more sensation from moving at 2000 km/hr than does his ground-based controller sitting comfortably in an armchair in Houston. Neither can an aircraft passenger sense the speed at which he is travelling.

Perception

Once information is sensed, it passes along various neural pathways to the brain. It can meet interference on this journey through lesions or chemical agents. But usually the message reaches its destination and is processed. The conclusion then reached about the nature and meaning of the message received is called perception and this interpretive activity provides a major breeding ground for error.

Basic principles governing how objects are organised and perceived were formulated in the early part of this century and are known as the *Gestalt* laws.

These are concerned with the structural or organisational arrangement of elements that influence the way we perceive things. For example, elements which are closer together tend to be seen as a group, as do elements which are similar. These laws of organisation begin with a distinction between what is seen as the figure or object and what is seen as the background. This principle is relevant to certain visual illusions, discussed in Chapter 5, when a picture or drawing is ambiguous and perception of the figure and background alternate. This separation of figure and background provides the basis of all perceptions.

During the perception process, account will be taken of the context in which the message is placed. For example, a 'cut-off date' will be perceived quite differently in a Middle East prison scenario than if it were quoted in the office of an electric power company considering unpaid electricity bills. If a newspaper captions a photograph of a person as 'best shot', other sources of information are employed to determine whether the description refers to the winner of a shooting competition or a terrorist serving a life sentence. The stimulus, whether aural or visual, is the same; the perception of the stimulus is different.

Sometimes perceptions are developed on the basis of inadequate or ambiguous information; when the information is insufficient we may unconsciously fill in the missing information ourselves. For example, it may be reported that a number of aircraft were lined up waiting for take-off clearance, with two in front, two in the middle and two in the rear. From this information it will probably be concluded without further question that there were a total of six aircraft. However, the report could also be applicable if there were only four aircraft.

Expectation, or what is called 'set', has an influence on perception. It is commonly said that we hear what we expect to hear and see what we expect to see. This effect in radio communication is discussed in Chapter 7 and its relevance to visual illusions is covered in Chapter 5. It is an important source of error. For example, this sequence of letters seems clear enough:

E D C B A

Yet in the interpretation of the fourth letter, expectation was applied in order to draw a conclusion as to what it represented. When placed in another context, exactly the same visual information will result in a totally different interpretation. Invert and check the following:

This may be seen as closely related to habit and is particularly applicable to the aviation environment. Flight and simulator training and the learning of Standard Operating Procedures (SOPs) are aimed at establishing a pattern of habitual behaviour. Because of the practice of standardising equipment and procedures within an airline it is also often possible for the operating habits formed on one aircraft to be carried over to subsequent aircraft.

This serves a useful purpose, as it avoids the necessity of each individual having to devise his own work practices and allows advantage to be taken of previous industry experience in optimising procedures. Design carry-over reduces confusion and minimises training costs. The standardisation and training also allows routine tasks to be performed with less attention and effort, leaving more time available for the accomplishment of other activities, and facilitates the exchange or substitution of individual crew members on very short notice.

Experience and habit, however, may not always be beneficial. Once a certain pattern of behaviour has been established, it may be very difficult to abandon or unlearn it, even when it is no longer appropriate. There have been many cases when a pilot has acted in a way which would have been appropriate to an earlier aircraft, but with different instruments, controls and systems, the perception and subsequent action were incorrect (e.g. CHIRP Feedback Nos 6 & 7). Aircraft are believed to have crashed from this type of human error resulting from a reversion to an earlier behaviour pattern (Rolfe, 1972).

Two situations are particularly vulnerable to this kind of reversion. Firstly, when concentration on a particular task is relaxed and secondly, in situations of stress. A characteristic of reversion is that the original experience or habit concerned may not have revealed itself for a very long time and the person concerned may be quite unaware of the danger lurking just beneath the surface.

The relationship of reliability or freedom from errors to experience is not a simple one. Physiological deterioration begins at a rather early age and then continues throughout life. Emotional stress from domestic and family origins also often increases as the years go by. Sometimes career frustrations only start to assume significant proportions in the second part of a career, when promotional prospects

39

begin to disappoint and other professional pressures are generated. Set against these adverse factors applying with increasing age is the favourable effect of experience. In general, experience tends to lead to a reduction in both errors and exposure to those situations in which errors can have serious consequences. Just where the crossover point comes, when the deterioration with age exceeds the improvement with experience, is debatable and will certainly vary significantly between individuals. This question is particularly relevant in connection with age limits applying to the issue of pilot licences and it has been argued extensively and vigorously, notably in the USA (e.g. Mohler, 1984).

Motivation is discussed in more detail in Chapter 6 but a short reference must be made to it in any consideration of human error. Motivation reflects the difference between what a person can do and what he *will* do in any particular set of circumstances. A fundamental weakness in traditional forms of proficiency checking and testing is that they generally only show a person's capability or capacity under test conditions. They do not necessarily reflect his performance or reliability when away from the test situation and supervision. This depends on motivation. If the motivation to perform a particular task is too low, then a reduced performance with more errors can be expected. There may be various reasons for such inadequate motivation. A person may be getting inadequate job satisfaction. He may be physically or psychologically unwell or suffering from emotional stress or fatigue, all of which may diminish motivation. He may be experiencing boredom which is often associated with monitoring automated systems.

Motivation has been the subject of much research. Yet it remains a complex and often unclarified aspect of human behaviour.

Not unrelated to motivation is arousal and this can be likened to alertness. A high level of arousal is usually accompanied by a high level of attention and is the opposite of drowsiness or day-dreaming. Various psychological and physiological parameters can be used to indicate arousal level. A high arousal level usually indicates either a high level of activity or preparedness for a high level of activity, either physical or mental.

Early this century, two psychologists produced a curve demonstrating a hypo-

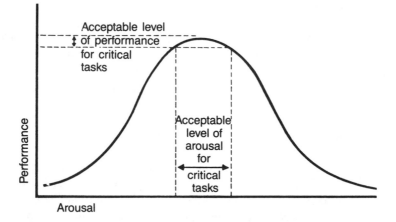

Fig. 2.4 A hypothetical relationship between arousal and performance (after Yerkes *et al.*, 1908).

thetical relationship between performance and arousal (Fig. 2.4). This diagram, called the Yerkes-Dodson or Inverted-U curve, has been used in discussions of human behaviour ever since (Yerkes *et al.*, 1908). The curve suggests that where a high level of performance is required then only a narrow band of arousal is acceptable. When the arousal is higher or lower, reduced performance can be expected. In the lower ranges could be included drowsiness and boredom; in the upper ranges, excitement, fear and panic. Certain psychological disorders might be located at the extreme ends. Human error can be expected to increase as the arousal level moves away from the optimum for a given task. It has been suggested that the optimum level of arousal for accuracy is lower than that for speed (Poulton, 1970).

Decision-making

After having drawn conclusions, rightly or wrongly, about the meaning of the message, information processing progresses from perception to decision-making. Once again, a source of error can be identified. Many factors may be weighed before a decision is made on the appropriate action to be taken in response to the stimulus. Sometimes the weight given to certain aspects of training or past experience can be excessive or inappropriate to the exclusion of more pertinent factors. Emotional or commercial considerations may play a role and distort the decision-making process. Factors such as fatigue, medication, motivation and physical or psychological disorder may apply influences at this stage and lead to an erroneous decision.

A very important contributor to human error is the false hypothesis or mistaken assumption. It has been involved in countless aircraft incidents and accidents, including Tenerife in 1977.

No civil pilot would deliberately make a take-off or landing without a clearance from Air Traffic Control (ATC). Yet there are many cases of uncleared take-offs on record; something had occurred which caused the pilot to make the faulty assumption that he was properly cleared. Similarly, no pilot would deliberately land at the wrong airport, yet every airline has such landings on its files. Certain information, such as runway direction, provided the basis for the false hypothesis and all later actions.

Perhaps the most dangerous characteristic of the false hypothesis is that it is frequently extremely resistant to correction. It may be very easy to adopt but very difficult to relinquish and one seems to be locked into it. A complete structure of false actions can then be built on the original false premise. Evidence or verbal input contradicting the false hypothesis is simply rejected until it becomes overwhelming and then it is often too late. This resistance to change has been discovered in many accidents, particularly since the introduction of the Cockpit Voice Recorder (CVR) which has revealed that doubts expressed by other crew members were overruled by the strength of the belief in the false hypothesis.

There are certain situations in which the false hypothesis is most likely to occur. These were derived from an analysis of a number of aircraft and railway accidents (Davis, 1958). Firstly, when expectancy is high. This occurs typically after long experience of a given event or situation. For example, long experience may have demonstrated that it always appeared to be safe to start a descent from cruising altitude on a particular route after passing a certain position. However, on an isolated occasion at night, a quite exceptional head-wind existed, of which the pilot was not aware; he descended and crashed into the mountains before reaching the airport. He had made a false assumption about his ground speed as a result of previous experience. The same kind of error can occur as a result of high expectancy in verbal communication, particularly that between ATC and the aircraft. ATC messages are sent in a form of verbal shorthand – this is worse in some countries than others – short-cutting standard phraseology. Sometimes part of the message is lost as a result of poor quality transmission or reception. In both cases, there is a tendency to fill in the missing parts of the communication, based possibly on past experience.

Secondly, when attention is diverted elsewhere. A number of accident investigations have revealed that when crew members were concentrating on a particular problem, they made false assumptions about other aspects of the flight – often aircraft altitude. The Lockheed 1011 accident in 1972 at the Everglades in Florida is a typical example of this phenomenon (NTSB-AAR-73-14) but there have been many more.

Thirdly, when it serves as a defence. People like to hear good news and not bad. They prefer to avoid anxiety and problems. As a result, incoming messages may be subconsciously modified to alleviate the anxiety and so provide the message which is preferred. One might say that they are ignoring or evading the truth. A pilot may place greater credence on evidence that the weather is likely to improve and reject clues which might suggest the contrary. Feelings associated with rank and status may increase vulnerability to this kind of influence.

Fourthly, following a period of high concentration. In these circumstances there is a reaction after the effort applied in maintaining a high level of concentration. A relaxed attitude takes over, the stressful and difficult part of the flight has passed and the scene is set for the adoption of a false hypothesis. This situation may occur after a period of bad weather or complicated procedures. Caution is set aside and, in clear weather, the pilot may land on the wrong runway.

And lastly, as a result of the effects of motor memory. If an action has been taken, such as operating a wrong switch but in a similar way to the correct one, then the occurrence of a false hypothesis may be anticipated. The operator will be convinced that he has done the right thing and will not easily be convinced otherwise. This might apply to pulling a wrong circuit breaker or even closing down the wrong engine or shutting off the wrong fuel valve.

One very fundamental characteristic of the information processing system requires mention. Although man has a vast capacity for sensing information, the decision-making stage of the process consists of just one single channel. In other words, although information may be being sensed from the approach lights, the altimeter, the airspeed indicator and ATC, the decision-making channel is being time-shared between the different inputs. This bottle-neck impedes the whole processing system (Broadbent, 1958; Poulton, 1971).

While one piece of information is being processed, the others are shunted temporarily into the notoriously unreliable short-term memory store to await available time in the single channel. When time becomes available, they are retrieved from storage and despatched on their way. Many factors can affect the efficiency of this storage and retrieval operation and, once again, a source of potential error is revealed.

While it is often felt that one can do more than one thing at a time, this subjective feeling results from the rapid scanning and time-sharing which takes place between one piece of information and another.

Because of its limited capacity, the decision-making channel is also vulnerable to overload conditions. Coping with overload, usually involving load-shedding, may be poorly managed and result in important information being discarded. When stress is high, concentration on a single stimulus may occur, with other important inputs being ignored. In the visual sense this is called perceptual tunnelling.

Action

Following the decision, certain action (or inaction) will be initiated. Muscles will be directed, possibly to manipulate controls or generate spoken instructions. This is another stage where error is revealed. If the control has been badly designed from a Human Factors standpoint, the action taken by the operator may be wrong. If a control on a new aircraft is similar but moves in a different direction from that in a previous aircraft flown, then an error in operation may be anticipated. Such an error will have originated earlier in the processing system and is often predictable.

It is also possible to predict, as proposed in what has been called Murphy's Law, that if equipment is designed in such a way that it can be operated wrongly, then sooner or later, it will be. If identical plugs are attached to 110 volt and 220 volt electrical appliances, then sooner or later the 110 volt appliance will be plugged into a 220 volt power supply, if this is available.

Motor-memory, which was explained in the discussion on the false hypothesis, is also relevant to this action or output stage of the information processing system.

Feedback

Once action has been taken, a feedback mechanism gets to work, and this is essential for the efficient operation of such a system. Inadequate or inappropriate feedback can interfere with the system in such a way as to generate errors.

Eye witness errors

Eye witnesses can play a significant role in the reconstruction of accidents and incidents and so contribute to flight safety. Their role may be in the investigation of the event itself or in the litigation which today so often follows it. They may be crew members or witnesses on the ground. Yet their testimony can be notoriously unreliable and so it is worth examining briefly some of the origins of the errors and distortions which are associated with this unreliability.

An eye witness may feel an obligation to defend a colleague or organisation, which can result in economical, selective or flexible use of memory. The huge sums now involved in post-accident litigation – particularly in the USA – can introduce diverse pressures. However for the purpose of this text we shall restrict comment to more conventional interpretations of human error, though the distinction is not always easy to establish.

A practical way to assess the quality and accuracy of eye witness reports may be to examine the possible influence of perception, communication, retention and the circumstances of the event itself, in the light of the previous discussion of error sources (Rolfe *et al.*, 1984).

Perception

Stress and attention selectivity can have an effect on perception but expectation is generally of greater significance in distorting eye witness recollection of events. If a car passed by in torrential rain, it is quite possible that a witness later would recall with conviction in testimony that the driver's window was closed and the windshield wipers were operating, even if this were not the case. It would normally have been expected in such a situation.

The August 1987 Detroit MD-80 accident provides a practical aviation example of the phenomenon (NTSB-AAR-88/05). The aircraft took off at dusk wrongly configured, with flaps and slats retracted. This was clearly established by the Digital Flight Data Recorder, the flight path reconstruction, the Cockpit Voice Recorder and a significant part of the wreckage examined. However, a First Officer of the same company, sitting in a parked aircraft on a taxyway about halfway along the take-off runway, testified that he specifically saw the flaps and slats properly positioned as the aircraft passed him. He also added that if they had not been extended the take-off would have been aborted following a take-off configuration warning, thus demonstrating the presence of a major influencing factor – expectation. This warning was, in fact, inoperative at the time. This First Officer as well as his captain said they were fairly certain that the gear was raised after lift off. In fact, the gear had not been raised, though expectation ruled that this would normally have occurred. Another First Officer also testified that she actually saw the flaps and slats being extended before the aircraft started to taxy, at about the time this might have been expected to occur. Yet definitive evidence demonstrated that this could not have been so.

Communication factors

Suggestion can be an additional source of error in eye witness reports. This does not have to be intentional, though even subtle implication and the sequence of questions can introduce errors in the recall of events. Leading questions and a desire to please the questioner (see also Chapter 10, page 234) may also influence the answers given. Questioning of a witness is a very skilled task and it is unfortunate that the investigation of routine incidents in air operations is normally conducted by staff with no such specialised knowledge or training. For example, 'how fast was the aircraft travelling when it smashed into the loading ramp' may elicit a different response from 'what was the speed of the aircraft when it came into contact with the loading ramp'.

Discussions between a witness and other people – particularly interested parties – before the report is made and recorded must raise concern as to the reliability of the report. The eye witness in the 1987 Detroit accident, quoted above, had such discussions. The US Airline Pilots Association (ALPA) policy on the sequestration of pilots after an accident is relevant here. The September 1989 La Guardia Boeing 737 accident (NTSB, 1989) created a stir in the media after the pilots were taken into ALPA sequestration after the accident and were not immediately available for questioning.

ALPA feels that the statements taken from a pilot immediately after an accident are often imprecise as a result of the trauma of the event and may not convey the pilot's actual intent. The Association therefore has a policy of isolating accident pilots from the official investigators for a period after the accident. During this period, the pilots are questioned about the accident by the ALPA representative, allowed to recover from the trauma, 'collect their wits about them and to accurately reconstruct their actions and the cockpit events'. As an example of the kind of discussion that may occur, the ALPA representative may say during sequestration to the pilot 'Well, is that really what happened, or did this happen?' (Martinez *et al.*, 1990).

To facilitate this protective process, ALPA has available to members a 24-hour 'hot line' for use in the event of an accident and aims to get an attorney to the scene of the accident and to the pilots immediately. In the UK, too, the British Airline Pilots Association (BALPA) arranges for a legal representative to be present at the time of the interview. With effect from December 1989, the effective time of sequestration in the USA was somewhat limited, as the pilot was required to submit to a drug test within 32 hours of the accident (FAR 121.457). A drug test taken so long after an accident is of limited value, but this is another matter. The NTSB requires that the rights of the crew members must be made known to them but that the interview should not be unnecessarily delayed (NTSB, 1983). In the USA and the UK – two countries with advanced accident investigation methods – there is no specific limitation on the time which may be allowed to pass between the accident and the interview.

ALPA, BALPA and others are genuinely concerned that pilots who have fully cooperated with accident investigators may have to face criminal charges based on the testimony which they have given voluntarily. This has occurred in several countries, including Egypt, Greece, Japan, Switzerland, the former USSR and Venezuela (Donoghue, 1989). In some cases it could be argued that the 'crime' was no more than an error of judgement but perceived differently for various reasons by local authorities. This threat to the participating eye witness does not encourage full and accurate recall of the event and is a factor in union protective measures.

There may be some conflict between the need to protect witness testimony from external influences and other biases and the requirement to protect the rights of the pilots concerned. Such a potential conflict, in a slightly different context, appears to have been resolved in confidential reporting systems such as ASRS and CHIRP (see Chapter 1, page 18). It has not yet been resolved in accident investigation.

Retention
Retention of the observed phenomena in the mind of the witness is also subject to influence. There may be a natural decay in memory, but to counterbalance this there is also a tendency to fill in gaps, particularly under unskilled or subjective questioning.

Repetition or restatement by an eye witness of his recollection of an incident will increase the certainty with which that recollection is held. A statement made in public or in officially recorded testimony is likely to increase future commitment to that position and increase self-confidence in it. This increasing confidence may quite mistakenly be seen by investigators or jurors as reflecting the quality or accuracy of the report. It may result from the more positive fixing of the event in memory or, more simply, from the embarrassment of later contradiction or weakening of the original position taken. There may be no correlation between confidence and accuracy (Rolfe *et al.*, 1984).

The nature of the event

The nature of the event itself can influence eye witness recall. It was perhaps significant that although the First Officer who witnessed the 1987 Detroit MD-80 take-off accident, noted above, claimed he could recall details of the accident aircraft well before the crash had occurred, he could recall nothing of other aircraft which had taken off immediately beforehand. Yet he was in a stationary aircraft and had not changed location. He could recall that the flaps/slats were set – of direct relevance to the crash – but could not recall whether the landing lights were on. The trauma of seeing an aircraft explode in flames may reduce the reliability of the subsequent recall, generating possible distortions, additions or subtractions.

The duration of an event and the ambient conditions – noise, light, visibility, etc. – may also affect the accuracy of recall.

Report evaluation

In assessing the reports of eye witnesses, several precautions must be considered and these are listed in Appendix 1.2A.

In the regular 33-day College of Aeronautics (Cranfield, England) accident investigation course $1\frac{1}{2}$ days are spent on witness interviewing and statement evaluation. The subject also plays a major role in the 5-day exercise during the course.

Authoritative texts are available on eye witness testimony to assist those who may be called upon to collect such information (e.g. Baddeley, 1982; Cohen, 1989; Loftus, 1979).

The classification of errors

The need to classify

In order to describe a person we may say that he is a man of distinction, well-dressed, tall and honest. Similarly we may describe an error as being one of substitution, system-induced, random and reversible. By classifying people we are better able to identify them. Similarly with errors. There are many ways of doing this. Classification may be in terms of source, as already discussed, cause, consequences or general nature. We will here simply review a few of the most commonly used classification terms.

Design-induced and operator-induced

An error which occurs at the *L–H* or *L–S* interface may result from a failure to design the *Hardware* or the *Software* properly taking into account the normal characteristics of the *Liveware* or the operator. This is often called a design-induced error as distinct from operator-induced.

As an example, there have been historically countless cases of confusion in handling the flaps and the gear controls on DC3 aircraft as they are in close proximity to each other and of comparable shape. Similar confusion between controls continues to be reported from time to time in the ASRS and CHIRP confidential reporting systems. Further examples of control and display designs which can induce errors are discussed in Chapter 11.

An operator-induced error can be attributed directly to inadequate performance on the part of an individual reflecting, perhaps, deficient skill, motivation or vision. An error may be to some degree both design- and operator-induced.

Sometimes a distinction is made between a design-induced error, referring to the *Hardware*, and a systems-induced error, referring to the *Software*.

Rather similar to these forms of classification is one which refers to situation-caused errors and contrasts these with what are described as human-caused errors. Some studies have pursued this grouping; one has reported two to four times as many situation- as human-caused errors (Kragt, 1978).

Random, systematic and sporadic

Another form of classification can be illustrated in the way rifle shots are distributed on a target (Fig. 2.5). When the shots appear to be scattered at random across

Random error   Systematic error   Sporadic error

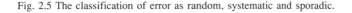

Fig. 2.5 The classification of error as random, systematic and sporadic.

the target without any discernible pattern, this is called random error. Many factors may influence the range of this variability. A second kind of error is characterised by a small dispersion which is offset from the desired point. This is called systematic error and is often caused by only one or two factors. A third type is called sporadic error and this is said to exist when, after a routinely good performance, an isolated error occurs. Sporadic errors are very difficult to predict.

If we relate these three error types to the flying situation, we could say that the pilot whose landing touch-down point varies without a recognisable pattern is committing random errors. The one who consistently undershoots is demonstrating a systematic error. While the one who normally lands the aircraft accurately, but then inexplicably makes a rare undershoot, is experiencing a sporadic error. The same kind of error classification can, of course, be applied to the performance of other tasks, such as the stewardess making coffee or the mechanic carrying out maintenance work.

Omission, commission and substitution

One of the most common errors recognised is failing to do something which ought to be done such as missing an item on a checklist. This is called an error of omission and introduces another way errors can be classified. An error of commission is doing something which ought not to be done, such as calling passengers to board the aircraft on the scheduled departure time when the flight has a one-hour technical delay. A third kind of error in this form of grouping is the error of substitution, which is taking action when it is required, but the wrong action. This kind of error has led to disaster on a number of occasions when a pilot closed down the wrong engine after an in-flight engine failure.

Reversible and irreversible

In terms of safety, the classification of errors as reversible and irreversible is a useful taxonomy when discussing strategies to meet the challenge of human error. A mountaineer climbing without a safety rope is likely to commit an irreversible error if he slips. With a safety rope, the same error is likely to become a reversible error. A pilot who miscalculates the amount of fuel to jettison in an emergency situation has made a reversible error if the calculation is to be rechecked before action is taken. But the actual dumping of too much fuel is an irreversible error.

Meeting the challenge of human error

Allocation of tasks between man and machine

Perhaps the first step to take in meeting the challenge of human error is to determine whether we need to meet it at all. One possible way to handle the problem of human error in a particular task is to take the task away from man altogether and give it to the machine or the computer. While there will still be scope for human error in controlling the machine or programming the computer, the execution of the original task will be free from the effects of human fallibility.

As a question of general principle in any system design, man should be given the tasks which he does best, and similarly the machine. While this distribution is not yet entirely resolved, a great deal is known about the relative characteristics of man and machine and much has been published on the subject since the so-called Fitts Lists (e.g. Fitts, 1951; Singleton, 1974; McCormick *et al.*, 1983). A few items

Table 2.1 Some characteristic differences related to the performance of various kinds of activity, which may provide guidance in the allocation of tasks between man and machine.

Property	Man	Machine
Monitoring		✓
Overload breakdown		
(gradual)	✓	
(sudden)		✓
Reasoning		
(deductive)		✓
(inductive)	✓	
Speed		✓
Power		✓
Consistency		✓
Complex activity		✓
Short-term memory		✓
Computation		✓
Error correction	✓	
Intelligence	✓	

illustrating characteristics which can provide a basis for task allocation are listed in Table 2.1. It has long been recognised that man's performance of tasks requiring him to monitor or detect brief, low intensity and infrequently occurring events over long periods is poor. In 1943, the Royal Air Force (RAF) asked if laboratory tests could be done to determine the optimum watch length for radar operators on anti-submarine patrol, as it was believed that targets were being missed. Mackworth later devised a test to simulate the task of a radar operator and this represented the first attempt to explain the phenomenon which has become known as the vigilance decrement or the vigilance effect (Mackworth, 1950). The RAF had found in Coastal Command that a marked deterioration in efficiency occurred after about 30 minutes and this has been confirmed many times since (Fig. 2.6). Much of the

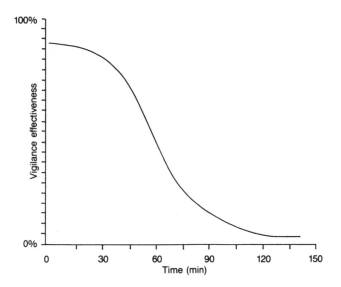

Fig. 2.6 An illustration of the kind of vigilance effect which can be expected in the performance of passive tasks with low signal rate. This shows a notable decline in performance after about 30 minutes.

research on the vigilance decrement has been done in laboratory studies. These have not always been representative of the operational environment, where many factors, not least, motivation, play a role, so that many such studies may have little operational application (Smith *et al.*, 1969). However, there have also been many industrial studies, particularly in Japan, and these have generally revealed such a decrement, although it has been shown to be task-dependent, and is subject to individual differences (O'Hanlon *et al.*, 1977). Practice does not seem to be effective in eliminating the decrement. It is necessary to recognise the existence of this potentially dangerous phenomenon in the design of equipment and procedures; it is relevant to the *L–H* and *L–S* interfaces in the *SHEL* model which was discussed in the previous chapter. Work on the vigilance decrement has been well surveyed and analysed (Davies *et al.*, 1981).

Man's short-term memory is also very susceptible to error and the transfer of some of this activity to data-link systems is likely to enhance overall system reliability.

49

Routine and repetitive operations create boredom with a loss of motivation and increased errors. The machine has no such problems with this kind of task. Man has difficulty in assimilating large amounts of information; the computer takes this in its stride.

On the other hand, machines are less adaptable than man. When they break down they usually do so abruptly and totally, while man's performance tends to become erratic and breaks down in lapses.

Man's performance and error rate are influenced by motivation, which is discussed in more detail in Chapter 6. The performance of the machine is free from motivational influences but, on the other hand, it does not really care whether it gets home safely. When tasks are allocated to man, then, particular attention must be paid to maintaining motivation at the highest possible level.

The two-pronged attack

Having established that man has a task to fulfil and having acquired some understanding of the nature of human error, the stage is now set to examine how this challenge to safety and efficiency can best be met. This involves a two-pronged attack.

Firstly, and as the initial prong of the attack, it is necessary to minimise the occurrence of errors. One way to tackle this is by ensuring a high level of staff competence through optimising selection, training and checking. It is important to recall here that a test or check establishes only how a person performs under the test or check conditions and does not say too much about how he will perform when not under supervision and when working in a non-test environment. Personality, attitudes and motivation play a vitally important role here. In tightening the selection process and increasing training and checking, there are clear economic limits. A point is reached where diminishing returns make further investment in this direction impractical.

It would obviously be an attractive solution if we could eliminate human error altogether but, as Cicero pointed out in his wisdom, errors are a normal part of human behaviour. Their total removal is an unrealistic dream. However restrictive is the staff selection process and however much is invested in training and checking, there will still be residual errors. The second prong of the attack, then, is to reduce the consequences of these remaining errors. In safety, we are primarily concerned with the consequences of error rather than the error itself, so this strategy is of the utmost relevance.

Throughout aviation history the emphasis has been on the refinement of man with the objective that errors will be totally avoided. It has been a commonly held philosophy that a person carrying as much responsibility as an airline pilot should not make mistakes. Even errors of judgement, as distinct from discrete errors involving incorrect procedures, usually suffer severe penalties from the authorities. That is, if the pilot should be lucky enough to survive. This is in contrast to attitudes elsewhere, such as in the medical profession, which is more tolerant of human error amongst its members than is the aviation community. A medical authority has said that 'physicians and surgeons often flinch from even identifying error in clinical practice, let alone recording it, presumably because they themselves hold ... that error arises either from their or their colleagues' ignorance or ineptitude' (Gorowitz *et al.*, 1976). When a protective blanket enshrouds human error, programmes to meet their challenge as applied in aviation and outlined in this chapter are effectively blocked. Such programmes cannot function without a recognition that errors are a part of normal human behaviour and that to conceal them is to encourage their repetition and all the consequences which follow.

The illusion that it is possible to achieve error-free operation has been a convenient concept in simplifying accident investigations and the allocation of blame but has tended to discourage proper attention to the second prong of the attack – living with errors but reducing their consequences.

From the discussion earlier in this chapter on the sources and classification of errors, many of the steps which should be taken to control them will be self-evident. Nevertheless, it may be worthwhile to review these steps referring to the sources and classification terms already quoted and within the general framework of the two-pronged attack.

Minimising the occurrence of errors

The *SHEL* model utilised earlier to locate the source of errors may also be useful in determining action which can be taken for their avoidance.

The quality and condition of the *L* component in the centre of the model will influence the performance of that component and the frequency of errors which can be expected. Tolerance to fatigue and other stress effects, various personality characteristics and the possession of skill, all influence liability to error. But, after a certain degree of refinement, we have to work with people as they are rather than as we would like them to be. Interfacing of the components of the system with man as he is, with all his rough edges, is a vital step in minimising the occurrence of errors.

The design of controls and displays is one aspect of the *L–H* interface. They must be so designed that they match human characteristics. Various forms of coding should be used for controls and their movement should meet human expectations. Displays must not only provide information but do this in such a way as to facilitate the human information processing task. While this seems like simply stating the obvious, it requires a knowledge of human behaviour and the way people process information, make decisions and act upon them.

Errors can also be minimised by improving the interface between the *L–S* components. This involves the non-physical aspects of the system such as the design of checklists, procedures, indexing of manuals, the design of maps and charts, airport and route guides. Strict adherence to SOPs reduces errors. However, SOPs can never be established for every possible condition and some flexibility must remain. The apparent contradiction between strict adherence to SOPs and the need for flexibility represents a dilemma which requires very careful consideration and skilful application.

Much progress has already been made in optimising the *L–E* interface through improved control of noise, vibration and temperature, thus reducing errors from these stressful factors. Many studies have been made on the effects of heat and cold on mental and physical performance and several reviews of these have been made (e.g. McCormick *et al.*, 1983). Incidentally, heat has a more adverse effect on mental performance than cold.

The last interface in the *SHEL* model was that between people, the *L–L* interface.

Concern, in virtually all aspects of operational aviation, has recognised that insufficient attention has been given to the interaction between staff in activities where they work as a team. In flight crew activities, the system of redundancy is designed to reduce the incidence and consequence of errors, but this has not been sufficiently effective. The *L–L* interface also encompasses leadership and command and it can be expected that weaknesses here will lead to misunderstandings and errors. Airlines are beginning to orientate training programmes towards enhancing more cooperation and communication between crew members and one such programme is briefly described in a later paragraph.

And so, as an important measure in the first of the three strategies to meet the challenge of human error, we can say that there must be an optimum matching of the components of the system with the characteristics of man. This is a concept at the very heart of ergonomics.

From Fig. 2.4 it was apparent that performance and thus errors vary with the level of arousal. The optimum level of arousal depends upon the nature of the task to be performed. A complex task does not require such a high level of arousal as a simple task – the Inverted-U curve is further to the right for the simple task. Boredom leads to a very low arousal level, when errors can be expected to increase. Aircraft design and operating procedures must aim to establish an optimum level of arousal throughout the different phases of flight. Vigilance and monitoring tasks are particularly sensitive to arousal level.

Training for a particular task will reduce the incidence of errors, though it will not totally eliminate them. Overlearning, which is discussed in Chapter 9, has a notable effect in making a skill more resistant to stress and error.

It is worth noting here that in attempting to reduce human error, that is, to modify human behaviour, on a long-term basis, exhortation alone is of little value. Simply issuing a bulletin asking that certain errors be avoided, or demanding that staff 'smarten up', is not likely to be productive unless it is accompanied by other appropriate measures as outlined here. In any case, such an approach seeks only to modify behaviour and not the situation in which the error occurred and which may have induced it.

Random errors can reflect a wide variety of origins. The scatter or range of variability can also be influenced by several factors. The most important of these must be determined and control over them tightened up. Personnel selection, training, checking and supervision may all play a role in this process. Systematic errors, usually caused by only one or two factors, are normally easier to correct once a proper analysis has been made. A good feedback to the operator will help to reduce this kind of error. Sporadic errors are usually difficult to predict or reduce as they do not occur frequently and often seem to be unrelated to other known factors in the work situation. They are usually resistant to correction through training or indoctrination. To determine the origin may require a profound analysis of all the circumstances of the case.

Self-pacing of a task, that is, doing it at one's own speed, contributes to error-free operation. This is particularly true when a person is suffering from fatigue or sleep disturbance. However, most flying tasks cannot be self-paced. The aircraft speed is largely pre-determined and this fixes a time-frame for the accomplishment of many tasks from navigation to the serving of meals. A proper workload distribution will enhance, though not guarantee, the possibility of self-pacing.

Reducing the consequences of errors

Having reduced error as far as is reasonable and practical, we must now determine how to live with the errors which, inevitably, remain. This is the second prong of the attack and the one which has been most neglected, partly as a result of the illusory notion of error-free operation.

In the field of equipment design we can make errors reversible. This concept is already being applied to visual display units (VDUs) of computer and navigation systems where the selected input first appears on a 'scratch pad' on the screen. It is then checked before being inserted. This might be called a 'select, check, insert' system which, while not lowering the original error rate, reduces the consequences of the error. The interlocking system employed in landing gear levers to prevent inadvertent gear raising on the ground does not affect the original error committed, but ensures that no undesirable consequence arises from it and permits the faulty action to be reversed. Wherever possible, the design of systems in which errors can have serious consequences should be such as to allow initial actions to be reversed.

A highly effective way to prevent the consequences of many errors from developing is through efficient cross-monitoring. Redundancy in flight crew and other critical team activity is provided to permit cross-monitoring; without it, the normal error rates discussed earlier in this chapter would never be acceptable. There is accumulating evidence, however, to suggest that this redundancy is not providing the expected protection. Increased attention must be given to this problem in the preparation of training programmes and operational procedures as well as in conceptual aspects of design. One factor influencing the effectiveness of redundancy is the 'trans-cockpit authority gradient' discussed earlier in this chapter. Most current CRM programmes deal with both cross-monitoring and the 'trans-cockpit authority gradient' with varying effectiveness. As would be expected, the problems appear to vary in their severity from one company to another and between countries.

When one crew member monitors and has reason to question the data used, the decision or performance of another, the effectivity of the monitoring depends on the response which it generates. The validity of such questioning must always be tested before it is rejected. In 1976, a 727 crashed at Ketchikan after it had approached too high and too fast and landed too far down the runway. The first officer had advised of both the excessive speed and height, but the captain failed to react effectively. In its report, the NTSB called for crew members to be more outspoken when they believe the flight is being conducted in an unsafe manner (NTSB-AAR-76-20). At Tenerife in 1977, the flight engineer expressed some doubts as to whether the runway was clear, but the captain brushed them aside (Spanish accident report). The unassertive copilot's concern in the 1982 Washington 737 accident mentioned earlier, also failed to generate a response from the captain (NTSB-AAR-82-8).

These accidents also raise the question of the role of personality and attitudes in crew cooperation and this is discussed in later chapters. Leaders in all walks of life have a tendency to reject questioning by subordinates. This enables them to continue with the performance of their task, as they perceive it to be, without disturbance or delay. But in flying, this is a luxury which is dangerous and cannot be tolerated.

Apart from people monitoring each other to limit errors or their consequences, equipment can also monitor human performance, as in the Ground Proximity Warning and the Altitude Alerting systems. There has been a dramatic reduction in altitude violations since the human limitation in vigilance and monitoring tasks was recognised and automatic altitude alerting and capture were introduced. But experience shows that when such protection is temporarily removed, the operation becomes particularly vulnerable to error (ASRS Callback No. 86). Human monitoring of aircraft and systems status has also been abandoned in many cases and passed over to the machine with the aid of electronics. Disagreement and asymmetry warnings are far more effective and free from error when discrepancies are detected technically and automatically.

Nevertheless, in spite of the assumption by the machine of many monitoring tasks once performed by man, his role as a monitor is still vital in meeting the challenge to safety and efficiency presented by human error.

One final remark should be made on the subject of minimising the occurrence of errors and reducing their consequences. As an overall philosophy, it is wise to use good judgement to avoid whenever possible situations in which superior skills must be applied to ensure safety.

Industry programmes

In some industries 'Error-Cause-Removal' (ECR) programmes have been introduced. These are particularly useful in identifying and eliminating situation-caused errors. These programmes emphasise prevention rather than cure and involve setting up an ECR team which consists largely of those working on the job. The teams have regular meetings and try to determine work situations which they believe will lead to errors and accidents. Proposals are then made to management and specialists are brought in to provide expertise in developing solutions (Swain, 1974).

Another approach is called a 'Zero-Defect-Programme' (ZDP). Such programmes are based on the assumption that it is possible to achieve error-free activity if a person has been adequately trained and his motivation is high enough. This type of programme is conceptually weak for two reasons. Firstly, it is difficult, if not impossible, to maintain motivation continuously at a very high level, however good the original training and incentives. And secondly, it tends to ignore the fact that many errors are situation- or design-induced and these are less likely to yield to motivational factors on the part of the operator himself. The result of such a programme can be a decrease in the errors reported by the employees, but not in the errors actually taking place.

Aircrew training and proficiency testing has traditionally been done on an individual basis. If each individual crew member were proficient, then it was assumed that the team consisting of these individuals would also be proficient and effective. This has not always proven to be the case, as was mentioned earlier. Some airlines many years ago introduced the concept of training and periodic checking as a crew as well as an individual. This is a central concept in all CRM programmes. Such programmes are encouraged by ICAO and several airlines and regulatory authorities. They are coming very close to becoming a regulatory requirement in some States.

By about 1975, Northwest Airlines (with FAA approval), and later NASA, began to formalise 'crew' as opposed to 'individual' training (Nunn, 1981). This led to a change in Federal Aviation Regulations (FARs) in the USA with respect to crew proficiency training. This type of training programme utilises a flight simulator and a highly structured scenario to reflect a total line operational environment. It is called 'Line-Oriented Flight Training' (LOFT). This is a training activity in which errors are allowed to occur as they would do on a real flight. The *L–L* interface which was discussed earlier is exposed in full measure as the crew work as a team to resolve problems with which they are presented.

The training is intended to improve the effectiveness of flight deck resource management and an important part of this is the management of human error. This is done by recognising the kind of stress situation in which error can be expected and learning how to reduce the probability of the error occurring and to limit its consequences if it does occur. This kind of training is only effective if errors are allowed to occur without penalising the trainee. In principle, the instructor in such sessions should remain unobtrusive and must not become involved in the operational situation. He must not instruct. An important element in the LOFT programme is the analysis, including self-analysis, which takes place in the debriefing, where the instructor acts as a moderator rather than a teacher (Lauber *et al.*, 1983).

Such programmes as LOFT can use video-taping of the training session, though this is not essential. Video-taping enables a more effective process of self-analysis to take place, as discrete as well as judgement errors are presented visually to the 'actors' themselves. This enhances motivation.

Living safely with human error

A prerequisite for the design of equipment and procedures must be that they should accommodate human error if safety is to be assured. This may involve the adoption or modification of attitudes by many in the aviation industry and governing bodies. Accepting the inevitability of error may be found difficult for some, particularly in the field of equipment design. Attitudes towards the relative merits of positive and negative reinforcement of behaviour and to motivation may require review and this is discussed in more detail in Chapter 6. Attitudes towards the main objective of accident and incident investigation may require adjustment, particularly in relation to questions of blame and cause, punishment and prevention.

It will involve looking more deeply into why the error occurred rather than accepting superficial descriptions which indicate only where in the system the error was located. It will certainly involve a better understanding of the nature and the mechanisms of human error than has generally been demonstrated in the past. And it will require recognition that meeting the challenge presented by human error is not an unskilled, Do-It-Yourself (DIY) activity but requires the utilisation of appropriate expertise.

In all the chapters which follow, background information will be presented which will aid in the formulation of an effective and constructive approach to the management of human error.

3

Fatigue, Body Rhythms and Sleep

Jet lag

The expression 'jet lag' is now well established in the English language as the description applied to the lack of well-being experienced after long-distance air travel. This terminology is unfortunate because the problem has nothing directly to do with jets and the lag description is, at best, debatable.

What is not debatable, however, is that most travellers take some days to recover from a long flight and during this time their performance, motivation, mood and behaviour may be expected to suffer to some degree. Sleep disturbance is experienced; bowel elimination and eating habits may be disrupted. Lassitude, anxiety, irritability and depression are often reported. Objectively, there is evidence of slowed reaction and decision-making times, defective memory for recent events, errors in computations and, highly significant in a critical operation such as flying aircraft, a tendency to accept lower standards of operational performance.

For the vacationer, recovery may simply mean taking the first couple of days relaxing quietly beside a hotel swimming pool. For the businessman, the condition could involve reduced performance in contractual negotiations, with financial losses as a result. For the flight crew member, disturbed biological rhythms may continue throughout a 30-year career and involve a public safety dimension, which cannot lightly be dismissed, and possibly a personal health risk.

The problem of jet lag also encompasses sleep disturbance and deprivation and these, too, have a different perspective when seen from the points of view of crew and passengers. If a passenger drops off to sleep during flight, the worst that can happen is likely to be that he misses a meal or the in-flight movie. In the case of flight deck crew the implications are more serious. Amongst crew members the possibility that one or more of them can occasionally drop off to sleep on long-range night flights has long been known. Cases are sometimes revealed where all on the flight deck have been asleep at the same time, with nobody awake to 'mind the shop' (e.g. CHIRP Feedback Nos 1, 2 & 9).

Various methods are used by crews to handle the problem. Fortunately, not many use that developed by an imaginative copilot in a large national airline. He uses a kitchen timer which he sets to the expected time of the next navigation reporting point, perhaps 30 minutes ahead. With the aircraft controlled by the autopilot, he then pushes his seat back and sleeps until the buzzer awakens him. He then makes his report to ATC, resets the timer and again slips off into the Land of Nod. A more imaginative idea is the German device consisting of a special kind of spectacle frame that can detect whether the eyes are closed and then give an alert. Another such alerting technique in the UK uses Galvanic Skin Response (GSR), or skin resistance, to detect low arousal. High skin resistance has a good correlation with sleepiness. Both devices have been under development for a considerable time. Perhaps the frequent and sometimes excessive use of coffee is the most common method used on board to stay awake and reference is made to this later in the chapter.

There have been many reasons why it was not palatable to recognise the existence of this crew performance problem. The elevated public relations image of the airline pilot is not supported by such ordinary human behaviour. Recognition may also have provided further ammunition in crew union bargaining, a prospect which would surely not be welcomed by most airlines which view such bargaining as having the sole objective of achieving more pay for less work. But with the confidential incident reporting systems now being operated on a national basis in several countries, recognition can no longer be withheld. Such information tends to leak to the media and attract public interest.

There is no doubt that cultural factors and airline policies play a role in the occurrence of unscheduled crew sleep at duty stations on board. Nevertheless, it remains a universal symptom of the overall situation associated with irregular work schedules and jet lag. Before considering effective measures to meet the difficulties associated with jet lag it is first necessary to understand some of the basic elements involved.

Fatigue

Recognition of the problem

The definition of fatigue is a source of difficulty which has tended to generate confusion and retard progress. Four interpretations of fatigue may be considered. Firstly, it may reflect inadequate rest, as covered extensively in this chapter. It may refer to symptoms associated with disturbed or displaced biological rhythms which are often described by the sufferer simply as jet lag. Thirdly, it may be interpreted as excessive muscular or physical activity – perhaps working too hard and long in the garden or skiing all day in deep snow. And finally, it could result when minimal physical activity has occurred but, as in preparation for an examination, excessive cognitive work has been undertaken. Sometimes more than one of these sources may be relevant to a particular condition described as fatigue. One must try to avoid using the term fatigue in a too generalised way when discussing the condition, its origins, effects and remedial action.

A major research programme by NASA on fatigue was initiated by the US Congress in 1980. This was aimed at assessing the psychophysiological effects on pilot performance and the operational significance for flight safety and efficiency of flying various types of flight and duty cycles. The NASA-Ames Research Center extended the original project, reported in 1985 and 1986, in view of the interest generated by the initial findings and their significance for flight safety (Foushee *et al.*, 1986; Gander *et al.*, 1985, 1986).

The long-distance traveller will probably say after his journey that he feels fatigued. In a survey in the USA, 93% of the pilots responding reported that fatigue was a problem and 85% said they had felt 'extremely tired or washed out during the previous 30 days' (Dexter, 1975).

In an analysis of data obtained in the USA from the confidential ASRS programme, NASA reported in 1981 that 'fatigue-associated decrements (in performance) resulted in substantive potentially unsafe aviation conditions. Considerations of duty and sleep are the major factors in the reported conditions' (Lyman *et al.*, 1981). The British CHIRP system announced in 1984 that the largest number of reports received concerned fatigue, sleep and the way in which work patterns are constructed (CHIRP Feedback Nos 6 & 9). With the background of these reports, an administrator of the CHIRP programme has concluded that the importance of disrupted sleep as a causative factor in accidents may have been underestimated (Green, 1985).

A far more dramatic demonstration of the possible effects of fatigue appeared during the investigation of the Challenger space shuttle disaster in 1986. Amongst a wide range of Human Factors deficiencies considered were the possible effects of sleep loss, excessive duty shifts, circadian or daily rhythm effects and the resulting fatigue on the decision to launch the shuttle in spite of concern about its safety (Presidential Commission, 1986).

The difficulty faced in trying to apply an enlightened approach to the problem of fatigue amongst crews is illustrated in the public comment of a senior executive of a leading British independent airline in response to such reports. He said that in crew scheduling, pilots should expect to 'take the rough with the smooth'. He felt that 'safety is not the real issue' and that 'any appreciation of crew workload must consider a period of at least 28 days' (CHIRP Feedback No. 6).

This policy, while recognising the existence of long-term cumulative and chronic fatigue, fails to recognise that short-term fatigue exists and can affect performance and thus safety. It further rejects too easily the sense of responsibility of airline pilots who report the adverse effect of fatigue they experience.

An airline pilot expressing the incredulity of many reading this official airline philosophy on fatigue could not resist subsequently going into print in response. 'By the same token', he wrote, 'if the senior executive breaks his leg, is he able to convince himself that he feels no pain having averaged it over the other 27 days when he didn't break his leg?' (CHIRP Feedback No. 7).

The facility now provided through confidential reporting schemes to bring forward airline and pilot views on such questions makes a useful contribution to understanding and thus progress in safety related to fatigue. Amongst the reports, however, there are occasionally depressing signs of ignorance and complacency, making the need for more formal education in Human Factors painfully evident.

Low humidity and noise

There are numerous factors which contribute to the subjective feeling of fatigue and one of these is low relative humidity (RH). A range of 40–60% is often seen to be an agreeable room RH (Grandjean, 1973), but on long-range flights it may fall to as low as 3%. The low cabin RH results from the low water content of the atmosphere at the altitude at which jet aircraft fly most economically. Discomfort of the air passages of the nose and throat is experienced and there may be an increase in risk of infection. Humidifiers can be installed on the aircraft but these incur certain economic penalties which make their installation the exception rather than the rule in civil aircraft. This question of low RH is discussed in more detail in Chapter 13 in connection with the cabin environment.

Noise is another contributor to fatigue and this is also discussed together with other environmental factors in Chapter 13.

Fatiguing procedures

Passengers and crew are both subject to fatigue resulting from pre-flight preparations. This is sometimes aggravated, too, by excessively long delays performing customs and immigration formalities at the end of a flight. Such officials, routinely performing within their eight-hour working day, seem to take little or no account of the fact that the crew may not have seen beds for approaching 24 hours and the passengers possibly much longer. Crew check-in times are usually in the order of one to one and a half hours before departure, but passengers are less fortunate. While some stations require only 20 minutes to handle an international passenger check-in (e.g. Manchester, Düsseldorf), others may possibly require as much as two hours of frustrating and fatiguing queuing, checking, security procedures, waiting and sometimes arguing in the face of over-booking (e.g. Bogota, Damascus, Delhi, Jeddah, Maracaibo, Tel Aviv). Clearly, there is scope for a substantial reduction in fatigue by improving passenger handling procedures at many airports.

But low RH, noise, aggravating and inefficient ground handling procedures, and some might add hypoxia and ozone, are not the primary causes of the feelings of fatigue or lack of well-being experienced by long-distance air travellers. To understand this phenomenon we need to understand something of the body's normal patterns of sleeping and waking and the rhythms which apply to most of the biological systems within the body.

Body rhythms

Chronobiology

We live in a universe of rhythms. Our galaxy rotates once every 200 years. Sun spots occur every 11 years and tides every $12\frac{1}{2}$ hours. But the most significant rhythm for the inhabitants of this planet is the earth's rotation once every 24 hours.

It has long been known that most biological processes are rhythmic or cyclic in character. That is, they consist of sequences of events which repeat themselves at regular intervals. The sequence may be simple or extremely complex and the time span or period of a particular cycle may be anything from a fraction of a second to many years. Such rhythms occur in all forms of life and man is no exception.

The first scientific reports concerning biological oscillations appeared in the literature more than 200 years ago. Since 1950 there has been an acceleration in research in this field which has resulted in the creation of a new sub-discipline, chronobiology. An International Society of Chronobiology serves as a focal point of such scientific activity and there is an international journal of chronobiology.

New terms, such as chronohygiene, chronotolerance, chronotoxicity, chrono-pharmacology and chronotherapeutics have now entered the language.

Sodium and potassium excretion, amino acids, cortisol and other hormone levels and blood characteristics, for example, all oscillate during the 24 hours to a rhythm. Even the sensation of pain varies at different times of the day. The toxicity of drugs varies according to the time of day when they are administered. As an example, an antihistamine may last 15–17 hours when taken at seven in the morning, but only 6–8 hours when taken at seven in the evening. Even the effect of aspirin shows a circadian rhythm.

The highest point in the rhythm curve for each of the body's systems is called the acrophase. The time when this occurs will not only vary between systems but differences may also be found between ethnically and geographically different populations, if a simple time-of-day criterion is used and different living patterns are ignored (Halberg, 1977).

The fact that the acrophase for the toxicity of a drug may be different from that for its therapeutic or healing effect is likely to have an increasing impact on medical treatment of diseases. By selecting the appropriate time of day for administering the drug, the maximal benefit can be obtained with the minimal toxic effect.

The search for the pacemakers or circadian controlling mechanisms in the body, showed little success until late in the 1970s. By then it had become clear that there was no single pacemaker but that control of a substantial part of the body's rhythms resided in the suprachiasmatic nuclei of the hypothalamus, deep inside the brain. When this part of the brain was removed in animals, the rhythms which tracked the rest/activity cycle ceased to oscillate. The source of control of other rhythms, such as body temperature, remained the object of further research (Moore-Ede *et al.*, 1983).

Rhythms and time cues

The most common of the body's rhythms is the circadian (*circa-dies*) or 24-hour rhythm, which can be related to the earth's rotation time. This periodicity is not quite true, however, because experiments carried out in temporal isolation, that is, where all time cues have been removed, have demonstrated that what is called the free-running cycle is, on average for humans, closer to 25 than 24 hours. This varies between individuals and is usually in the range of 24 to 27 hours.

In the real world, the cycle is maintained at 24 hours by what are called entraining agents or *zeitgebers*. The most powerful are light and darkness, but there are others such as meals and physical and social activity and these also have an influence on the oscillatory functioning of the body's systems. Sleep is also cyclic and while it can itself be seen as an oscillating system it also has an influence on other body systems. For example, the peak growth hormone concentration occurs about 60–90 minutes after sleep onset, but this phase can be shifted immediately with shifting sleep time. A sleep disorder known as Delayed Sleep Phase Insomnia results from limitations in a person's ability to accelerate his free-running cycle to keep it synchronised with the actual 24-hour day.

A typical example of a rhythmic system is oral temperature (Fig. 3.1). This temperature rises during the day to reach a peak during the evening, with extraverts and evening types tending to reach the peak a little later than introverts and morning types. It will be seen that sleep normally occurs when the temperature is falling

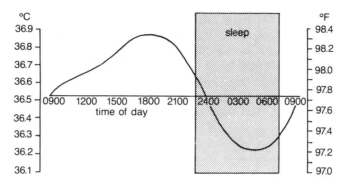

Fig. 3.1 The circadian rhythm of oral temperature.

and waking occurs when it is rising.

The 24-hour rhythm is the most commonly recognised of the body's rhythms, but there are others. The female 28-day menstrual cycle is one. A 90-minute cycle, called the ultradian cycle, has been detected and this is rather close to the hourly Basic Rest and Activity Cycle recognised in infants. It will also be seen in the pattern of sleep stages discussed later in this chapter.

Birthdate Biorhythm Theory

In passing, a brief diversion from reality should be made into what has been called the Birthdate Biorhythm Theory. This is based on the assumption that there are three long-term body rhythms, viz 23 days (physical), 28 days (emotional) and 33 days (intellectual). By plotting a person's own cycles, based on his date of birth, it is claimed that the low points in the cycles, said to coincide with accident proneness, can then be forecast and appropriate precautionary measures taken. The theory has been applied industrially in a number of countries, notably the USA and Japan, with some 'success' claimed. However, no causal relationship has been established and it cannot be concluded from these cases that the reduction of accidents occurring confirms the validity of the theory. On the contrary, statistical analysis of a very large number of aircraft accidents has failed to reveal any correlation between these three cycles and the timing of the accidents. Other extensive research has also failed to detect any validity in the theory and considerable literature is now available on the subject (e.g. Englund *et al.*, 1978; Khalil *et al.*, 1977; Wolcott *et al.*, 1977). While scientific evaluation of the theory is considered closed, commercial activity in the field is not.

Rhythm of performance

Since body temperature and other physiological processes exhibit such a striking circadian rhythm it might be expected that the brain, which is supported by at least some of these processes, will also demonstrate a rhythm which will be reflected in measured performance.

Numerous experimental programmes have now been carried out which show this to be the case (e.g. Higgins *et al.*, 1975; Klein *et al.*, 1972). The actual form of the curve is task-dependent, meaning that it will vary according to the task (Fig. 3.2). It is important to understand that this reduction in performance during certain parts of

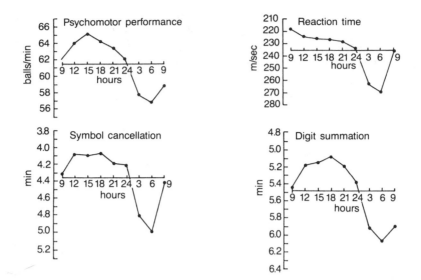

Fig. 3.2 Human performance of various tasks varies during the day with a rhythm that often tends to correspond with that of body temperature (Fig. 3.1). This variation is in addition to any effect from sleep deprivation (Klein *et al.*, 1972).

the 24 hours is not the result of sleep deprivation, which will be discussed shortly, but is an entirely separate performance-reducing factor. While this is a natural rhythm, this is not to say that it is immune from external influence. Practice will raise and flatten the curve, as will heightened motivation or increased effort.

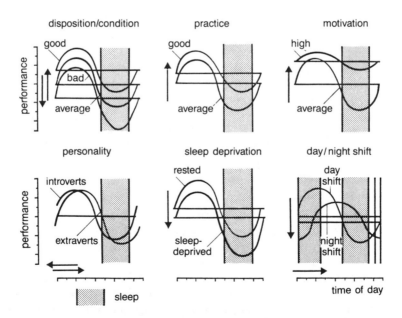

Fig. 3.3 Behavioural performance rhythms can be modified through various factors (Klein *et al.*, 1976).

Personality differences of introversion and extraversion will shift the curve to the left or the right (Fig. 3.3). The range of oscillation, that is, the difference between the maximum and the minimum performance scores within a cycle, is also task dependent. In tests with certain shiftworkers, it has been shown to be as high as 30–50% (Klein *et al.*, 1976). It tends to be greater with complex than with simple tasks.

The loss of performance arising from this natural rhythm may be greater than that arising from the loss of a single night's sleep. Any flight crew member can testify that on a long flight which encompasses the normal night's sleeping period, a task may be done more efficiently after the normal 'waking' day period begins than during the previous 'sleeping' night phase. This is experienced in spite of the sleep deficit which is accumulating.

One characteristic of the daily rhythm of performance that is not apparent from Fig. 3.1 is the existence of a post-lunch dip. This appears to occur regardless of whether or when lunch is actually taken (Colquhoun, 1971). This suggests a time of day when it might be preferable to avoid critical tasks which require optimum performance, if one has such a choice.

The significance of the performance circadian rhythm is now being increasingly recognised. Fig. 3.2 (page 62) was included as an appendix in the report on the 1978 Beaulieu Heath accident in the UK, which occurred in the early morning hours (AIB AAR 7/80). Even before this, in the report of the near-disaster at Nairobi in 1975, reference was made to the fact that the event took place at the time when the biochemical, physiological and psychological functions were at the lowest point of their circadian rhythm (AIB AAR 14/75). The report on the 1979 Three-Mile Island nuclear power station accident in the USA also cited performance rhythm as a contributory factor (Payne *et al.*, 1979). It was also considered in the 1986 Challenger shuttle disaster investigation when assessing the decision inadvisedly to launch (Presidential Commission, 1986).

Disturbance of biological rhythms

It was only with the advent of electric light allowing routine shiftwork, and air travel permitting rapid crossing of time zones, that the disturbance of biological rhythms became a serious problem. It influences both efficiency and safety and attracts public attention.

A disturbed pattern of biological rhythms is now the foundation upon which the life of the long-range flight crew member is based. The condition is sometimes called circadian disrhythmia, or desynchronosis. Objection to the inference of -nosis led to the term transmeridian dyschronism, and when the disturbance is related to shiftwork, metergic dyschronism. In long-range aviation, the disturbance results from irregular work schedules superimposed upon time zone changes, so perhaps another term will be created. Perhaps we shall finish up where we started with jet lag.

Whatever the condition is called, from jet lag to transmeridian and metergic dyschronism, it is important that its nature is understood and its consequences properly faced.

When the pattern of one's living environment changes, as occurs in shiftwork or irregular working schedules or when travelling across time zones, the rhythm of all

Table 3.1 Shift rates after transmeridian flights for some biological and performance functions (Klein *et al.*, 1980).

	Westbound	Eastbound
	min/day	
Adrenaline	90	60
Noradrenaline	180	120
Psychomotor performance	52	38
Reaction time	150	74
Heart rate	90	60
Body temperature	60	39
17-OHCS	47	32

the cyclic systems in the body must readapt to the new environment. This is a more complex process than once thought.

Four factors play a major role here. Firstly, systems shift their phase at different rates, so while they are shifting they are not only out of phase with local time, but also out of phase with each other (Table 3.1). In view of the delicate balance maintained by the body's homeostatic, or self-regulating, mechanisms, it is not unreasonable to hypothesise that this unbalanced state of the body's chemistry may be a challenge to the overall well-being of the traveller and contribute to his fatigue. We have, after all, thrown a heavy spanner into the homeostatic works.

A second factor is that resynchronisation occurs at a different rate depending upon whether the phase must be delayed, as on westbound flights, or advanced, as on eastbound flights. Although the time of day of the flight has a major impact on the sleep pattern and so also the feelings of fatigue, most travellers recognise that recovery from eastbound flights is more difficult than from westbound flights. Man has a free-running circadian rhythm of about 25 hours and a range of entrainment averaging about $23^{1}/_{2}$–$26^{1}/_{2}$ hours. Clearly the entrainment capability eastbound is therefore less than it is westbound. In some cases of, say, an eight-hour time zone difference, it has been found that a person's systems may elect to delay 16 hours rather than advance eight hours, resulting in a very long readjustment period (Mills *et al.*, 1978). Whether the system phase is to be advanced or delayed is only one factor affecting the rate of resynchronisation (Table 3.2). A third confusing factor is that resynchronisation does not appear to occur at a constant rate. Data obtained by averaging physiological and performance parameters suggest a non-linear recovery which can be mathematically represented by an exponential function. The studies showed that the out-of-phase condition, or time zone difference, is halved each 48 hours and a model has been created for predicting the recovery period required after transmeridian flights (Fig. 3.4). This model has certain limitations; it is a mean of eastbound and westbound flights and takes no account of individual differences in entraining capacity. It should not be applied to eastbound flights with more

Table 3.2 Some factors affecting relative rate of resynchronisadon of disturbed biological rhythms (Klein *et al.*, 1980).

Faster	Slower
Evening types	Morning types
Extraversion	Introversion
Low neuroticism	High neuroticism
Younger	Older
Low pulse & respiration	High pulse & respiration
Labile rhythms	Stable rhythms
Strong time cues	Weak time cues
Low amplitude	High amplitude
Delay shift (W)	Advance shift (E)

than about a nine-hour time difference (Wegmann *et al.*, 1983).

Finally, there are substantial differences in the ability of individuals to adjust their circadian rhythms to repeated transmeridian shifts. This further complicates management of the problem.

Chronohygiene

The pharmaceutical industry has recognised the demand which might exist for a chronobiologic drug which could accelerate the resynchronisation of disturbed body systems. Although experimental work has been done in this area using, for example, mepiprazole hydrochloride, results have been disappointing. Drugs such as theophylline and pentobarbitone have been demonstrated to shift the phase of the body rhythms in animals, but both have undesirable side effects. Nevertheless, by 1975 it was possible for scientists working in this area 'to propose the feasibility in the near future of writing rational circadian phase control prescriptions for the

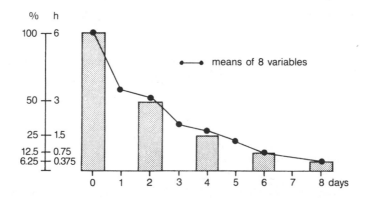

Fig. 3.4 The shaded bars represent a model for predicting the average resynchronisation of eight variables for eight post-flight days starting with an out-of-phase condition of six hours. The curve represents actual values from a group of eight subjects (Wegmann *et al.*, 1983).

transmeridian traveller or shiftworker' (Ehret *et al.*, 1975). Other work has been done using electrical currents to influence circadian rhythms, but here, too, little of practical value has emerged. Work with the pineal hormone melatonin shows promise and this is discussed later in this chapter.

Following the suggestion that poor chronohygiene was a contributory factor in the Three-Mile Island nuclear power plant near-disaster in 1979, renewed attention was given to the problem in the USA. Shiftwork practices were reviewed and a programme of chronohygiene involving timing of meals, use of drugs, exercise, sleep and social cues was proposed to accelerate the phase shift (Ehret, 1981). At about the same time, the US Army carried out a programme aimed at inducing a more rapid phase advance of the biological rhythms following the deployment of military personnel from the USA to Germany. The model of Ehret, developed at the Argonne National Laboratory in Illinois, was used in conjunction with the manipulation of social cues, light/dark cycles, and rest/activity patterns which are known to be effective synchronisers of human circadian rhythms (Cuthbert *et al.*, 1979; Graeber *et al.*, 1979). It was demonstrated that a self-imposed living schedule following transmeridian flight may modify the rate of physiological and psychological adjustment to the new time zone. This schedule gave emphasis to light/darkness, meal times, meal content (carbohydrate or protein), controlled use of caffeine and social cues (see Appendix 1.3). The efficacy of this dietary programme is, however, open to debate. Hypnotics (sleeping drugs) are not chronobiotic drugs and do not seem to be effective as entraining agents.

Sleep

The nature of sleep

The most common physiological symptom associated with long-range flying is disturbance of the normal sleep pattern. This may or may not involve an overall sleep deprivation. The problem arises from having to work, or at least travel, during the normal sleeping period and from difficulties in being required to sleep when the biological rhythms are not in a sleeping phase. It has been established that the major influence on total sleep duration is the time of day of sleep and the relation of the sleep phase to the body temperature curve shown in Fig. 3.1 (Akerstedt *et al.*, 1981; Czeisler *et al.*, 1980).

Most adults and certain animals take sleep in one long period a day (monophasic) while infants and many animals whose activity is not largely controlled by the light/dark cycle, take sleep in several periods during the day (polyphasic). Once the monophasic pattern has been established it becomes a natural rhythm of the brain, even when prolonged waking is imposed. In cultures where afternoon siestas are the custom, the evolution from polyphasic to monophasic sleep, usually occurring before the age of ten years, is never finally completed.

The description of the 'golden age' of sleep research has been applied to the period from the early 1950s until the mid-1960s. Although it was already apparent as long ago as 1935 that there were different stages of sleep, it was not until nearly 20 years later that serious study began, particularly by Kleitman and his colleagues in the USA. In 1957 a coding of these stages was established using primarily electroencephalogram (EEG) electrical brain wave characteristics of each stage of sleep. It was another decade before the terminology was standardised (Rechtschaffen *et al.*, 1968) and this has been used since in the reporting of sleep research, though some anomalies still remain. In view of the relatively recent start of serious sleep research and the pace at which it progresses, one must be cautious in going back too far in the literature when seeking information.

During those years of research much was learned of the nature of sleep. Yet the precise function of sleep and its different stages remains very much obscure. We should first then review what is known.

Sleep patterns

Sleep has been divided into two basic kinds, orthodox and paradoxical (or REM). Paradoxical was so named due to the paradox of an almost waking brain wave pattern combined with largely paralytic muscles. The acronym REM is derived from the initials of Rapid Eye Movement, one of the earliest characteristics discovered in paradoxical sleep. Orthodox sleep is sometimes called Non-REM sleep.

Each kind of sleep has its own characteristics (Table 3.3). When a child is born,

Table 3.3 Some characteristics of orthodox and paradoxical (REM) sleep.

Characteristic	Orthodox Sleep	Paradoxical Sleep
EEG (Brain Waves)	Big slow waves	Low voltage waves, high frequency
EOG (Eyes)	Quiescent	Rapid Eye Movements
EMG (Throat)	Tensed Muscles	Relaxed Muscles
ECG (Heart)	Regular	Irregular
Dreaming	No recall normally	Recall
Sleep Walking	Yes	No
Major Body Movements	Less frequent	More frequent
Stomach Acids	Steady	Increase

sleep is divided equally between orthodox and paradoxical (REM) sleep. On reaching adulthood, the proportion of REM sleep has fallen to about 20% and continues to fall with increasing age so that by 70 years it represents only about 15%.

In addition to this basic division of sleep into two kinds, a further four subdivisions of orthodox sleep have been determined (Table 3.4). It will be seen that a person enters sleep with a dominant brain wave frequency of about 7–8 Hz and this decreases to about 2–4 Hz in Stage 4. It will be apparent why various techniques which tend to bring the brain wave frequency down to the alpha range (8–12 Hz) are recommended to the insomniac. Stages 3 and 4 are sometimes known as Slow Wave Sleep (SWS). Brain wave frequency ranges are often quoted as beta (above 12 Hz), alpha (8–12 Hz), theta (4–8 Hz) and delta (2–4 Hz).

During a normal night, sleep shifts from one stage to another about thirty times; it has been described as climbing up and down the staircase of consciousness

Table 3.4 Sleep stages, with four stages of orthodox and one of paradoxical (REM) sleep.

Sleep stages

Stage	0 :	Awake; 8 - 12 Hz (alpha, relaxed)
Stage	1 :	Low voltage mixed frequency; 2 - 7 Hz, decreasing alpha, light sleep. About 1 - 10 minutes. Transition stage.
Stage	2 :	Spindles, bursts 12 - 14 Hz. No alpha but theta and delta. About 10 minutes.
Stage	3 :	20% - 50%, 2 - 4 Hz. About 5 minutes. Transition between stages 2 and 4.
Stage	4 :	50%, 2 - 4 Hz.
Stage	REM :	Stage 1 EEG, REM. Low amplitude EEG, dream recall.

(Fig. 3.5). REM sleep recurs about once every 90 minutes (an ultradian rhythm) and increases as the night progresses. As dreams are only recalled from REM sleep, this explains why they are often remembered upon waking in the morning. It is clear that an early call which shortens a night's sleep will proportionately reduce the amount of REM sleep obtained.

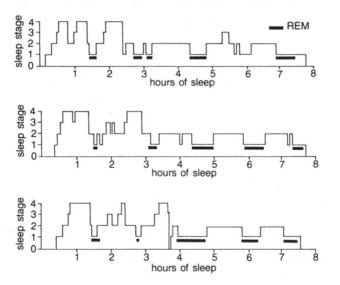

Fig. 3.5 The shifting of sleep stages over three nights for one subject. This illustrates the predominance of Stages 3 and 4 in the early part of the night, and the increasing REM sleep as the night progresses (Webb, 1971).

Naps and microsleeps

For those whose normal night's sleep is necessarily broken, the ability to nap would seem to be an asset. The nap is not simply a miniature version of a normal sleeping period; the stages of sleep which occur during naps depend on several factors, one of which is the time of day at which the nap is taken.

It appears that the restorative effect of a nap varies between individuals with those who habitually take naps appearing to obtain more benefit than non-habitual nappers, who sometimes perform worse after a nap (Lubin *et al.*, 1976). Other research suggests three major factors affecting the recuperative effect of naps – hours of prior wakefulness, the time of day and the nap duration (Naitoh, 1981). Nap duration appears to be particularly important, with time of day becoming very significant only when there is a state of sleep deprivation.

The application of naps to the life of the long-distance air traveller is important. Before a long night flight many will try to obtain some sleep, though the success achieved varies widely between individuals. After arrival, many will also try to take a nap to recover some of the sleep loss, though this often results in an unpleasant hangover feeling for the rest of the day. The ability to obtain such a prophylactic nap before a flight would appear to be a great asset and techniques to enhance this capability are a source of conversation amongst sleep-deprived crew members. One of the most promising aspects of the NASA-Ames research programme on fatigue has been a project to explore the use of scheduled naps by flight crew members having long-range flights. The FAA granted NASA and the cooperating airlines a special exemption to permit NASA to study the effectiveness of scheduled naps under real-world operating conditions (Henderson 1990 and Graeber *et. al.*, 1991).

Microsleeps are very short periods of sleep lasting from a fraction of a second up to two or three seconds. Although their existence can be confirmed by EEG recordings, the person is not generally aware of them. This makes the phenomenon particularly dangerous. They have been shown in tests to correlate with periods of low performance and they occur most frequently during conditions of fatigue. Microsleeps are not helpful in reducing sleepiness. A nap of not less than ten minutes appears to be necessary for sleep to be restorative (Bonnet, 1987).

The quality of sleep

It is a common experience that one night's sleep appears to be more restorative than another, yet there is no simple way of quantifying this quality. Length of sleep by itself is not enough – we ask 'how did you sleep?' and not 'how long did you sleep?'. It is possible to analyse sleep stages using EEG recordings, but then the meaning in terms of restoration of the different stages is uncertain. Various performance tests can be carried out, but these have often been shown to be insensitive to sleep changes. As a result, measurement of the quality of sleep is often left simply to subjective reporting.

The depth of sleep, like its quality, is an elusive characteristic. Even its definition is open to confusion and perhaps it would be better if it were abandoned as an explanatory concept. Usually, it is considered that Stage 4 is deep sleep, displaying the lowest brain wave frequencies. It is often thought that Stage 4 is the most restorative part of the sleep period, yet the sound intensity required to arouse a sleeping person is not much different in Stage 4 from what it is in REM sleep, with almost a waking brain frequency. In these confusing circumstances it may be wiser to concern ourselves with the restorative quality of sleep rather than its depth. Researchers Minors and Waterhouse of Manchester, England initially developed the concept of 'anchor sleep'. It consists of at least four hours of uninterrupted sleep at the individual's home domicile sleep time. It is an intriguing concept, but unfortunately, there is little convincing evidence that 'anchor sleep' has practical applications for the operational demands associated with airline scheduling (L. Connell, personal communication, June 1992). The area of concern is the performance deterioration that accompanies the crew sleep deficits that are an inevitable part of transport aviation.

Sleep and memory

Periodically, claims are made, particularly in Eastern Europe, that learning takes place from tapes played during sleep, though research in the West has been able to show only very limited absorption of material. However, there seems to be some increase in information retention just before dropping off to sleep, provided only a limited amount is attempted. It is known that a person who is asleep can respond to certain stimuli and can discriminate between names and tones (Williams *et al.*, 1964).

Research into the effect of sleep on memory has been carried out for many years. Early work suggested that sleep and night-time appeared to have a beneficial effect on memory compared with wakefulness and daytime (e.g. Jenkins *et al.*, 1924). Reference was made in later research to what was called the prior-sleep effect on memory (Ekstrand *et al.*, 1977). At about the same time it had also been demonstrated with children that there was a better retention of learning in the afternoon than in the morning (Folkard *et al.*, 1977). These studies raised a number of questions and later work attempted to determine whether the phenomenon resulted from sleep as a restorative process or whether it could be associated with circadian factors. Sleep is a time when restorative processes are enhanced and there is much evidence to suggest that this may be due to an increase in the net rate of protein synthesis during sleep. This restorative hypothesis of the function of sleep seems to be a useful predictor of the effect of sleep on memory (Idzikowski, 1984). It is not clear whether REM is important for memory as has been suggested by various researchers.

With the knowledge currently available, it may be preferable to study before a period of sleep rather than during the first couple of hours after sleeping. Flight deck crew members are involved in studying periodically for checks and examinations throughout a career and so these and any subsequent findings could be particularly relevant.

Insomnia

There are two kinds of insomnia and they must not be confused when seeking relief. Clinical insomnia describes the condition when a person has difficulty in sleeping under normal, regular conditions and in phase with his body rhythms. In other words, an inability to sleep when the body's systems are calling for sleep. Situational insomnia, on the other hand, refers to difficulty in sleeping in a particular situation, in this case, when the biological rhythms are disturbed. This often occurs when one is required to sleep when the brain and body are not in the sleeping phase.

It is likely that someone already suffering from clinical insomnia will also report situational insomnia when travelling or as a result of shiftwork, and the two conditions become confused.

Clinical insomnia is usually one of three types; an inability to get to sleep (sleep-onset insomnia), waking up in the night and then being unable to return to sleep and early waking in the morning. In seriously afflicted insomniacs the three conditons tend to become indistinguishable. Reference was made earlier to the special disorder known as Delayed Sleep Phase Insomnia.

It must be understood that insomnia is rarely a disorder in itself; it is normally a symptom of another disorder. For this reason, the common, symptomatic treatment of insomnia with hypnotics (sleeping drugs) or tranquillisers is inappropriate unless treatment for the underlying cause is also undertaken.

There are wide differences between individuals in their ability to sleep out of phase with the biological rhythms and in their tolerance to the sleep disturbance which is inevitable in long-range air travel. There are no doubt in some cases emotional stress factors involved, but it is likely that usually the differences are physiological and related to body chemistry and the characteristics of the individual's own body rhythms. Certain rhythm characteristics appear to be correlated with adjustment to shiftwork, suggesting the possibility of predicting the degree of tolerance of an individual to disturbing influences on biological rhythms (Reinberg *et al.*, 1980).

The need to obtain sleep when the body and brain chemistry is not in the sleeping phase is a fundamental requirement of the long-range flight crew member. A few have no difficulty, but at the other end of the scale there are those who find the situation intolerable and are forced to withdraw from long-range flying. Perhaps the most adverse condition is that which may be called the 24-hour ground-stop syndrome. This occurs at a crew 'slipping' station when the airline is operating a daily service. With such a schedule, if the crew goes to bed shortly after arrival, following perhaps a 12-hour working day, they will be out of bed again probably 14 hours or so before their next duty period starts. In other words, they will be starting their next working day at about the time their next sleeping period should be starting. This new working period can be more than 12 hours, the whole of which may then be carried out in a sleep-deprived state. Situational insomnia further aggravates the situation.

Drugs and sleep

Drugs of one kind or another can be used to induce sleep or to counteract the drowsiness and low performance associated with sleep deprivation. We should first look at those drugs commonly prescribed for clinical insomnia and, often erroneously, for situational insomnia.

There are two common groups of drugs used to induce sleep, the barbiturates and the benzodiazepines. The barbiturates can be fatal if taken in overdose, are seriously addictive and have an adverse effect on performance. There is no place for these drugs in the aviation situational insomnia case and even in clinical insomnia their use is now very limited. The benzodiazepines, best known under trade names such as Valium, Dalmane, Mogadon, Librium and Normison, are not toxic to the same extent as the barbiturates and for years have been spoken of by physicians as 'safe' medication. However, they are far from being harmless. They have an adverse effect on performance (in some cases for almost as long as certain barbiturates), they can be addictive, combined with alcohol the effects are unpredictable and they can sometimes cause a hangover effect.

In using any such drugs it is important to be aware of two characteristics of the drug; its half-life and its effect on performance. The half-life of a drug is the time it takes for the drug level in the blood to fall to half of its peak level. This should be known so as to recognise the risk of possible accumulation of the drug within the body. For instance, nitrazepam (Mogadon) has a half-life of 25–30 hours; for diazepam (Valium) it is about two to three days. In both cases a daily dose would lead to accumulation. Furthermore, there is a wide variation in the metabolic rate of individuals – a two-fold and sometimes even a four-fold variation may be possible. This will inevitably alter the effects of the drug on an individual basis, so it cannot just be assumed that a drug with a short half-life will never accumulate in the body. There is no simple way to predict this reliably without doing an individual test to determine a person's own rate of drug elimination. The effect on performance is more difficult to ascertain. This is partly because the effect is task-dependent, partly because of individual differences and circadian effects and partly because the pharmaceutical industry is not required to carry out such experimental work before marketing a drug. Nevertheless, independent studies have been made of drug effect on performance by, amongst others, the RAF Institute of Aviation Medicine in England (e.g. Nicholson *et al.*, 1982).

All performance test results should be evaluated with caution. They may apply to one specific task only which may or may not be truly representative. Furthermore there is probably a circadian effect which may or may not be apparent in the test data. Individual tolerance to the drug can be expected to vary due to age and other reasons. There may also be several other real-world variables which are not reflected in the laboratory situation. Often a small number of test subjects is used, which may not properly reflect the full range of the target population.

In the Falkland Islands conflict in 1982, temazepam (Normison), a metabolite of diazepam, was used to ensure adequate sleep for flying staff on what were very demanding flying schedules. Some transport crews accumulated 360 hours within three months, while other reconnaissance crews were reaching 100 hours in two weeks. Individual flights lasted up to 20 hours, with extensions up to 28 hours if a landing in the Falklands was not possible. These flight schedules are about double those found acceptable in normal civil flying and far exceeded what would be compatible with a tolerable sleep pattern. In these exceptional circumstances, temazepam, with its rapid absorption rate and its short half-life, was said to be effective in 20 mg doses for assisting military crew members to obtain sleep at various times of the day (Baird *et al.*, 1983; Lader, 1983). A temptation to extrapolate from this military application of drugs to the civil aviation environment should be tempered with a due understanding of the fundamentally different objectives of wartime military flying and routine peacetime civil flying.

Later studies were conducted using another short-acting hypnotic called brotizolam. In the research, three test subjects were used in the UK and three in the USA in a cooperative programme. When the drug was used before a full overnight sleep of eight hours, no residual effects were found on performance the following day. However, some residual effects were found when the drug was used for shorter four-hour sleep periods. Some individual differences were also found in the performance effects (Nicholson *et al.*, 1985).

The half-life of a drug and the duration of its effect on performance should not be confused. As an example, nitrazepam has a shorter half-life than diazepam but a much longer detrimental effect on performance. This again emphasises the need for caution when hypnotics are used by flight crew members or others engaged in skilled activity. And, of course, individual responses differ.

A number of precautions are therefore necessary, particularly for flight crew members, before using drugs in the aviation environment. The medical practitioner prescribing them should be well aware that they are to be used for the management of situational insomnia and not clinical insomnia. These precautions are listed in Appendix 1.4.

The fact that a sleeping tablet can be bought without a prescription means that it is not highly toxic but it does not mean that it is free from adverse effects on performance and can safely be used in the flying environment.

Alcohol is a general, non-selective Central Nervous System (CNS) depressant and must be considered as a drug. It is widely used in the aviation environment, as elsewhere, to induce relaxation and sleep. Alcohol may have some soporific effect, depending upon the phase of the body's chemistry rhythm when it is ingested. It may also initially sometimes appear to have a stimulating effect, but this may be due to the loss of behavioural inhibition. What is certain, however, is that even when it does induce sleep, the pattern is not normal, with a suppression of REM sleep. This disturbance may persist after all alcohol has disappeared from the blood. The sedative effect of alcohol is transient in the case of alcoholics, as tolerance increases and a self-sustaining disregulation ensues.

Beverages containing caffeine such as coffee, tea and various soft drinks increase alertness and normally reduce reaction times (Table 3.5). However, it has also been shown that caffeine disturbs sleep, depending on the amount taken and the time interval before sleep. It has also been accused of a relationship with heart disease and diabetes and of being addictive (Goldman, 1984). Caffeine can already be detected in the blood five minutes after ingestion and its half-life is at least three and a half hours. Individuals differ in their metabolic rate and this may be reflected in their different tolerance to caffeine. This drug, like others mentioned earlier, disturbs the pattern of sleep, reducing Stage 4 and REM. When caffeine is removed from a drink, the sleep-disturbing effect is also removed (Karacan *et al.*, 1976).

While amphetamines have been demonstrated to help maintain the performance of even complex tasks during sleep deprivation, they only postpone and do not remove the effect of sleep loss, and some lack of efficiency may still remain. The sleep period following the use of amphetamines can be expected to be disturbed (Higgins *et al.*, 1975).

There are some who claim that aspirin helps them to sleep. Research into the use of aspirin for this purpose has been scant, but the data available suggest that aspirin, like most hypnotics, tends to reduce Stages 3 and 4 of sleep and increase Stage 2 in compensation. It also disturbs the normal pattern of sleep while being used and for some days afterwards (Horne *et al.*, 1980).

Other drugs, such as antihistamines which have drowsiness as a side-effect, have sometimes been incorrectly used by crew members to help induce sleep. An earlier anti-malarial drug (Daraprim) which had an arousal effect, was used by some to counter the sleepiness felt when required to work during the body night. This drug was also reported to have a disturbing effect on sleep. The abuse of drugs to

Table 3.5 Some common sources of caffeine.

Source	Approximate Amounts of Caffeine per Unit
Beverages	
brewed coffee	85 mg per 150 ml*
instant coffee	60 mg per 150 ml
decaffeinated coffee	3 mg per 150 ml
brewed black tea	50 mg per 150 ml
brewed green tea	30 mg per 150 ml
instant tea	30 mg per 150 ml
cola drinks	32-65 mg per 12 oz
cocoa	6-42 mg per 150 ml
	mg/tablet
Prescription medications (USA)	
APCs (aspirin, phenacetin, and caffeine)	32 mg
Cafergot	100 mg
Darvon compound	32 mg
Fiorinal	40 mg
Migral	50 mg
Non-prescription (USA)	
Anacin, aspirin compound	
Bromo-Seltzer	32 mg
Cope, Easy-Mens, Empirin	
compound, Midol	32 mg
Vanquish	32 mg
Excedrin	60 mg
Pre-Mens	66 mg
many cold preparations	30 mg
many stimulants	100 mg

* Based on 150 ml average cup size.

counter the effects of the irregular work/sleep pattern in long-range flying must be discouraged. Its occurrence may be seen as reflecting the need for more educated overall attention to the subject.

Sleep functions and requirements

While much is known about what occurs during sleep, the precise function of sleep and its different stages remains largely in the realm of hypothesis. It has been shown, for example, that following deprivation of one kind of sleep a rebound of that kind of sleep will occur, suggesting a need for it. Yet such a need cannot be easily demonstrated or confirmed in research studies. For instance, after SWS deprivation there will be an excess of SWS in the next sleep period; after REM deprivation, there will be an excess of REM sleep. After total sleep deprivation, there will be a rebound of SWS and REM but not Stage 2, though it would be incautious from this to dismiss Stage 2 as without function.

There have been many theories proposed on the restorative function of sleep, particularly for the repair of body tissues suffering from the wear and tear of the waking hours and for some form of brain restitution. In order to obtain some insight into just why we sleep, many studies have been carried out depriving test subjects of sleep over periods of several days.

Research studies which have reduced regular sleep by smaller amounts, such as an hour or so a night over several months, suggest that it is possible to learn to live with less sleep than we considered normal (Johnson, 1982). While such programmes have not revealed physiological damage as a result of this deprivation, it cannot be concluded that such damage would not ensue when applied over long periods. The possible effect on life-span cannot be measured and individuals may be expected to react differently. Furthermore, there could be psychological effects such as a degradation of motivation. The post-lunch dip in performance can expect to be aggravated by the effect of sleepiness. Research using rats has demonstrated severe pathological conditions such as stomach ulcers and internal haemorrhage and even death in some of the animals which were sleep deprived (Rechtschaffen *et al.*, 1982).

It has been difficult to determine the role of sleep in human body restitution, a process which also continues during daytime. However, caution is again necessary. While the Olympic athlete takes no more total sleep than the sedentary office worker, his sleep pattern will show more Stage 4 when the core body temperature has been raised. Heavy exercise outside athletics also results in more Stage 4 sleep. The possible relationship between body temperature and sleep is mentioned later in this chapter and again in Chapter 4.

While the physiological effects of sleep deprivation are less than clearly defined, sleep is clearly essential for psychological performance, which shows rapid deterioration with sleep loss. Changes which take place suggest some impairment in the CNS, from which it may be deduced that the brain needs sleep. The impairment, however, does not seem to last long after normal sleep patterns are restored.

After a period of total sleep deprivation it is unusual for the full amount of sleep to be reclaimed. In experimental work only about a third of the lost sleep seems to be recovered and this is mostly SWS – REM and Stage 2 being abandoned. Fig. 3.6 indicates how much sleep could be lost by a crew on a line flight.

The first part of a night's sleep contains most of the SWS and it seems that this period, about the first five hours, may be the most important. However, the fact that it has not been possible to establish the function of REM sleep does not mean that

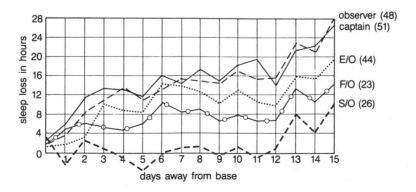

Fig. 3.6 An example showing how sleep deficit on a long flight can accumulate as the flight progresses. This was a London–San Francisco–Hong Kong flight on a Boeing 707 (Preston, 1975).

it has none. For instance, much has been written on the self-regulating function of dreams, which are only recalled from this stage of sleep, and this aspect cannot be easily dismissed (Lavie *et al.*, 1984).

Total sleep time amongst healthy humans conforms to a normal distribution (see Fig. 12.2) within any age group. In young adults this would show a mean of about seven and a half hours with a standard deviation (see Appendix 1.5 for definition) of one hour. In studies of short (less than six and a half hours) and long (more than eight and a half hours) sleepers it has been found that the amount of SWS, that is, Stages 3 and 4 combined, remained about the same. Studies with even wider extremes of normal sleep have produced similar results, with SWS remaining roughly constant and the variation occurring largely in Stages 1 and 2.

Not only is it possible to learn to sleep somewhat less than considered normal, but the reverse is also true. Many can return to sleep after waking in the morning, adding an hour or so to the full night's sleep total and thus establishing a new norm. It is again significant that this extra, probably superfluous sleep, contains largely REM and Stage 2 and little or no SWS. This again suggests that SWS plays a more essential role in the maintenance of well-being than other sleep stages.

One of the most significant conclusions from research is that sleep appears to play a role in the maintenance of motivation. So that even if the sleep loss may not have reduced significantly the actual capacity to perform a task, it may have reduced the motivation to do so. This loss of motivation can be at least partly restored by extra effort, though an individual must be aware of the phenomenon to take corrective action. This has a major relevance to safety.

Conclusions on sleep functions and requirements are not as definitive as one might wish but there are indications that sleep may be more important for the brain than for the rest of the body. It also appears that the human system is more willing to accept the loss of REM and Stage 2 sleep than SWS. It is also apparent that there are individual differences in the amount of sleep needed to perform normally and that sleep loss refers to the individual norm rather than to some nominal requirement. Research into these aspects of sleep is a continuous process and updated reviews are published from time to time (e.g. Horne, 1985).

Human performance and sleep

Sleep deprivation means an overall loss of sleep, while sleep disturbance refers only to the disruption of the normal sleeping schedule. In long-range flying, sleep disturbance is virtually inevitable as aircraft fly round the clock and the irregular working periods break into the normal body nights. Avoidance of sleep deprivation and minimising sleep disturbance must remain the objective of those planning such activities.

It is generally thought that the main problem facing crew members is that of disturbance but it is clear that loss of sleep can accumulate on long flights. While there are large differences in tolerance to the stress of long-range flying and generalisations cannot be made from the experiences of one crew or one crew member, the consequences of such sleep loss cannot be ignored (see Fig. 3.6).

Much research has been done on the effect of disturbed sleep. Vigilance and calculation tasks have been shown to be significantly impaired and mood adversely affected, for example, simply by displacing the sleeping period by two to four hours (Taub *et al.*, 1973).

It is difficult to demonstrate consistent results in studies using partial sleep loss due to the almost limitless variations in conditions surrounding the case and a complex interaction of the task, the situation and personal factors. However, clear reduction in performance of certain tasks has been shown following loss of a night's sleep. In one such laboratory study, the sleep-deprived performance decreased progressively during the experiment (Hockey, 1970).

Even small periods of sleep deprivation have been shown to affect performance. In one experiment it was demonstrated that more signals were missed and there were occasional lapses in performance following a sleep loss of five hours on one night or two and a half hours on two consecutive nights (Wilkinson, 1969).

In fact, lapses and inconsistency in performance are characteristics of the sleep-deprived person. An analogy may be that he is not like an aircraft which performs perfectly until it runs out of fuel and then suddenly stops flying or like a mechanical toy which slows down steadily and gradually before it finally ceases to perform. He is more like an old car which runs, falters, runs again, and so on; one moment it works fine, the next it misfires and its performance plummets dramatically only to recover again.

The performance decrement resulting from sleep deprivation increases with altitude and is further increased with higher workload. This has a particular significance in long-range flying which is typified by sleep disturbance and deprivation and high cruise altitudes (Mertens *et al.*, 1985).

Laboratory studies often suffer from some lack of credibility when viewed from industry and this was mentioned earlier relative to drug research. Sometimes the task and the environment appear to be too remote from the real working situation. Frequently the test subjects, usually as a result of economic constraints, are not representative of the people actually involved with the job in industry in terms of age, skill or motivation; they are often young students, nurses or military personnel. Sometimes they appear too few to encompass the range of individual differences which can reasonably be expected to be encountered operationally.

In studying sleep and circadian rhythm problems, subjective and objective data have been obtained in numerous studies in the field, covering world routes such as Tokyo–Moscow, Brussel–Rio, Dublin–New York, Amsterdam–New York and London–Hong Kong. In all such work an almost insurmountable difficulty is faced in trying to control tightly the many variables involved in the operation. This does not mean that the work is without value, only that interpretation calls for much care.

One study reported in 1985 which was sponsored by the CAA in the UK, attempted to avoid many of the pitfalls noted here. The study investigated the effect of one night's sleep loss on the performance of 16 pilots in actual flight, in a flight simulator and in a variety of laboratory tests. This comprehensive programme demonstrated that flying performance deteriorates considerably with sleep deprivation. The performance loss was somewhat greater in the simulator than in the actual aircraft, possibly due to motivational factors. All the laboratory tests (with one exception) also revealed a performance decrement. There appeared to be some correlation between many of the experimental measures and the introversion/extraversion dimension of personality (Farmer *et al.*, 1985). The findings of this important study suggested that loss of sleep was profoundly disruptive of the performance of pilots. The study provided some scientific and objective data to support the numerous subjective confidential reports from pilots submitted to the ASRS and CHIRP centres relating reduced flying performance to disturbance in the waking/sleeping cycle, as well as earlier surveys (e.g. Hawkins, 1980).

In addition to the effect of sleep loss on the performance of specific tasks it has been shown that subjective attitudes are also affected; mood, appearance and behaviour all appear to suffer. Even small amounts of sleep loss seem to affect motivation. As motivation represents the difference between what a person can and will do in a particular set of circumstances, this has a direct relevance to performance.

The effect of sleep loss is task-dependent. The more complex task tends to suffer more than the simple task but the more interesting task suffers less than the monotonous or duller task. Vigilance and monitoring tasks, increasing with the trend towards automation of flight deck activities in modern aircraft, are therefore particularly vulnerable to sleep loss and this fact should be an important consideration in the design of equipment and procedures.

One of the most dangerous aspects of performance degradation with sleep loss is that a person is unlikely to be aware of the manner and extent of his deteriorating performance. We might compare this lack of awareness of performance deterioration with hypoxia and alcohol consumption. Subjective feelings of sleepiness correlate well with periods of reduced performance and so a recognition of these feelings can provide a warning signal.

Research has been conducted to determine the level of performance which can be expected just after awakening. An early study done in connection with the US space programme showed that performance of a number of tasks was reduced for several minutes after waking (Seminara *et al.*, 1969). This and other studies showed the major performance loss to occur within the first three to four minutes with a measurable loss, depending on the task, extending for some minutes longer. Work in France has shown it possible to measure performance decrement as much as 20 minutes after waking (Angiboust, 1970). Later work suggested that performance after waking may be better if the person expects to stay awake than if he is to be allowed to return to sleep after the test (Robson *et al.*, 1974). This again suggested the importance of motivational factors in task performance. Careful attention to this sleep inertia effect must be given when considering whether to permit staff engaged in critical activities to nap at their duty station. The other side of this coin is that consideration must also be given of the possibility that flight crew performance would decrease even more without the scheduled naps. Tha NASA special project involving naps is taking a very careful look at these problems.

Facing the problem

The sources of the problem

At the beginning of this chapter, reference was made to the possible financial impact on business travellers and the safety aspects on crew members of the disturbance of sleep and biological rhythms. Specifically, this chapter has discussed four major sources of performance degradation in this environment:

- circadian rhythms; doing a daytime task in the body night and working when the biological rhythms are desynchronised one with another
- sleep disturbance and deprivation; working while suffering from loss or displacement of sleep
- drugs; working while under the influence of certain medication or drugs
- sleep inertia effect; performing a task just after waking

With these sources in mind it is now necessary to examine how best to face the problem which is presented to safety, efficiency and personal well-being by long-range flying.

Work and rest scheduling

As the problem for crews is created primarily by the irregular working and resting pattern, enlightened crew scheduling or rostering is a crucial factor. Analysis of personal sleep records over long periods has demonstrated the relationship of the route schedule to crew fatigue resulting from sleep deprivation and disturbance (Hawkins, unpublished data). Computer scheduling, increasingly used in airlines, lacks flexibility and usually takes insufficient account of natural circadian variations and certainly of personal tolerance. Currently used criteria which usually assume that the human is some kind of machine which can automatically obtain restorative benefit from a given number of hours off duty, regardless of when these hours are given, are inadequate. They ignore scientific knowledge and inevitably lead to lowered performance during aircraft operation. A first requirement, then, is enlightened design and supervision of crew scheduling.

For the businessman, there is much scope for planning a trip taking into account the adaptation characteristics of the body as outlined in this chapter. Modifying bedtime a few days before the trip and adapting meals when away and on return will help to minimise sleep loss and accelerate body readaptation. This programme depends on the phase shift required.

For crew members, the stationing for a period of months at an overseas base and flying just one part of the network, is likely to have a favourable influence on reducing the sleep disturbance. The ending of such stationing by one major airline caused the senior medical officer to remark that 'medically speaking, the cancellation was a retrograde step and physiologically it has done much harm' (Preston, 1972). Where there are crew unions, work and rest patterns together with other conditions of employment are usually the subject of management/union negotiations, sometimes seen as horse-trading, without input from specialists in the life sciences. Such work and rest patterns are, however, probably less deleterious in general than when no unions are involved.

Virtually all of the problems associated with crew members' rest and sleep have been exacerbated by the increased range in both the miles and the time aloft of our latest aircraft. For example, one of the recent inaugural flights of a latest transport took almost 17 hours. A great many time zones were crossed and the normal circadian rhythms of everyone connected with this flight were severely disrupted. An attempt to meet some of the problems associated with the crews on these aircraft has been the provision of aircraft bunks or sleeping quarters for both flight and cabin crews – an improvement over simply reserving a first-class seat for resting crew members. While in-flight bunks may provide far from ideal rest facilities, there seems little question that they can help if they are used judiciously with their ultimate use determined by both regulatory and collective bargaining requirements. It is hoped that such use can be based upon solid research involving a real-world problem.

Regulations

Anyone flying in the belief that the International Civil Aviation Organisation (ICAO), the aviation branch of the United Nations, is controlling these matters will be disillusioned. ICAO Annex 6 (paragraph 4.2.9.3) requires only that an operator shall formulate rules limiting the flight time and duty time of crew members, but gives no guidance as to exactly what limits should be applied. The current issue of ICAO Circular 52-AN/47 contains a compendium of actions by different states to implement this requirement and a summary of the limitations which are applied by states and operators.

Individual countries all have their own regulations and these differ widely. For example, in one country with a large national airline, a pilot is legally permitted to fly as active crew for 20 hours and in addition an unlimited number of hours as non-active, dead-heading crew or passenger. Non-active crew flying is not considered in these particular national regulations as duty time. In an advanced European country, even in the 1980s, a pilot could legally start a $16\frac{1}{2}$–hour duty day some 12–16 hours after the end of the last proper sleep period, little account being taken of circadian dysrhythmia and effects on human performance. The rules usually represent absolute limits to provide the operator with flexibility in operations and are not normally designed for continued, routine application. However, in many states they are routinely applied over long periods as the norm rather than the exception and this often gives rise to warnings from affected pilots about the impact on safety (e.g. CHIRP Feedback No. 9). They sometimes place within the discretion of the captain a decision to override the duty time limitations in the event of unscheduled delays. The argument that certain flight time limitations, established for safety, can be set aside if the cause happens to be unscheduled and their application would cause commercial penalties or inconvenience, is not entirely a rational one.

Formulae

Formulae for assisting in work and rest planning have been published from time to time. One for pilots and flight engineers was published in the USA by a former head of the FAA aviation medicine branch (Mohler, 1976). However, a subsequent accident report showed that the limits had been grossly but quite legally exceeded, and the formula remains no more than a recommendation (Glines, 1976). For passengers, the late head of the ICAO aviation medicine branch devised a formula to calculate the minimum rest period which should be taken before commencing work after a flight (Buley, 1970). This is not suitable for flight crews and was in any case developed before much of the relevant research on circadian rhythms outlined in this chapter was carried out. Later, an attempt was made to compare the models of Mohler and Buley, together with those proposed by Gerathewohl and Nicholson, when applied to an airline route from Europe to South America. The Mohler formula appeared to provide the most realistic index of load, but a new model was proposed which provided further refinement to that of Mohler (Wegmann *et al.*, 1985).

A fundamental decision must be made by the long-range traveller in this respect. He must determine whether he wants to synchronise, that is, to shift his phase, to his new destination or whether, assuming a rapid return, he wishes to remain as far as possible synchronised with his place of departure. Upon this decision will rest his personal scheduling strategy. There is clearly much scope for both passengers and crew members for the application of more enlightened scheduling.

Diet adaptation

For those long-distance travellers who make a single journey followed by a stop of a week or more in the new place, the initial phase of the body and its shifting pattern is rather predictable. With greater time zone changes some confusion can arise as a result of the reverse adaptation syndrome, that is, resynchronising by delaying instead of advancing the phase. But for crew members the situation is much more complicated. A return polar flight from, say, London westbound to Tokyo (a 15-hour time difference) with an intermediate crew stopover in Anchorage (a 9-hour time difference) can lead to total circadian rhythm chaos when some rest time is spent in each place. It would be impossible to predict exactly the phase of the biological rhythms on return to London.

The US Army programme mentioned earlier utilised staff who were first stabilised in US time, made the single trip to Europe and then became stabilised in European time. This kind of schedule applies to many passengers but not to many intercontinental airline crews, who are constantly on the move. There are even times when a crew member does not want to accelerate resynchronisation with local time but prefers to maintain the phase associated with the place of departure.

It is unrealistic to suggest that there is a simple dietary plan which crews can always use to enhance circadian rhythm adaptation. However, a basic understanding of the importance of meal times and the various other chronohygiene measures which can be taken related to light/darkness, rest/activity schedules and social interaction will be helpful. For those who wish to apply some of the concepts which have been developed, notably by Ehret and his colleagues at the Argonne National Laboratory, a selection of these is listed in Appendix 1.3.

In general, it has been said that a carbohydrate meal in the evening, particularly when combined with a food high in tryptophan (a serotonin precursor) such as cheese or milk, tends to increase the availability of serotonin to the brain. Serotonin is a neuro-transmitter associated with sleep. However, this dietary interaction with sleep remains open to debate.

Drugs and hormones

No totally effective and acceptable drug has been developed for routine and long-term application in the control of these problems. Nevertheless, research has concentrated on developing short-acting drugs to produce sleep in special situations and they have been applied effectively in a military environment. Other work continues to find a chronobiotic drug which would shift the circadian phase in the required direction. Development in this field is continuing and so specific recommendations here would not be appropriate. In any case, it would be rare to find a drug which in the long term did not have some adverse side effect when used on a routine basis and so caution should be applied in relying on pharmacology to solve the problem.

Caffeine as a drug performs a certain role in shifting the circadian phase of body rhythms but is otherwise generally sleep disturbing and should not be used within several hours of sleep by those experiencing sleeping difficulties.

Alcohol is perhaps the most common drug used to induce relaxation but it has other adverse effects and is not recommended, except in very small quantities, in this environment. Other methods of relaxation should first be learned and applied.

Perhaps the most promising substance yet employed for the relief of the symptoms of disturbed circadian rhythms is melatonin. This is a methoxyindole secreted by the pineal gland. Following experiments at the University of Surrey with animals where melatonin was found to promote the resynchronisation of disturbed biological rhythms, controlled, double-blind tests were conducted on a group of 17 human subjects. These were performed on the London–San Francisco flight, with an 8-hour time zone difference. The results demonstrated a significant reduction in the symptoms of jet lag amongst those using melatonin (Arendt *et al.*, 1986). At Monash University in Australia, other related studies have been conducted with similar favourable results (Monash, 1985).

Environmental optimisation

The long-range traveller, whether he happens to be a crew member on duty, a businessman or scientist on his way to a conference or simply a vacationer, should ensure that the sleeping environment is optimal. Where contracts are made with hotels, these requirements can be specified in the contract. In other cases they should be requested when booking a room. A number of other steps can also be taken at a personal level to enhance the prospect of obtaining satisfactory sleep. These are listed in Appendix 1.6.

Relaxation and other techniques

The similarity between the restorative effect of sleep and deep relaxation is not precisely known. A number of techniques which induce relaxation have been used with success to encourage sleep in cases of clinical insomnia. Sleep latency, or the time it takes to fall asleep, has been reduced and in many cases sleeping patterns return to normal (Bootzin, 1981; Nicassio *et al.*, 1974).

Furthermore, it has been demonstrated that certain of these techniques have restorative characteristics which can partly make up for lost sleep. While little systematic work has been done in the long-range aviation environment to research the application of such techniques there have been airline programmes on a small scale. One such technique, Autogenic Training, has been made available to staff of three European airlines. These programmes have confirmed that the most common response of trainees is improvement in sleep both at home and in the travelling environment of disturbed biological rhythms (Hawkins, 1984a). Autogenic Training is a method which has been well documented and applied clinically for more than 60 years. It involves the use of passive concentration on certain formulae which are aimed at producing a state of psychophysiological relaxation known as the autogenic state. This facilitates enhancement of certain homeostatic, self-regulating mechanisms in the body. It requires skilled teaching and is not a learn-it-yourself method. However, once learned it is totally self-administered, requires no equipment and needs no further assistance from the therapist for lifetime use. It is easy to learn and has no ritualistic, religious, mystical or folkloric connotations and is thus more palatable to people with technical orientation than certain methods with oriental backgrounds. A more detailed review of the various methods available, with their advantages and disadvantages, has been made elsewhere (Hawkins, 1980).

Experiments have demonstrated that the use of a non-specific self-relaxation method is not normally effective in producing the desired results (e.g. Bootzin, 1981; Nicassio *et al.*, 1974). In fact, it may be counter-productive and worse than no technique at all, possibly as a result of the frustrations arising from the lack of a desired result.

There is some experimental evidence to support the commonly held notion that a warm bath may help to promote sleep. It has been demonstrated that after prolonged sitting in a warm bath in late afternoon, there were significant increases in sleepiness experienced later at bedtime. Furthermore, there appeared to be an increase in SWS and some decrease in REM sleep. No significant effects on sleep were found when using a cool instead of a hot bath (Horne *et al.*, 1985a).

Exercise and sleep has also been the subject of some experimental studies. It appears that exercise must be vigorous rather than of long duration and may have a greater effect on physically fit individuals than on the unfit. The most common finding in these cases is an increase in SWS (e.g. Bunnell *et al.*, 1983; Torsvall *et al.*, 1984). Body heating effects may play a similar role in the increase in SWS which follows a warm bath, a sauna and vigorous exercise (Horne *et al.*, 1985b).

Another factor relating exercise to sleep is that tension and anxiety are known to be associated with insomnia and it has been demonstrated that exercise can have a tranquillising effect on these (Greist *et al.*, 1979). This is mentioned again in the following chapter.

Evidence leads to the conclusion that the problems outlined in this chapter can at present best be met through proper chronohygiene, including enlightened work scheduling, and a scientifically developed relaxation technique aimed at enhancing the body's own self-regulating mechanisms. Chemical interference with these mechanisms through drugs of one kind or another should be seen as a last resort to be applied only when other more natural processes have failed, and then only as a temporary measure. Subject to further research, the hormone melatonin may provide a means of accelerating the resynchronisation of disturbed biological rhythms.

4

Fitness and Performance

Health and performance

Incapacitation

In the first chapter, reference was made to the confusion which often exists between the scope of Human Factors and medicine in aviation. This is at least partly due to the fact that both draw information from the discipline of physiology. Medicine is concerned with the prevention, diagnosis and cure of sickness, and in aviation, the influence of the environment of flight on these. Human Factors, on the other hand, is certainly concerned with individual well-being but is also particularly involved with human performance and factors which cause variations in performance.

It will be clear, then, that the question of physical and mental fitness is of concern to the physician as well as to the Human Factors specialist. This is one area where there should be effective communication and cooperation between the two. An ergonomically badly designed seat can aggravate the effects of natural spinal deterioration. The result can be pathological, requiring medical treatment, but it can also cause pain and discomfort, which in turn may lead to distraction, reduced motivation and lowered performance. The input into seat design, concerning well-being and effects on performance, is one of ergonomics.

Total

In the extreme case, pathological conditions occurring during flight can lead to a total loss of performance. Certain cardiovascular disorders, such as heart attacks or strokes, can cause sudden incapacitation and in a few cases may have contributed to accidents.

More frequent, however, are incapacitations resulting from gastrointestinal disorders. An analysis in one US airline showed that these represented nearly half the total in-flight incapacitations of pilots, while heart attacks accounted for only one seventh (Mohler, 1984). The onset of incapacitation following eating of contaminated food can be rapid and total. Such cases amongst flight crews occur sufficiently frequently for some airlines to insist that the two pilots eat different meals and some forbid the consumption by crews on duty of seafood altogether. Although food poisoning is usually temporary in nature, death can follow, as occurred in 1984 to a passenger of a major airline several months after contracting salmonella poisoning from a meal in flight. While these conditions are medical in nature, the risk of an unpremeditated and sudden loss of part of the operating crew must be taken into account in the operational design of procedures, systems and flight deck layout.

Many airlines now include sudden incapacitation of a pilot in flight simulator training programmes. Experience has shown that the first time a copilot or first officer is confronted with the simulated sudden demise of the captain, the actions which follow may be unpredictable. Obvious incapacitation presents little problem in the actual take-over, though the assumption of command may create difficulties. Control transfer can usually be expected to take place in a very short time – one analysis of training sessions has shown an average of 3.5 sec (Orlady, 1975). However, subtle incapacitation presents another problem and when this has been introduced into training exercises in flight simulators, it has sometimes resulted in a crash (NASA, 1981). This reflects the difficulty of recognition of and response to a serious performance failure of a pilot flying the aircraft. There is a natural reluctance to take away control from another qualified pilot. Nevertheless, such an intervention even by a junior member of the crew may at some time be necessary.

The prevention of total in-flight incapacitation lies primarily in the field of aviation medicine. Although certain aspects of this prevention, such as the interface between medical staff and crew members, contain elements of Human Factors, this area will not be pursued further within the scope of this book.

Partial

We have referred here to total incapacitation as being the total removal from the system of a person as an effective component. This can be seen as an area of medical responsibility. But there are many non-medical reasons which can cause a reduction in capacity and which would not be apparent in routine medical screening or examination. Total incapacity is often detected quickly by other crew members. Reduction in capacity, on the other hand, may be far more insidious as it is frequently not apparent to others – and sometimes not even to the person himself. Such partial incapacitation or reduction in capacity may be the result of factors such as fatigue, stress, sleep and biological rhythm disturbance or medication. As was discussed in the previous chapter, the capacity to perform may also be influenced by the circadian effect.

In spite of evidence which suggests that reduction in capacity to perform occurs frequently, it is debatable to what extent this reduction in capacity is the direct cause of accidents. It is likely that in most cases it could be shown, if the crew had survived, that they had at the time adequate capacity to perform the task properly had they utilised that capacity to the full. The cause of the reduced performance, then, may possibly often be found in a lowered motivation rather than in a lowered capacity *per se*.

Reduction in capacity may also occur as a result of pathological conditions. One such condition usually causing only temporary and partial lowering of performance is hypoglycaemia, or low blood sugar, but this can often be satisfactorily controlled by dietary care (Harper *et al.*, 1973). Many suffering from mild hypoglycaemia will not have sought medical advice and may be quite unaware of the origin of the occasional symptoms of lack of well-being which they experience. The possibility of hypoglycaemia having played a role in poor pilot performance was cited in the report of the 727 accident at Ketchikan in 1976 (NTSB-AAR-76-20).

Definition of fitness

When we refer to fitness it is reasonable to ask, 'fitness for what?'. And so once again we must look for a suitable definition. Certainly we are not concerned here with the fitness needed by competition-class athletes. For some purposes, physical fitness may be used to mean simply aerobic work capacity or as a measure in a specific performance test.

But for our purpose, fitness might be defined as a condition which permits a generally high level of physical and mental performance. It suggests an ability to perform with minimal fatigue, to be tolerant to stresses and to be readily able to cope with changes in the environment.

Beliefs and evidence

More than 2000 years ago in the dialogues of Plato it was advised that we should 'avoid exercising either mind or body without the other, and thus preserve an equal and healthy balance between them. So anyone engaged in mathematics or any other strenuous intellectual pursuit should also exercise his body and take part in physical training'. In more recent times, President John F. Kennedy said that 'physical fitness is not only one of the most important keys to a healthy body, it is the basis of dynamic and creative intellectual activity'.

Closer to aviation, Ross A. McFarland wrote in 1953 in his 'Human Factors in Air Transportation' that a pilot in poor physical condition is more subject to error and poor judgement. He also declared that a pilot in good physical condition is more apt to be mentally alert and have a greater capacity for arduous mental work.

The belief that there is a direct connection between physical fitness and mental performance is widely held and is clearly not new. Though scientific evidence to support this belief is scarce, the brain does depend for its functioning on a number of body systems and body chemistry. As these are influenced by physical exercise, the idea that there may be an association is not unreasonable.

Although any relationship between physical fitness and cognitive and psychomotor performance remains speculative, there does indeed appear to be an established relationship with emotion, so providing a physiological–psychological link (Morgan, 1984). There also appears to be an association, though not necessarily a causal one, between physical fitness and mental health. A reduction in depression has often been demonstrated to follow aerobic exercise programmes and it has appeared that these could be more effective than psychotherapy in the treatment of this disorder (Greist *et al.*, 1979). The improvement does not appear to occur, however, until late in the programme and as many who enter such programmes tend to drop out early, this conclusion may be very significant. Vigorous physical activity has also been shown to reduce tension and anxiety to a degree comparable with that achieved by the tranquilliser meprobamate (de Vries *et al.*, 1972).

In controlled studies, various researchers have found increases in extraversion, self-confidence, self-awareness, and in the elderly, improvements in memory following physical fitness programmes (Fentem, 1978). It is not surprising, therefore, that physical fitness is likely to lead to improved self-esteem.

We may have already established a link between physical fitness and cognitive and psychomotor performance, albeit an indirect one. It is known and has been well demonstrated that physical fitness can have favourable emotional effects. It is also known that emotions can affect motivation, which is addressed more fully in Chapter 6. Motivation is imposed upon capacity to determine how a person will perform in a given situation. Thus, even though it may be difficult to establish a direct link with cognitive and psychomotor capacity in laboratory conditions, in practice, an indirect link via motivation may exist. It is also believed, of course, that physical fitness provides resistance against fatigue, which also has an adverse effect on the performance of cognitive and psychomotor tasks.

While scientific evidence relating mental performance with physical fitness is scarce, this is not so with respect to the performance of physical tasks and tolerance to environmental stresses. There is strong evidence of physiological improvements when those with a sedentary life-style take up regular exercise. They report an improvement in their physical work capacity and find less effort is required to carry out given tasks. They suffer less fatigue and recover faster (Fentem, 1978).

We should look now at six factors which are known to have an influence on fitness, as defined here, and so may also have an influence on performance.

Factors affecting fitness

Exercise

Benefits
Three general benefits can be attributed to fitness achieved by a suitable exercise programme. Firstly, it is normally reported by those who have achieved physical fitness that they feel better. This sensation improves morale and motivation and can be expected to influence performance at work and harmony at home. Less irritability can favourably influence the *L–L* interface discussed in Chapter 1. The improvement in postural tone of the skeletal muscles should be of assistance to those who are required to sit for long periods in an aircraft seat, reducing discomfort and distraction from this source. Moderate exercise improves the appetite and physically fit people often say they enjoy their food more. Once physically fit, eating does not cause such an increase in weight. This is due to the changed activity of the body enzymes.

The second benefit of exercise is related to better health. The body deteriorates if it is not used. Overweight contributes to arthritis, backpain, breathing problems and cardiovascular disorders. Optimum weight charts are available in numerous publications (e.g. DHSS, 1979). There is much evidence that regular exercise reduces the chance of having to face two of the big killers of our time, heart attacks and high blood pressure. And, for those with flying licences to consider, it also reduces the risk of premature retirement on medical grounds. Exercise improves the appetite. More specifically, physical conditioning results in an increase in the stroke volume of the heart which results in a lower heart rate and thus a rapid improvement in the efficiency of the heart. It also results in better extraction of oxygen from arterial blood, which may be related to a more effective redistribution of blood to the working muscles. Peripheral blood circulation regulation also improves and this can be observed by a decrease in systolic blood pressure for a given workload. Tolerance to heat and cold also appears to be greater (Myles *et al.*, 1974).

One of the major enemies of the cardiovascular system is fat. This is deposited in the bloodstream as a result of noradrenaline production which can arise from emotions such as anger, aggression or frustration. High cholesterol (fat) levels in the blood increase the risk of atherosclerosis. Noradrenaline is also produced by exercise, but in this case, the fat it produces is burned up by the physical activity, thus protecting the system.

The consequences of the tensions and stresses of daily life are increasingly seen as damaging to health and performance. Continual emotional stress without overt activity can have a cumulative adverse effect. Exercise tends to dissolve the tension and stress.

The third of the benefits of exercise is that the fit person is likely to look better and this acts as a reinforcement to feeling better. This effect influences performance through an enhancement of motivation.

Types
An effective physical conditioning programme should contain three sorts of exercises (Fig. 4.1):

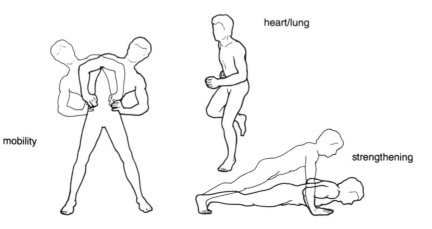

Fig. 4.1 One example each of mobility, strengthening and heart/lung exercises (Health Education Council).

- mobility exercises; to assure that all the major joints and muscles move freely through their full range
- strengthening exercises; to provide some cushion of reserve for a special occasion, such as lifting a heavy piece of luggage or pushing a car
- heart and lung exercises; to improve the condition of the heart muscle and increase the amount of oxygen that the body can process in a given time.

These exercises are sometimes called aerobics.

It is rare for normal work to provide those kinds of exercises necessary to ensure fitness. The calorific cost of piloting an aircraft, for example, is only about half of that of driving a train and one tenth of that of cross-country skiing. Yet even in the cases of work involving high calorific cost, the regime of exercise is unlikely to be a properly balanced one. The best way to achieve this kind of fitness is through one of the systematically designed and medically approved programmes which are now available. It is possible to learn all that is necessary from a book (e.g. Carruthers *et al.*, 1976; Cooper, 1970; Steen, 1979). In the 1970s, the Douglas Aircraft Company in Long Beach sponsored a physical conditioning study programme, using pilots as test subjects. Within two to three months of starting the programme, clear subjective and objective improvements in physical condition and tolerance to stress of all kinds were reported (Gaume, 1981).

In addition to these overall fitness programmes, special exercises have been recommended by Finnair for use in flight where movement is restricted. A problem experienced by flight deck crew members and passengers is discomfort in the lower legs, which has been correlated with foot swelling, following long periods of sitting. It has been demonstrated that these problems can be significantly reduced by using modest leg activity while sitting (Winkel *et al.*, 1986). Exercises for the eyes have also been recommended (see Appendix 1.7).

It is desirable that any programme to enhance physical fitness should be enjoyable and not simply a chore. There will certainly be some who find no direct pleasure in the systematic fitness programmes involving more formal training sessions. Many possibilities are available for those favouring less organised methods, though it must be recognised that these are unlikely to produce such a balanced degree of fitness.

Programmes

Modification of life style. At a modest level, it is very easy to modify one's approach to transportation and personal movement. It is possible to take the staircase instead of the lift, at least part of the way. We can take the bus for only part of the journey, and then walk the rest. We can take the dog for a walk more often, even without the dog. And make part of a long lunch a long walk.

Walking and jogging. Walking is good as an introduction to more effective physical activity, but unless it is very vigorous and extended, it is unlikely to provide the intensity of effort needed to achieve and maintain full fitness. Jogging is the next step and much evidence is available to support its claim to be a very effective contribution to fitness and health. It should be approached, however, intelligently. There are certain heart, lung, blood vessel and other physical conditions which would make vigorous exercise, including running up a flight of stairs, dangerous. Anyone aware of such a medical problem should first consult a physician familiar with exercise programmes before embarking upon such activity. As a general principle, it is not a bad idea to have a medical check within a few months before starting a fitness programme and recommendations are available for such a check for different age groups (Cooper, 1970). There are a number of conditions when it is advisable not to jog at all (see Appendix 1.8).

When engaged in vigorous exercise such as jogging, the heart rate should be monitored occasionally. A good rule of thumb is 200 less one's age, less an unfitness handicap which is taken as 40 for the very unfit down to zero for the very fit. It is important to note the time it takes for the heart rate to return to normal after the exercise as this can also be seen as a measure of fitness. As a general rule, it should be below 100 within ten minutes but details can be found in relevant fitness books. Breathing rate should also return in a few minutes to normal.

Cycling. As a non-programmed form of exercise, cycling also has much to recommend it though, like jogging, it has to be vigorous enough to keep an elevated heart rate for an extended period. In urban areas this is not always easy to achieve.

Swimming. This is an excellent form of exercise and has many advantages over other sports. Movements are smooth and rhythmical. The body is supported. It is often available in tropical countries where high temperatures discourage other kinds of vigorous activity. But as with other physical activities, it must be vigorous enough to generate a sustained rise in the heart rate; simply paddling or floating is not enough.

Ball games. Generally, these are less effective, as it is only the running which provides the significant exercise. This is often in just short bursts, with short-term, high stress placed on the muscles. Games such as golf and cricket are no doubt good for morale but have no significant effect aerobically, that is, to provide protection for the heart and lungs. However, they do contribute a little as mobility exercises.

Hatha Yoga. This is probably the oldest existing physical culture system and is the form of yoga best known in the West. It is based on a subtle coordination between physical activity and breathing. Certain of the exercises are also suitable for use on board an aircraft. Some flying staff using the method report improvement in general fitness and an increase in tolerance to stress.

Proper exercise is an important element in achieving physical fitness. There are a number of processes by which fitness can be destroyed and we should look briefly at these.

Smoking

Probably none of the fitness-destroying processes causes more emotional response than smoking. In the aircraft cabin, disputes with smoking as their origin are common and sometimes lead to acts of personal assault. On the flight deck, too, disputes due to smoking are increasing. These aspects are discussed in Chapter 13; here we should just look at some facts as now available from research and statistical analysis.

There are three significant components of tobacco smoke from a health and performance standpoint. Tar, which has been demonstrated to cause cancer in animals. Nicotine, which is associated with cardiovascular health and has an effect on performance. And carbon monoxide, which is a toxic gas created by incomplete burning of any organic matter in the cigarette and is also associated with heart troubles. The presence of carbon monoxide also reduces oxygen distribution to the brain and body and thus directly affects performance. There are numerous other possibly harmful constituents of tobacco smoke (DHEW, 1979).

The study of Human Factors teaches that statistics should be used judiciously and this caution must be applied to the mass of statistics published in connection with smoking. In particular, a distinction must be made between a causal relationship, when a cause and effect have been directly demonstrated, and a statistical relationship, where one element is common to two components but where it has not been demonstrated that one results from the other. In the case of smoking, most of the evidence shows a statistical relationship between smoking and ill-health rather than a causal one. However, the statistical relationship has become so overwhelming that few now doubt that a causal relationship exists. Nevertheless, this does not rule out other factors such as genetics and a person's constitution playing a role in the relationship.

In the UK, USA and elsewhere, major studies have been published linking smoking to cancer and cardiovascular disease. They have also demonstrated the effects of filters and what occurs after giving up smoking as well as the different effects on men and women (e.g. Royal College of Physicians, 1983). We should look a little more closely here at the effect of smoking on performance. The effect on physical performance as a result of the damage to the heart and lungs is well-known. In the UK alone, the cost to the National Health Service of treating diseases attributed to smoking was quoted for 1981 as about £155 million (Finsberg, 1982). Perhaps less well appreciated are the psychological effects.

Nicotine

Nicotine is the source of satisfaction and addiction in smoking. It results in increased adrenaline and noradrenaline output and a raised level of physiological arousal, even though the smoker may at the same time claim to feel more relaxed. This phenomenon (Nesbit's paradox) has not been adequately explained. It has been suggested that it may arise from the increased feeling of self-confidence to cope with current tasks, thus leading to a reduction in anxiety associated with the performance of the task. Various psychological explanations have been offered for this paradox (e.g. Schachter, 1973). The claim that smoking favours an ability to concentrate may arise from the fact that cigarette-induced arousal seems to be associated with increased selectivity of attention. This gives the impression to the smoker that he is focusing attention on relevant rather than irrelevant aspects of his current task. It has been demonstrated that a smoker's reaction times increase when he is deprived of his tobacco. Also that because of the raised arousal level generated, the gradual decrease in efficiency which occurs in monotonous or boring tasks can be counteracted to some extent by smoking (see Fig. 2.4). However, there is also evidence that when complex tasks are involved, performance may be worse in the smoking condition than in the deprived condition (Elgerot, 1976).

There is some evidence to suggest that the increase in arousal has an influence on memory with short-term memory being somewhat worse and long-term memory somewhat better. While various interpretations can be placed on these findings, they do tend to support some earlier theories on arousal and memory (Myrstan *et al.*, 1978).

When cigarettes and alcohol are used together, the arousal level is higher than when using alcohol alone. The depressant effects of alcohol seem to be counteracted by smoking.

These influences can be used to predict the effect of short-term abstinence of tobacco in the habitual smoker. Such effects are in addition to emotional aspects of enforced abstinence such as irritability, annoyance and frustration.

Carbon monoxide

A second component of tobacco smoke which is related to performance is carbon monoxide. Although cigars and pipe tobacco produce more carbon monoxide than cigarettes, the levels in the blood are usually higher with cigarette smokers because they generally inhale more. While filters in cigarettes reduce the level of tar in the smoke inhaled, they appear to increase the amount of carbon monoxide which is allowed to pass into the lungs and thence the bloodstream.

A little physiology is relevant here. Oxygen is usually collected from the lungs and transported through the body and the brain by a component of the blood called haemoglobin. Haemoglobin is attracted to carbon monoxide 210 times more than it is to oxygen. Consequently, when both are present in the lungs the carbon monoxide will be distributed in preference to the oxygen resulting in an oxygen deficiency in the brain and body. When carbon monoxide is combined with haemoglobin it becomes carboxyhaemoglobin (COHb). It is likely that carbon monoxide is the factor causing the less efficient exercise observed in smokers. This is because the main determinants of a person's tolerance of heavy physical activity are the maximum oxygen intake and the ability to transport oxygen from the air to the working tissues. Aerobic performance of smokers is significantly worse than non-smokers (Cooper *et al.*, 1968). Performance of the non-smoking endurance athlete could be adversely affected by exposure to the smoke of others. It must be concluded that any exposure to carbon monoxide will cause a deterioration in tolerance of exertion in proportion to the increment in COHb (Shephard, 1982).

As a baseline, a non-smoker has about 1–3% COHb in the blood. A smoker has about 4–10% depending on how much he smokes and whether or not he inhales. When reducing smoking, the COHb levels fall as the number of cigarettes is progressively lowered.

The COHb of a non-smoker may rise to about 5% when he is sitting in a badly ventilated, smoky room. He is then exposed to Environmental Tobacco Smoke (ETS) and becomes what is known as a 'passive smoker'. The smoke drawn in by the smoker, called mainstream smoke, is largely retained and only a small part is exhaled. When the cigarette is not being drawn, so-called sidestream smoke is generated and passes directly into the air and is breathed by passive smokers. Although the constituents of mainstream and sidestream smoke are essentially the same, they are present in quite different proportions, mainly as a result of the different temperatures of burning. Exhaled mainstream smoke also has the constituents in different proportions from the inhaled smoke. It is noteworthy here that some irritant gases, such as ammonia, and some cancer-forming substances, such as dimethylnitrosamine, are present in far higher proportions in sidestream than mainstream smoke (DHEW, 1979). Evidence is accumulating that ETS may pose a health hazard to the non-smoker (Royal College of Physicians, 1983). And he frequently suffers discomfort, annoyance and sometimes an allergic reaction.

Another passive smoker who is often forgotten is the unborn child. The mean birth weight of babies born to smoking mothers is less than normal and the stillbirth rate is much higher. It is not clear whether the damage results from the nicotine or oxygen deprivation as a result of the carbon monoxide. Pregnant women and the newborn also have an increased vulnerability to cigarette smoke (Shephard, 1982). A reduction in the baby's weight has also been shown to occur when the mother is only a passive smoker but the father is a smoker. A Danish study using 500 women found that on average the baby's weight was reduced by 120 g for each packet of cigarettes (or cigar or pipe equivalent) smoked by the father a day (Rubin *et al.*, 1986).

A wide variety of studies have been made on the effects of carbon monoxide on CNS function. Reductions have been shown experimentally in visual discrimination and judgement of time intervals with 4–5% COHb (Beard *et al.*, 1970); manual dexterity and memory effects with 7% COHb (Bender *et al.*, 1972); vigilance task performance, which is the earliest detectable effect of carbon monoxide with 2–4% COHb (Beard *et al.*, 1967, 1970). Although there is some discussion on the experimental basis of a number of studies, most researchers are prepared to accept that some deterioration of psychomotor function caused by carbon monoxide can be expected at blood COHb levels from about 3%. Similar levels can also adversely affect ischaemic heart disease. At high altitude, even a COHb level of 1.5%, easily reached with passive smoking, can present some danger to a cardiac patient (Shephard, 1982).

Safety

Oxygen deprivation resulting from smoking-induced COHb levels can be compared with the hypoxia effect of altitude. It can be seen that a cabin altitude of 5000 ft and a COHb of 5%, the equivalent physiological altitude is 10 000 ft (Fig. 4.2). In 1976, the Airline Pilots Committee of the Public Citizens Research Group petitioned the FAA in the USA in connection with smoking on board. The petition aimed to amend part 91 of the FARs so as to prohibit smoking on the flight deck of commercial aircraft and prevent any crew member from flying within eight hours of smoking. Arguments were presented related to visual and other performance decrements resulting from smoking, both with respect to the smoker and the non-smoker or passive smoker. The FAA Civil Aeromedical Institute in Oklahoma City reviewed the literature on the subject and reported its findings in 1980. It concluded that it was possible that smoking had contributed to some general aviation

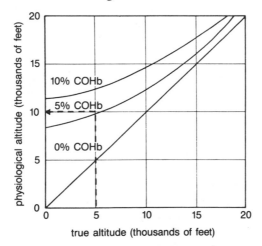

Fig. 4.2 The effects of carbon monoxide on altitude tolerance. For example, a person with 5% COHb at 5000 ft experiences the physiological effect equivalent to 10 000 ft altitude (McFarland, 1953).

accidents in the USA. However, the factors involved in these accidents seemed far less likely to apply in commercial aircraft operations. The FAA felt that additional studies of smokers' performance of realistic tasks, the effects of withdrawal and of aircraft accidents and incidents were needed before conclusions could be drawn. It nevertheless also endorsed anti-smoking programmes for what it described as 'improved health, maintained levels of performance and safety'. The petition was based on the possible effects of smoking on performance and not health and so the latter was not considered in the FAA study (Dille *et al.*, 1980). Later the FAA prohibited smoking in the passenger compartments of airline flights within the domestic US. The pilot compartments were specifically excluded because it was feared that withdrawal symptoms could adversely affect the performance of some pilots.

Although smokers sometimes report unpleasant symptoms such as depression, irritability and difficulties concentrating when deprived of their tobacco, psychological tests have not shown the deterioration in performance to the degree smokers themselves expected in this situation. However, there is some experimental evidence to suggest that as a result of the general arousal increase from smoking, performance of the smoker may possibly be reduced if deprived of smoking when carrying out certain boring or monotonous tasks (Myrstan *et al.*, 1978). It should also be remembered that irritability and frustration caused in the smoker who is deprived of his drug or in the non-smoker who becomes a passive smoker against his will, are not conducive to harmonious and effective working relations.

Trends

The long-term solution is the gradual eradication of smoking on all airline flights. As has been previously noted, non-smoking is no longer a voluntary exercise in the US. One study reported that in the UK four out of five pilots were non-smokers (Sloane *et al.*, 1984). In the short-term, the most reasonable objective may be to allow the smoker when possible to indulge in his chosen habit. But under such conditions that the non-smoker is not required to breathe any of the polluted air generated by the smoker exercising this freedom. Paradoxically, although this might suggest the aircraft toilet as a suitable smoking location on board, this is in fact designated specifically as a non-smoking area. This is due to a demonstrated fire hazard with the risk of total aircraft destruction (see Chapter 13).

Finally, in case it is thought that anti-smoking campaigns are new phenomena, it might be recalled that King James I of England (1603–1625) reacted with some strong words about the smoking made popular in the country by Sir Walter Raleigh – apart from having him beheaded. In his 'Counter-blaste to Tobacco', the King said that smoking was 'lothesome to the eye, hateful to the nose, harmful to the braine, dangerous to the lungs, and in the black stinking fume therof nearest resembling the horrible Stigian smoke of the pit that is bottomless'. But his denunciation appeared to have been ineffective and only in the last part of the 20th century, as smokers became a minority of the population in many developed countries, did non-smokers again become more vociferous. Backed very belatedly by prestigious medical organisations, they began to repeat the King's strictures in a somewhat more modest language but with an intolerance which seemed to grow with each success they achieved. His Majesty would have been well pleased with the current trend of events.

Alcohol

Another socially acceptable drug in most parts of the world is alcohol. Many cultures throughout history have co-existed with alcohol as a harbinger of merriment and relaxation. Yet it must be recognised as a drug which is as powerful and complex in its actions as many drugs which are available only on the prescription of a physician. It is used socially to induce an altered state of consciousness.

In the USA it was said in 1971 that the cost of alcoholism in industry was about $10 billion a year and the overall cost of alcohol abuse was estimated at about $25 billion a year (Royal College of Psychiatrists, 1979). But here we are concerned not so much with financial cost but with well-being and performance and thus safety. And not so much with alcoholism as a disorder but with the routine use of alcohol in daily life, though the distinction may be a little blurred. People who take just a few drinks a day socially risk becoming alcoholics within some years; particularly those who have a drink because they feel they need one. In a survey of British airline pilots, 98.7% reported drinking alcohol. This occurred on average four times a week, on each occasion averaging four drinks. About 12% reported drinking as a way of coping with stressful situations (Sloane *et al.*, 1984).

The extent to which the origins of alcoholism lie in genetics, the socio-cultural environment or in personality is not clearly defined. In fact, in some cases psychiatric illness is an individual predisposing factor. Nevertheless, a mass of research information is available about why people drink harmfully, but as reflected in Bertrand Russell's quotation in Chapter 1, the major problem is to get the message across. After all, abolition of the rum ration in the British Royal Navy was recommended in 1834 but not implemented until 1968. Good, easily readable and authoritative information is now available on the subject (e.g. Mohler, 1980; Royal College of Psychiatrists, 1979).

Performance

The use of alcohol is of particular concern in an industry where a high level of performance is mandatory. Alcohol impairs discrimination and perception in the visual and auditory systems. It disrupts short- and long-term memory, the thinking and decision-making processes and coordinated hand–eye movements. It slows reaction times. It lowers inhibitions and increases recklessness; in Rome 2000 years ago Seneca wrote, 'drunkenness is nothing but a condition of insanity purposely assumed'. Even at social drinking levels, thought processes are adversely affected (Parker *et al.*, 1977). After a substantial drinking session, higher mental and reflex functions can be affected for two to three days after stopping the drinking (Ryback *et al.*, 1970).

A particularly dangerous aspect of alcohol consumption is that the drinker feels his performance is improving, but, in the words of Porter in Shakespeare's Macbeth in a slightly different context, 'it provokes the desire, but it takes away the performance'. As a result, risk-taking increases and errors of judgement pass unnoticed.

There is a striking similarity in the behaviour of a person suffering from hypoxia (lack of oxygen) and one under the influence of alcohol. This applies not only to the manner in which performance is reduced but also in the increasing self-confidence which accompanies this degradation and the insidious manner in which the effects are gradually manifested. In aviation, it can be said that altitude in-

95

creases the performance-degrading effect of alcohol.

Metabolism

About 20% of the alcohol drunk is absorbed in the stomach and 80% in the intestines. It is then carried around the body in the bloodstream. As soon as it enters the bloodstream it begins to affect judgement, behaviour and skills.

The rate at which alcohol is absorbed depends on various factors such as total body weight, the food eaten previously, the speed of drinking and the type of drink. It is absorbed much faster from the duodenum than from the stomach so the rate at which it passes the stomach has a fundamental effect on the rate at which it is absorbed in the blood and circulated in the body. It appears that drinks of medium alcohol concentration have a higher absorption rate than those with either a high or low concentration. Other constituents of the drink also affect the rate of absorption, probably as a result of their effect on gastric motility and perhaps also on regional blood flow. When diluted to the same alcohol concentration, that in beer is absorbed slower than whisky, whisky slower than gin and gin slower than red wine (Becker *et al.*, 1974). Sugars in sweet drinks slow down absorption while the carbon dioxide in drinks such as sparkling wines and spirits with soda water, accelerate the absorption. The faster the rate of absorption the higher the peak blood alcohol concentration (BAC) achieved. Taking a meal before a drink may reduce the peak alcohol level by almost 50%.

BAC is measured in milligrams of alcohol present in 100 millilitres of blood and this is often abbreviated to mg%. After stopping drinking, the BAC level can be reasonably expected to fall by about 10–15 mg% an hour, though the metabolic rate varies widely between individuals. Furthermore, not only is the toxic effect of alcohol dependent on the time of day when it is taken, but the rate of elimination, or more accurately, conversion in the liver, also appears to vary with a circadian rhythm. The rate of metabolism has also been reported to vary with other factors such as the use of oral contraceptives by women (Sturtevant *et al.*, 1979).

Safety

As the BAC rises, the brain and nervous system are increasingly affected and changes in behaviour occur. Deterioration in brain and body functions can be readily detected, depending on the task, from a BAC of about 20 mg% upwards. Many studies have been conducted, notably in transport and industry, on the relationship between BAC and accidents (e.g. DOT, 1985; Hore *et al.*, 1981; Metz *et al.*, 1960; Observer *et al.*, 1959). From different studies it has been demonstrated that in industry those with more than 30 mg% have considerably more accidents than others and that the accident rate of those who abuse alcohol is three times that of other workers. French studies concluded that 15% of accidents serious enough for work to stop were caused by alcohol; afternoons were a particularly bad time of day. In road transport it has been shown in the USA that those with 80 mg% had double the sober accident risk, with 150 mg% ten times and with 200 mg% twenty times. Another estimate quotes the accident risk doubling with 0.85 litres (1½ pints) of beer, increasing to 25 times with 2.85 litres (5 pints). In the UK in 1984 more than 30% of drivers killed and 30% of those involved in an accident had above 80 mg% BAC. In France the percentage was higher still.

In the aviation environment there is less 'drinking and driving' than on the roads and the sample of information from accidents is too small for meaningful statistical analysis. Some experimental work has been done, however, using a Cessna 172 aircraft. The study required experienced and inexperienced pilots to fly ILS approaches with various levels of blood alcohol. Glide slope and localiser deviations, as well as procedural errors were recorded (Fig. 4.3). It is interesting to note that of

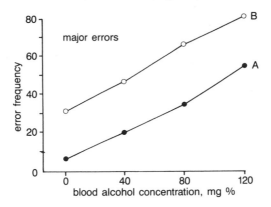

Fig. 4.3 Major errors in procedures made on ILS approaches in flight by experienced (A) and inexperienced (B) pilots in a Cessna 172 with various levels of blood alcohol (Billings *et al.*, 1973).

the 32 flights performed at each BAC (40 mg%, 80 mg% and 120 mg%) it was necessary for the safety pilot to intervene and take over control once (40 mg%), three times (80 mg%), and sixteen times (120 mg%) (Billings *et al.*, 1973). It has also been demonstrated that drinking hangover effects can cause a seven-fold increase in flight planning errors (Wise, 1980).

In spite of all that is known of the impairment of performance due to alcohol, some 50% of general aviation pilots in the USA responding to a survey, stated that in their view it would be safe to fly four hours after drinking alcohol. Analysing the answers further, it was estimated that 27–32% of the respondents considered flying after drinking within a time period which would result in a 15 mg% BAC or higher to be a safe behaviour. The relationship between these attitudes and actual behaviour can only be hypothesised. However, it was noted that about 20% of general aviation pilots killed in accidents in the USA were found to have a BAC of 15 mg% or higher (Damkot *et al.*, 1978).

In commercial aviation, evidence of working under the influence of alcohol is usually difficult to obtain and in any case it is no doubt less than in general aviation. Nevertheless, dramatic reminders occur from time to time to demonstrate that the problem cannot be ignored. One such case was the DC8 which crashed after take-off at Anchorage in 1977 killing all on board. The pilot was found to have three times the BAC permitted for driving a car (NTSB-AAR-78-7). Occasionally, violations of alcohol regulations are brought to the attention of the FAA. In each of the three consecutive years 1980–1982 a civil airliner crash occurred where the influence of alcohol on the pilot's performance was cited in the report, though these did not involve major operators.

ICAO decrees that a history of alcohol addiction is a permanent bar to a flying licence (Annex 1, paragraph 6.2.4.3). However, licensing authorities revalidate a licence if treatment appears to have been effective (Schwartz *et al.*, 1978). The situation facing flight crews is more severe than in many other occupations. Duty-free alcohol is available on most international flights. It is customary for crew members to drink socially at night-stops, away from home and family, and these sessions can sometimes develop into heavy drinking. In the previous chapter, reference was made to the use of alcohol by many crews to 'wind down' after a flight and to facilitate sleep – alcohol is a CNS depressant. Because of the habitual use of alcohol in this way, some will inevitably become addicted. The first sign that a person is becoming addicted is likely to appear to personal friends and working colleagues rather than to a medical department or licensing authority. They therefore carry responsibility to apply whatever personal influence they have to achieve a change of course and encourage acceptance of counselling or medical intervention; guidelines are available for aiding recognition (HEC, 1974) and a summary is provided in Appendix 1.9.

Drugs

The 20th century's tool for the management of stress is drugs, that is, tranquillisers, sedatives, hypnotics. And then, sometimes, amphetamines to counteract the resulting drowsiness. Furthermore there is a tendency to look to drugs to relieve the symptoms of a wide range of minor ailments such as headaches, colds and allergies. In road transport a highly significant statistical association has been shown to exist between the use of tranquillisers and accidents (Skegg *et al.*, 1979). No comparable data are available in aviation though it is known that such drugs are used extensively by flight crews to obtain sleep in the disturbed biological rhythm environment of long-range flying, as was discussed in the previous chapter. A survey carried out in the 1970s in one European airline revealed that 25% of the flight crew members interviewed admitted to using such drugs sometimes or frequently in the working environment while nearly a further 21% admitted to rare use. At home, away from work, the comparative figures were 2% and 8%, respectively, confirming that the problem was an occupational one (Hawkins, 1980).

A 'safe' drug for a general medical practitioner is primarily one with low toxicity. Even in the 1970s one major airline was counselling its pilots that nitrazepam (Mogadon) was a 'safe' drug for routine use by flying personnel. As one of the benzodiazepine group it is certainly less toxic than a barbiturate, though when it is in the body together with alcohol its effect can be potentiated to an unpredictable degree. But as with all such drugs, it has an adverse effect on the performance of skilled tasks and so safety is given another and less quantifiable dimension. This was discussed in Chapter 3 and reference should be made to Appendix 1.4 for precautions which should be taken before using this kind of drug.

In the USA alone, there are some 1300 basic prescription-only drugs on the basis of their generic or medical name. However, when marketed by the pharmaceutical industry they are given trade names, and there are about 7000 of these, some of which are the same drug under different names and some are combinations of the generic drugs. But the over-the-counter, non-prescription drug field is even more extensive. In the USA there are some 300 000 of these. While these are reasonably safe under certain circumstances, the fact that no prescription is required for their purchase does not imply that they do not affect the performance dimension of safety. In particular, antihistamines are known to cause drowsiness and these are used in anti-cold, anti-allergy and anti-sickness remedies. Some influence coordination.

Drugs may be classified in six major groups with respect to their effect on performance and their acceptability in flying (Mohler, 1983). In any case where doubt exists, reference should be made to a physician familiar with performance requirements in the flying environment.

We are not concerned here with the socially unacceptable drugs, such as lysergic acid diethylamide (LSD) and the opiates like heroin, opium, methadone, and morphine; there is simply no place for these and other psychotropic drugs in aviation.

Stress

Any authoritative work on stress must inevitably involve setting forth into the semantic jungle surrounding the word itself; this in turn leading to similar hazardous expeditions with motivation and fatigue. Such an adventure is beyond the scope of this book and in any case has been bravely undertaken elsewhere (e.g. Murrell, 1978). For our purpose, a stressor may simply be defined as an event or situation which induces stress. This pressure and the resulting stress may have adverse results, such as job dissatisfaction, reduced work effectiveness, behaviour changes or health damage; one might say, the human system is no longer able to cope and begins to break down.

The extent to which this breakdown follows, depends on the individual's response to the stressor; on his adaptive capability. More than a century ago people were writing of the 'stresses of modern living'. At that time aviation was confined to balloons, so one must be cautious in identifying stress with aviation, or any other particular industry or occupation, for that matter. Or even the nature of modern society.

In aviation

Nevertheless, the aviation industry does involve a cocktail of stressors which is unique when combined with a critical need for a high level of performance. Flight crews, particularly long-range flight crews, face an unusual collection of ingredients in the cocktail. Early examination of flying stress goes back a long way, though stress then was interpreted in physiological terms only (Flack, 1918) and as noted in Chapter 1, this resulted in the involvement of physicians. Perhaps the earliest stressors recognised were those created by the immediate environment such as noise, vibration, temperature, humidity extremes and acceleration forces. Many of these now have less significance than formerly and have been replaced by more complex factors, the implications of which have not always been fully understood; not least, their impact on motivation and other behavioural characteristics. Working/resting patterns for crews while away from base in long-range flying are now totally irregular; not even the degree of regularity involved in shiftwork is maintained. Upon this irregular working/resting pattern is superimposed the disturbing influence of transmeridian flying on the circadian rhythms. Cases of pilots, sometimes both pilots, dropping off to sleep during cruise flight at night are certainly not unknown, as officially revealed from confidential incident reporting and as long recognised as occurring by airline crews (e.g. CHIRP Feedback Nos 1, 2 & 9).

To the physiological circadian rhythm disturbances is also added the affective (emotional) stress associated with family separation for nearly one third of the working life. This can be particularly disturbing when the children are growing up and a wife needs support. In a UK survey of the wives of airline pilots, domestic role overload was cited as the aspect of life causing most dissatisfaction (Cooper *et al.*, 1985). Domestic stress on the pilot has been shown to have an effect on flying efficiency (Haward, 1974).

Other affective psychological stresses are also present, such as the feeling of insecurity induced by six-monthly medical and proficiency checks. The pilot, unlike most other professional people, is routinely required to perform, in perhaps hostile conditions, with a company check pilot or state aviation inspector looking over his shoulder, constantly posing a psychological threat to his security. Job security, notably in smaller airlines not cushioned from competition by state protection, and particularly under deregulation policies, is a worry reported by pilots and their wives. Role overload and other role pressures are occasionally experienced by flight crews, as by other staff with responsibility, and some degree of fear of flying may also exist in certain cases. Even with the young and enthusiastic pilots in military aviation, affective stress is also present. One study cited 71% of pilots as admitting to being worried by personal and domestic problems during the previous year (Aitken, 1969).

It has long been known that affective stress can produce certain characteristic types of pilot error (Davis, 1949). While earlier studies were concerned with the wartime environment, long-range commercial flying has introduced its own stressors. Later studies also revealed impairment of flying skills with affective stress (Haward, 1984).

At certain times a crew member may suffer from cognitive stress, as distinct from affective stress. It might be described as the mental workload becoming excessive. This is most likely to occur at the more critical and high workload phases of flight such as take-off and landing, particularly in adverse weather conditions. It may also arise in emergency conditions in any stage of flight.

The typical personality of a pilot may not be an optimum one for handling emotional problems. It has been suggested that he is a person who typically denies his internal emotional life. Under stress he is likely to seek a constructive solution and acts out his frustrations if he does not succeed in achieving a particular objective. If he should be unavoidably confronted with his emotional life, he may possess inadequate strategies for coping with the situation (Ursano, 1980). Studies of psychosocial stress in pilots, conducted in the UK, concluded that a relatively high percentage of pilots were identified as having a health risk, as categorised by their overall mental ill health assessment (Sloane *et al.*, 1984).

The pilot lives a life of deadlines. He is under constant pressure to maintain a public relations image. He is exhorted endlessly to be disciplined, responsible, vigilant, dedicated; to avoid complacency and to be economically conscious. He works under threat of immediate media spotlight in the event of an incident or accident. Yet to admit to suffering from the pressures upon him may be seen, in a society which extols achievement and competitiveness, as an admission of failure. And so, too often, the existence of the related symptoms are denied by the individual and ignored by his company. Until, that is, they become apparent through behavioural changes such as aggressiveness or a reliance on alcohol, or through sickness, such as hypertension, heart disease or peptic ulcer or reduced performance. This might be too late to avoid an accident and maintain operational and commercial efficiency. It may also be too late to save the individual's career.

Domestic influences
In discussing occupational stress it is not possible to ignore the interaction between work and non-work factors. It is delusory and unhelpful to instruct staff to leave their domestic problems at home; a person has but one brain and body to suffer from the effects of stress at home, and these cannot simply become healed on arriving at work. Frustrations at home can usually be expected to be reflected in attitudes at work; frustrations at work can frequently be expected to influence harmony at home.

Changes in the pattern of life can create pressure and some correlation has been shown between these changes, and accidents and incidents. In this connection, attempts have been made to weight various life changes in terms of so-called Life Change Units (LCUs). These weightings are based on the personal importance attached to events by a large number of people surveyed in the USA (see Appendix 1.10). Although each event may not, in isolation, significantly affect health and performance, they are cumulative. Even positive events, such as a wedding, can induce stress in normal life. As the total number of LCUs increases, it has been shown that the risk of becoming involved in an accident increases. Using this concept, a person with 300 LCUs may have double the risk of being involved in an accident compared with someone with only 150 LCUs. An awareness of this general relationship is the first step to applying protective measures (Holmes *et al.*, 1967). The hypothesis that pilots who contribute to aircraft mishaps are more likely to be identified with having troubles in interpersonal relations and experiencing life change events has also been supported by other studies (e.g. Alkov *et al.*, 1982).

Management
We referred earlier to the two most commonly applied means of managing stress, alcohol and drugs. But these both have disadvantages which make them unsuitable. It must again be emphasised that damage from a stressor arises from an individual's response to it, rather than the stressor itself. While one finds managerial responsibility stressful, another finds it stimulating and pleasurable. One finds flying in monsoon conditions challenging and interesting; another finds it simply exhausting and unpleasant. The stressor is the same; the response is different.

It would be a happy solution if we could all simply avoid having to face situations which we find stressful, and there is indeed some merit here. Generally speaking, we all have to face such situations domestically and at work and the remedy must rest with a modification of one's response to them.

A number of techniques have been developed to induce a more relaxed and healthy response and these have been outlined elsewhere (Hawkins, 1980). One in particular, Autogenic Training, has been applied systematically in the aviation environment and this has appeared to be effective in providing a tool to manage stressful situations and to reduce the adverse consequences, such as sleep difficulties, which often arise from them (Hawkins, 1984a). Sportsmen, schoolchildren, industrial workers and businessmen, have all claimed improved performance after using this method. Whichever method is used, as was mentioned earlier, it has been determined that non-specific, self-relaxation attempts are usually ineffective for the management of stress symptoms and the technique employed should be a properly developed and learned one (e.g. Bootzin, 1981; Nicassio *et al.*, 1974).

Increasing recognition of the problem of stress within the pilot community, and the apparent lack of adequate supporting measures from elsewhere, have led to the formation in some national pilot organisations of a Pilot Advisory Group (PAG). This group consists of peers in the piloting profession. Ideally, members of the group should have been specially trained, though in fact, they are likely to be high in piloting experience, with a caring personality, but possessing only modest qualifications, if any, in social or occupational psychology. They should be fully equipped, nevertheless, to know when and to whom to refer colleagues seeking or accepting advice; realistically, this is the most effective contribution they can make.

This group has the primary task of providing guidance for those colleagues with stress-related difficulties, sometimes reflected in drug or alcohol abuse. But it also has a role in providing psychological and motivational support for those pilots who may for one reason or another be off flying for an extended period. The group may also have a role to play in the trauma which may confront a pilot following an incident or accident. It has been said that the traumatic effect of a hijacking on a crew member has been generally neglected (Johnston, 1985).

Whether or not PAGs are introduced in pilot communities depends on numerous factors such as attitudes towards stress, human behaviour and performance and the extent to which the link between stress and safety is recognised. Their success depends on the confidence which they are able to command, which in turn is related to the personality and skill of members of the group and the links they have established with specialists engaged in stress management and therapy. Introductory notes on the PAG system were circulated to member pilot associations by IFALPA (document L84b168).

Some time before the PAG system was introduced, the Association of Flight Attendants (AFA) in the USA, the world's largest representative body for cabin staff with more than 21 000 members, established an Employee Assistance Programme (EAP) with similar objectives. The AFA has pointed out that flying staff have little routine contact with their supervisors – perhaps only once in a three-month period – so that timely recognition by them of stress-related problems is unlikely. Contact with peers – working, eating and night-stopping together – is much closer and there is frequently a supportive relationship between crew members. The peer-referral EAP initiative capitalises on this relationship.

The EAP committee members are full-time flight attendants some of whom have previous AFA committee experience and some have backgrounds in the caring occupations. They must be respected, trusted and well liked by their peers. Many of these committee members have received specialised training averaging over 100 hours each from AFA national staff and consultants. This training covers a comprehensive range of topics including recognition of behavioural problems, alcoholism, drug abuse, care of rape victims, eating disorders, domestic violence and various other common sources of stress. The results of the first few years of the AFA EAP have been encouraging (Feuer, 1984).

At a more specialised level, a comprehensive programme can be put together to tackle the problem of stress at work. This also encompasses stress which may be social or domestic in origin (Fig. 4.4). The keystone of the programme is the test which provides for each individual an analysis of his own psychological and bio-chemical stress profile. It can be used as an early warning system to reveal signs of an unfavourable stress response so that timely remedial or protective action can be taken. The three-part test utilises a comprehensive computerised questionnaire, a personal medical/psychological interview and a blood analysis (a single finger-prick sample). From the blood analysis, levels of stress-related components, such as cholesterol, glucose, uric acid, GGT and free fatty acids, are measured. The psychological, biochemical and interview information, including blood pressure readings, are then professionally analysed and a comprehensive and confidential report provided for the participant. The analysis indicates the individual's stress level and potential for stress-related disorders (Harris, 1986).

While this analysis can be conducted in a normal medical environment, in the UK a mobile stress monitoring service can also be made available to visit industrial or business organisations and provide the analysis in-house. The analysis will reveal the various sources of the stress condition. The programme then makes available follow-up facilities to deal with any potentially adverse indications. Here, two different channels are provided, each under specialised guidance. Firstly, on an individual basis, a personal counselling service, a stress response modification programme (such as Autogenic Training) and in more severe cases where immedi-ate remedial action may be indicated, treatment perhaps involving medication.

The second remedial channel relates more directly to the working environment

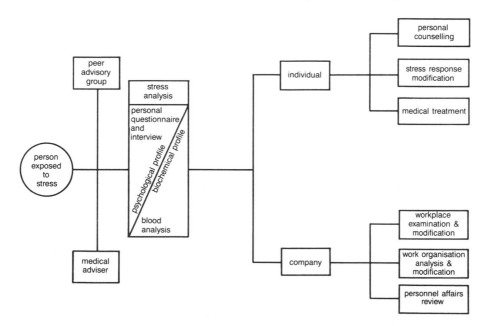

Fig. 4.4 A comprehensive programme to tackle latent or active personal stress conditions. The programme encompasses stress with origins in the domestic, social and work environments and draws upon expertise in psychology, biochemistry, medicine and ergonomics or Human Factors.

and this will be conducted by an ergonomist or Human Factors specialist. It will be recommended when the stress profile suggests that work factors are contributing to the stress condition. This phase of the programme will examine the workplace, including environmental factors such as noise, lighting, temperature, humidity, space, and other ergonomic aspects of the equipment and furnishing. It will study the work organisation, including aspects such as the role of the individual with the company, authority, responsibility, communications, personal interactions and duty periods. And thirdly, it will examine personnel affairs such as questions of company support, behaviour reinforcement, pension and sickness benefits and job security, to determine whether any of these may represent a source of stress.

Such a programme, applying the skills of applied and clinical psychology as well as medicine, provides a comprehensive and effective approach to the problem of stress in aviation or any other occupation. However, its acceptance depends upon a degree of enlightened response from management, staff and possibly unions and must be operated on the basis of full personal confidentiality.

Diet

The primary function of food is to provide energy to build and repair the tissues of the body. There is much evidence to associate a badly planned diet with ill health and low performance (DHSS, 1977). There has been a shift away from complex natural carbohydrates such as fruit, vegetables and grains, towards an excess of fats and refined sugar and refined carbohydrates. This has been described as 'incredible nutritional folly' (Airola, 1977). A knowledge of the necessary constituents of food and the proportion of the total food intake which they should represent is necessary if a balanced diet is to be achieved.

Carbohydrates are absorbed rapidly and provide one of the chief and most immediate sources of energy. They are the main fuel used by the muscles and the only fuel which can be used by the CNS. All carbohydrates are converted in the body to glucose. Sugar is a carbohydrate and is a rapid source of energy. However, the quantity eaten generally provides far more calories than are required and this may lead to obesity, heart disease and dental decay, amongst many other disorders. Refined sugar is a standard ingredient in a huge variety of processed food. In the USA it is estimated that only one third of the refined sugar is bought as such, the rest being eaten as a component of processed food. A British adult consumes about 43 kg of sugar a year; in the USA the per capita consumption is about 10 kg more than in the UK.

Fats provide the most concentrated source of heat energy in the body. They can be stored in the body in larger quantities than any other food substance and, because they take longer to digest, defer the sensation of hunger. They contain the bulk of fat-soluble vitamins and contain twice as many calories as carbohydrates. It is widely believed that in general people eat too much fatty food. The shift in eating habits towards meat and dairy products has notably unbalanced the diet; even a lean sirloin steak can be 10% fat and some cream cheeses are about 50% fat. Fats which are not consumed are stored in the body and lead to obesity and increase the risk of heart disease.

Proteins are needed for the building and repair of body tissues and are composed of smaller units known as amino acids. Proteins are broken down relatively slowly and provide a more lasting source of nourishment than carbohydrates. There are different sources of protein and it has been suggested that vegetable protein (e.g. in potato) may be preferable to animal protein (e.g. meat, milk, eggs). Furthermore, protein eaten raw, is preferable to that cooked.

Minerals have no energy-producing value but are essential for maintaining health. For example, iron is needed for the production of haemoglobin, which was mentioned earlier in this chapter and which transports oxygen from the lungs to the brain and body tissues.

Water is a very important part of the diet. It prevents dehydration of the intestinal contents, helping to prevent constipation, often reported by flying staff on irregular flight schedules. The amount of water needed depends on the rate of dehydration and this, in turn, depends on the aircraft cabin relative humidity and the level of personal physical activity. This is discussed in Chapter 13.

Fibre is a complex mixture of mostly indigestible plant substances, including what used to be called 'roughage'. Throughout man's evolution, his daily diet came mainly from plants such as unrefined cereals, pulses or legumes (e.g. beans, peas or lentils), root vegetables, fruit and nuts, and his bodily systems developed upon this basis of dietary fibre. The general shift towards refined foods, meat and dairy products has reduced the fibre intake to the extent that this deficiency is now suspected of contributing to numerous disorders from constipation and haemorrhoids to diabetes and certain forms of cancer. Fresh fruit and vegetables are high in fibre. Wholemeal bread has three times the fibre content of white bread.

Vitamins are complex substances which the body requires in very small amounts. Deficiencies can cause symptoms such as a feeling of malaise, proneness to infection and slow healing of wounds. In rare cases, more serious sicknesses can occur, such as scurvy (vitamin C), rickets or osteomalacia (vitamin D), beri-beri (vitamin B1) and anaemia (vitamin B12).

Each vitamin is found in certain foods; some are richer sources of vitamins than others though none contain the entire range. Food processing can reduce vitamin content, so in this respect it is wise to use the least processed foods. Vitamins are either fat-soluble (e.g. A, D, E and K) or water-soluble (e.g. B and C). An excess of water-soluble vitamins is probably not harmful as these will be excreted from the body. However, an excess of the fat-soluble vitamins can be dangerous as these will be accumulated in the body with a potentially poisonous effect. Water-soluble vitamins are removed from food with boiling or soaking of the food. Major sources of vitamins in food are listed in Appendix 1.11.

A balanced diet should provide in proper quantities for the body's needs with respect to:

● nutrients (proteins, fats, carbohydrates, vitamins, minerals and water) – the raw materials needed to build and maintain the body. It is most unlikely that a person eating a normally varied diet will be short of proteins, vitamins and minerals. It is probable that there will be an excess of fats and carbohydrates. In special cases, such as pregnancy, some supplements are advisable.

● energy (calories) – a source of power for the body. Usually, there is an excessive intake of calories.

● dietary fibre (a complex mixture of plant substances) – a component of food which is essential to proper functioning of the digestive and bowel processes and its lack is increasingly believed to be associated with other disorders. It is generally deficient in modern diets.

This general overview of dietary factors is not comprehensive and excellent small, authoritative and easily readable books on the subject are available elsewhere (e.g. 'The Manual of Nutrition' and 'Eating for Health', published by the British Government, HMSO, London).

5

Vision and Visual Illusions

Light

Characteristics and measurement

Light is formally defined as radiant energy which is capable of exciting the retina of the eye and producing a visual sensation. In other words, it is visible radiant energy. Light waves are a narrow band somewhere near the middle of the total radiant energy spectrum. Immediately adjacent to the visible spectrum are the areas of ultra-violet radiation (which causes sunburn) and infra-red (causing heating), both biologically important.

Measurement of light is called photometry and there are several concepts and terms associated with it. One problem is the multiplicity of different units used for measurement, some based on US units and others on the Système International d'Unités (SI). For the purpose of this chapter, the simplified data in Table 5.1 are sufficient. Intensity, illumination and luminance are the three basic terms.

Table 5.1 Three ways in which light can be described and the SI and US units associated with each description.

Origin	Point Source	Falling on Surface	Reflected from Surface
description	intensity	illumination	brightness or luminence
photometric units	candela candela (US)	lux footcandle (US)	candela/m² footlambert (US) millilambert (US)

The original standard was the standard candle, which produced an intensity of 1 candle-power. The present standard, the candela, is about the same, but is more precisely defined related to platinum at 1773°C. Illumination is calculated by the inverse square law, that is, if the distance from the light source is doubled, the illumination becomes one quarter. As the SI units are based on the metric system, and 1 m^2 is about 10 ft^2, it can be said that approximately 10 lux = 1 foot candle.

After light strikes a surface some of the energy will be absorbed; each different surface will absorb a certain combination of wavelengths. Reflectance of a surface is the ratio of the amount of reflected light (brightness) to the amount of light striking it (illumination). The panels and structure on the flight deck should obviously be treated with a low reflectance paint so as to minimise glare. Paintwork for the lettering of certain signs may require a high reflectance paint.

In relating different kinds of lamps to the illumination provided, it is not enough simply to compare the wattage of the lamps. For example, a 30 watt fluorescent lamp can typically provide 30% more light output, last 20 times longer and consume 40% less energy than a 75 watt incandescent lamp. Even incandescent bulbs of the same wattage can emit quite different amounts of light. Selection of the optimum kind of lighting, then, can affect workplace efficiency, use of energy and cost.

Another common term related to illumination is contrast. This is the relationship of brightness or luminance of a target to its background or surround. A contrast of 100% (positive) means the background brightness is twice that of the target.

A further word which we may come across in discussing vision is refraction. When light passes from one medium to another of a different density it is bent, or refracted. Each medium has its own refractive index. The angle of refraction is also a function of the wavelength of the light, so that it is broken into spectral colours when passing through a prism.

Colour

Variations in the wavelength of energy within the visible spectrum give rise to the perception of colour. A light source which includes all wavelengths in about equal proportions is called white light. However, most light sources have more of one wavelength than another and this creates a more blue, yellow, green or red appearance. The colour of a light source is measured by what is called colour-temperature. This relates the colour to that of a heated piece of iron which gives a similar colour appearance. For instance, the midday sun may have a colour-temperature of about 5500K. When cockpit lighting changed from red to white, a low colour-temperature white was generally specified. This was slightly reddish so as to preserve to some extent dark adaptation of the eye.

The light waves entering the eyes consist of different wavelengths which are later resolved as different colours. In the USA about 6% of healthy adult males have a notably reduced sensitivity to colour, though only about 0.003% are completely colour-blind. Less than 1% of women suffer from this condition, which is a congenital, inherited, sex-linked defect. It is not a disease and cannot be cured. It results from a defect in the structure of the colour-sensitive cones in the retina of the eye. The most common form it takes is red/green blindness, which afflicts about 4% of the US male population. Red and green are seen as shades of yellow, yellowish brown and grey. This phenomenon is obviously important in the selection and use of colour coding for purposes such as warning systems, displays, electrical engineering, and documentation (Van Cott *et al.*, 1972). However, it generally does not create a severe handicap in normal life, as visual acuity is not affected and a person is often not even aware of the defect until undergoing a special test (Ishihara colour-blindness cards or a computer graphics technique). It may prove a bar to certain occupations where safety or fine colour discrimination are factors.

The eye

Physiology

The eyes are often considered as man's primary gateway to knowledge. From a structural point of view the eye can be likened to a camera (Fig. 5.1). There is an aperture or pupil which controls the amount of light entering the eye and this is

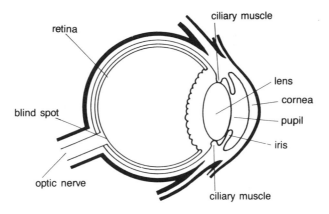

Fig. 5.1 The principal features of the human eye.

adjusted by the muscles of the iris. The iris comes in a variety of colours which are of greater significance to the poet and song-writer than to the physiologist. Light then passes through a lens which, together with the cornea (the transparent window on the front of the eye), focuses light on to the back of the eye, the retina. The retina corresponds to the film in the camera. The image on the retina is reversed and inverted.

Focusing on an object is called accommodation, which is the process by which the refractive power of the eye is modified for viewing nearby objects (Fig. 5.2). This functions in a different way from a camera, where the lens is moved backwards and forwards. With increasing age, the lens tends to harden making the task of the ciliary muscles more difficult and so accommodation becomes less efficient.

The retina consists of a complex layer of nerve cells. From the retina of each eye an optic nerve passes to the brain and where this nerve leaves the back of the eye there is a blind spot, devoid of any light receptors.

The light-sensitive cells of the retina consist of two kinds of receptors called rods and cones, each having quite different characteristics. The centre of the retina is called the fovea and the receptors in this area are all cones. Moving outwards, the cones become less dense and are gradually replaced by rods, so that in the periphery there are no cones.

The cones are used for direct vision in good light and are colour-sensitive. This is

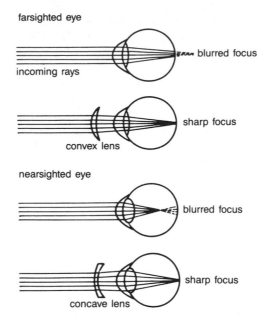

Fig. 5.2 Farsightedness and nearsightedness, the former being corrected with a convex lens and the latter with a concave lens.

called photopic vision. The rods are used in dim light and have no sensitivity to colour. This is called scotopic vision (Fig. 5.3).

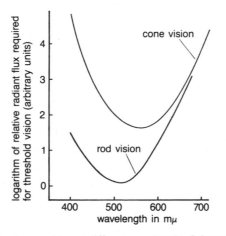

Fig. 5.3 Threshold rod and cone vision at different wavelengths. It is apparent that for detection of very low brightness levels, rod vision is used.

Each eyeball is held in position by six extrinsic muscles and these control movement of the eye under direction of the brain. They also provide for convergence of the eyes for depth perception or stereopsis. These muscles are under constant tension and form part of a finely balanced system which, when disturbed, can give illusions of movement.

110

The visual system at work

When the amount of light entering the eye changes, an adaptation takes place. This takes a little time, which varies with age (Fig. 5.4). This adaptation takes place in

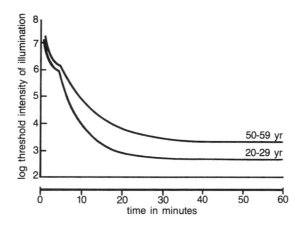

Fig. 5.4 Dark adaptation of the eye in subjects of different ages. A steady value, that is, full adaptation, is only reached after about 40 minutes. It can be seen that younger subjects are much more sensitive than older subjects and dark adapt more quickly.

two different ways. Firstly, a coarse adjustment is made by the pupil to allow more or less light into the eye. The pupil has a limited capacity of varying the light to which the eye is exposed by only 30 times, while the eye works efficiently over a total range of light intensity of more than a million to one.

A second and more significant process involves the action of the rods and cones. As the light decreases, the sensing task is passed over from the cones to the rods. This is apparent in the break in the curve in Fig. 5.4. The sensitivity of the rods involves a pigment called visual purple (rhodopsin) which is bleached in bright light and takes a certain time to be reconstituted as the light decreases. By this time, the light is inadequate for the cones, which are no longer functioning. As the light decreases, so sensitivity to colour is also lost – a garden of colourful flowers gradually becomes various shades of grey as dusk descends. The rods, it will be recalled, are not sensitive to colour.

As the fovea contains no rods, which would be required for vision in very low brightness levels, the centre part of the eye becomes blind in dim light. It is then necessary to look somewhat away from the visual target so that the peripherally located rods can perform their sensing task.

In order to reduce the time taken for the eyes to adapt to low brightness levels, wartime pilots used red light in their crew room. The rods are less sensitive to the red, longer wavelengths (Fig. 5.5). Red was also used in their cockpits so as to maintain their dark adaptation and also reduce the risk of being seen. In the early post-war years it was concluded that maximum rod dark adaptation, which in any case takes some 30–40 minutes, was not a civil aviation requirement. Furthermore, that the use of low colour-temperature white light resulted in only slightly inferior dark adaptation to red, and that the advantages of white were very significant (see Appendix 1.12).

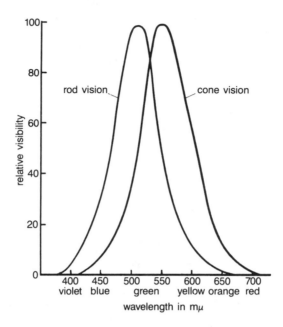

Fig. 5.5 Relative visibility of rod and cone vision at different light wavelengths. These curves show the relative visibility of different wavelengths that have the same amount of energy. Note that the eyes are more sensitive to longer wavelengths in daytime, viz when the cones are working. At night, when the rods are working, the shorter wavelengths are more visible (after Troland, 1934).

There has been a tendency in post-war civil flying to use higher brightness levels in flight deck lighting. Laboratory research suggests that little improvement in visual acuity, the ability to visually discriminate between adjacent objects, can be expected above 0.05–0.10 footlamberts. However, experience indicates that for long night flights, fatigue and drowsiness seem to be less with higher brightness levels. Furthermore, with increasing age, visual acuity can only be maintained by utilising increasing brightness. Up to the age of about 40, both static and dynamic (that is, when the observer or target is moving) visual acuity remain reasonably constant. However, after this age, both decrease more rapidly with dynamic visual acuity being only about half at the age of 65 of what it was at 40 (Burg, 1966).

Visual acuity is expressed in terms related to the smallest letter which a subject can read on a chart at 20 ft compared with the distance at which a normal person can read it. For example, 20/20 is normal vision and 20/40 would mean that the subject can read at 20 ft what a normal person can read at 40 ft.

Visual acuity can be affected by a number of factors such as brightness, brightness ratio and contrast, time to view the object and glare. Not only is glare a source of discomfort – it sometimes makes people feel tense and restless – but experiments have shown that visual performance is reduced. This depends on how close the glare source is to the line of sight. One study showed that with the glare source at 40° from the line of sight, the loss of visual effectiveness was 42%, while when only 5° away, the loss increased to 84%. The glare was half the level of the general illumination (Luckiesh *et al.*, 1927). Sensitivity to glare increases with age. One of the most prevalent beliefs is that glare causes visual fatigue. Other effects claimed include psychological stress, fixation and disruption of accommodation. However, very few studies have attempted to assess these effects. In fact, there are no objective means for measuring visual fatigue and an experimenter would have to rely on subjective opinion. While there may be some substance in the many claims made, further studies are needed before definitive conclusions can be reached. Literature reviews of the subject are available (e.g. Koffler Group, 1985).

Lighting specifications establish certain photometric brightness levels for displays. However, the features of a particular display, such as the spacing of graduations or the exposure of tapes, may result in an apparent brightness which makes it appear different from other displays which have the same photometric brightness. This introduces the question of the relationship between the information sensing function of the eye and information processing, the perception function, which is discussed in the next section.

To see clearly at different distances, two basic adjustments of the eye are needed. These adjustments change the refractive power of the lens, which we already noted is called focusing or accommodation, and change the convergence of the eyes, which is called binocular vergence. When the visual cues are weak, such as with empty field visual conditions in the daytime or at night, the muscles controlling these adjustments tend to take up an intermediate or resting state. This resting state – empty field or dark focus – is not at infinity, as was for long assumed, but on average at a distance of just under one metre, though there are wide variations between individuals. This is highly significant in searching for distant targets when visual cues are weak, as the eye will not be adjusted to detect them. The condition is aggravated by the fact that when there are other objects close to the eye, focusing is drawn closer to the observer. This is called the Mandelbaum effect and is more severe the closer the other objects are to the dark focus distance (Mandelbaum, 1960).

Conventional eye tests are generally adequate for predicting visual performance under normal conditions, with distinctive targets. Yet two individuals who have been rated as having 20/20 vision may differ greatly in their visual performance when it comes to tasks such as seeing signposts or pedestrians when driving at night. It has been known since the early days of jet flying that some pilots have more difficulty than others in detecting other aircraft at a distance either in the daytime or at night (Whiteside, 1952). The ability to detect outside objects when there is rain or other contamination on the windscreen can also vary considerably between individuals. The windscreen of cars and aircraft is often close to the dark focus distance for many people.

There may be several factors involved in these unpredicted differences in visual performance, but the difference in resting state or dark focus of the eyes may be a major element. The development of the laser optometer has allowed a person's dark focus to be measured easily. One study using student subjects showed the mean dark focus to be 67 cm (1.5 diopters) with the pattern amongst subjects approximating to a normal distribution (see Chapter 12). About 1% focused at infinity while others focused as near as 25 cm (4 diopters). Such wide dissimilarity could not be forecast from conventional vision tests (Owens, 1984).

Visual acuity appears to be optimal at the observer's dark focus. This suggests the use of individually adapted viewing distances for displays, where this is feasible. It has been shown to be possible to prescribe corrective glasses to improve vision when performing in conditions when the eyes tend to revert to their dark focus or empty field state of rest. Such corrective glasses are also effective for those with 20/20 vision.

Visual acuity can cause a real problem for ageing pilots. As a part of ageing the lens gradually loses some of its elasticity usually starting about age 40. This is an entirely natural phenomenon but the loss interferes with near vision. Near vision is important for pilots because they need the ability to see clearly at different lengths. Their visual tasks require three primary types of visual accommodation. The basic distances that are important are: near – 33–41 cm (13–16 in) which is normal reading or overhead panel distance; intermediate – 75–90 cm (30–35 in) which is the instrument panel distance; and infinity, i.e. the outside world. All three are important for safe flight.

At about 40 years of age most eyes need help in the 33 cm (13 in) range. This can be in the form of a lower reading glass segment (the so-called Ben Franklin glasses). Unfortunately, eyes continue to age and somewhat later accommodation at the instrument panel begins to deteriorate. This loss of accommodation is such a gradual process that it is often not obvious. The result, however, is increasingly fuzzy instruments, less accurate scanning and an increased potential for misreading some of the gauges or other indicators.

Many pilots have used trifocals to deal with these problems and have adjusted to them well. Others have secured different types of special corrective glasses. The important point is that help, if needed, is available through knowledgeable flight surgeons or opthalmologists.

The brain

Perception

Visual perception involves the eyes, the balancing mechanism of the ear (the vestibular apparatus) and the brain. It is also influenced by past experience. Stepping onto a moving walkway or an escalator when it is stationary gives a strange sensation. A conflict has arisen between the visual message, which says it is stationary, and past experience which says it should be moving and that an acceleration should be sensed in the vestibular apparatus and muscular action taken to maintain balance.

Perception is also forward looking and predictive. Fast ball games would not be possible without this element of what is expected to occur in the future, anticipating the curving service of a good tennis player or the movement of a flight instrument needle in the cockpit. When prediction is not possible, response suffers a delay. And so perception is greatly influenced by what we have come to believe about the properties of the object we are viewing and also what we expect the properties to be by the time we have to respond to them.

Different components of the body's sensing and processing systems are involved in this activity (Fig. 5.6). The interaction is such that the eye has been described as an outgrowth of the brain. The retina contains typical brain cells between the light receptors and the optic nerve and these modify the electrical activity of the receptors themselves. Some of the processing of visual information, then, is occurring in the eye itself. It is as though a specialised part of the brain surface has been transferred a short distance away to the rear of the eye.

Objects themselves are not uncertain or ambiguous and the image of the objects is usually reproduced faithfully on the retina – inverted and reversed. Uncertainty and ambiguity only occur when this correctly sensed information is processed by the brain. The visual input is then combined with other information based, for example, on learning, experience and expectation. Emotional factors also play a role in perception.

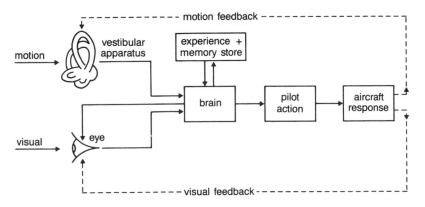

Fig. 5.6 A simplified model illustrating some of the components involved in visual illusions.

There is a blind spot in each eye, as noted earlier, the missing information from which is filled in by the other eye when using binocular vision. But even when one eye is closed, the blind spot is not apparent unless a specific test is made to demonstrate it (Fig. 5.7). This has relevance to visual search in flying.

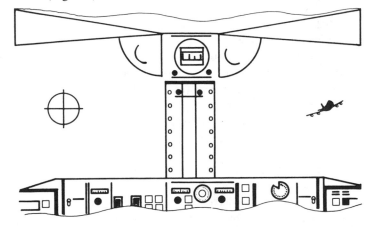

Fig. 5.7 For a demonstration of the blind spot, hold the drawing at arm's length, close the left eye and keep the right eye open. Now move the picture towards the face keeping the right eye focused on the cross. The aircraft will disappear and then reappear as it gets closer.

115

If vision of an object by one eye is obstructed, say, by a windshield post, then a visual target could be in the blind spot of the other eye and remain undetected.

Sometimes the information we require cannot be obtained from visual input alone. If we wish to determine a person's mood or attitude, his facial expression provides some basic clues but these need integration with other information before a conclusion is drawn. On the flight deck or in a flight simulator, a warning light can announce that a system is faulty, but previous experience may suggest that the warning is probably based on a faulty signal and that it can be ignored.

Another influence in connection with the performance of visual tasks is called fascination. This refers to the failure to respond to a stimulus even though it is within the normal visual field. This phenomenon generally arises from one of two causes. Firstly, the observer may be concentrating so much on one part of the visual task, say, the flight director, that he fails to respond to another more important or contradictory information source. Secondly, the observer may adopt a trance-like concentration on a display, receive the message, but not respond to it. This phenomenon of fascination may be aggravated by fatigue or anxiety.

When perception is influenced by expectation this is sometimes called set. In flying operations it can result in landing at the wrong airport, misinterpreting a light on the ground or in the aircraft or be the cause of other operational errors. The common saying that one sees what one expects to see is a reflection of this phenomenon. The principle also applies to the processing of aural information as we shall see in Chapter 7. Reference was also made to this in the chapter on error.

When the frequency at which a light source is flickering increases, at a certain point it is perceived as a steady light. This is called the Critical Fusion Frequency (CFF). For light levels where cone vision is operative, colour of the light has no influence on the CFF. It is important for the designer of a display to know the CFF so as to try to avoid the sensation of flicker for the user of the equipment.

In Chapter 3, reference was made to environmental time cues or *zeitgebers*. These are responsible for maintaining and resetting the biological rhythms which are disturbed in long-range flying. A primary *zeitgeber* is the natural light/dark cycle, even though the advent of artificial light adds a complication to the entrainment processes of the body's systems. Recent research strongly suggests that the judicious use of artificial light can significantly and constructively modify entrained biological systems. It is not known just how to use this knowledge in air transport aviation.

In the base of the brain are the suprachiasmatic nuclei, which appear to play a major role in entrainment of the body's systems to a 24-hour rhythm. The most important route by which these time cues reach the brain is the optic nerve. There appears to be a direct path from the photoreceptors in the retina, designed to sense ambient light, to this rhythm control centre in the brain. Experiments on animals have shown that when the optic nerve is severed, cutting off all signals from the retina, then the animal's biological systems lose their circadian rhythm and 'free run'. While this particular pathway is very significant, it is thought that other neural pathways may also play some role in this very complex process (Moore-Ede *et al.*, 1983). This function of the eye, which is quite distinct from seeing objects, has an important bearing on personal well-being in the physiologically disturbing environment of long-range aviation.

Depth and distance assessment

Judgement of depth and distance is an important perceptual ability which people possess. A number of cues are used to aid in accurate depth perception, such as binocular vision, perspective, apparent movement of the object, superposition, relative size, height of the object in a plane, and texture gradient. As pioneer pilot Wiley Post had only one eye, depth perception is demonstrably not derived from stereoscopic vision alone. In fact, this only plays a role up to about 12 m distance, beyond which the other static and dynamic cues are employed.

Primitive art used no perspective. It was not until the Italian Renaissance that reality began to be translated into rules of geometry for application to painting and drawing. Leonardo da Vinci did much to introduce added effects other than geometry to give impressions of depth and distance. In looking at such art, binocular vision is of no value and the artist must make use of a knowledge of monocular vision cues. The artist, in fact, is not trying to draw what exists; he is trying to reproduce what he believes would be the image of the actual scene on the retina, such as distant people being made smaller and less clear than close ones or the far end of a road looking narrower than the near end. Applying a knowledge of the world about us, different sizes, shapes and shadows are interpreted in terms of depth and distance.

Hypoxia and smoking

Hypoxia, or anoxia, is the condition resulting from oxygen deficiency. Generally, oxygen reduction with altitude up to about 8000 ft (a common cabin altitude upper limit on long flights) is unlikely to have a significant effect on performance. Above this, the effect becomes increasingly significant. However, reaction is not consistent; individuals suffer differently in performance degradation and acclimatisation is also an influencing factor. A reduction in night vision capacity and other adverse effects occur insidiously as the oxygen supply is depleted.

Smoking has the effect of introducing carbon monoxide into the bloodstream and thus reducing the flow of oxygen to the brain (see Chapter 4). A heavy smoker may experience an oxygen level in the blood equivalent to 10 000 ft altitude when only flying at a cabin altitude of 5000 ft, so some adverse effects on vision from such smoking may be anticipated. The oxygen depletion effects of altitude and smoking are cumulative.

Experimental work has shown statistically significant deterioration in various visual functions, such as visual acuity, brightness threshold and reaction to visual stimuli when COHb levels rise above about 5% (Shephard, 1982).

In the previous chapter, reference was made to the FAA literature study of the effects on performance of smoking on the flight deck. This was carried out in response to a public petition to have smoking banned in a similar way to alcohol. The possibility of impairment of vision was one of the aspects examined. The FAA felt that although smoking may have had an association with some general aviation accidents it was less likely that it would have been associated with commercial aircraft accidents. It nevertheless endorsed anti-smoking activity 'for improved health, maintained levels of performance and safety' (Dille *et al.*, 1980).

Visual illusions

Geometrical optical illusions

Although all pilots will have suffered from visual illusions of one kind or another when flying, the majority of the illusions will probably have passed undetected unless they led to a noticeable event. It is therefore important to recognise that we are all vulnerable to being misled by the perception and interpretation process which is applied to visually sensed information. One way to assist in that recognition is through the demonstration of common and well established geometrical optical illusions.

For more than 100 years, psychologists have been proposing theories to explain the distortions of these illusions. Many theories have been discounted, others remain, but little of a definitive nature has emerged. For the purpose of this volume we need only to be convinced of our own vulnerability; more profound studies are available elsewhere (e.g. Gregory, 1977; Robinson, 1972).

A number of such illusions have been researched in considerable depth and a selection of these is given in Figure 5.8.

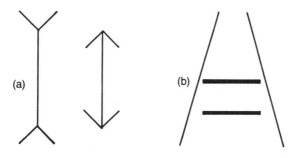

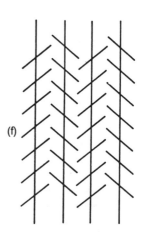

Fig. 5.8 A selection of common geometrical illusions.
a) Muller-Lyer illusion; the figure with the out-going
   fins looks longer than the other.
b) Ponzo illusion: the horizontal bar in the narrower part
   of the converging lines looks longer than the other.
c) Hering figure; the vertical lines look bent.
d) Poggendorf figure; the crossing line looks displaced.
e) Orbison figure; the square and circle look distorted.
f) Zollner figure; the vertical lines look tilted.

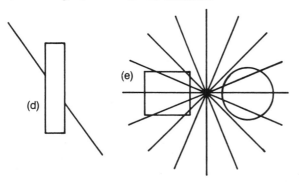

Depth and distance illusions

Of a somewhat different character are figures whose depth is ambiguous (Fig. 5.9). Looking at these figures it is possible to see changes taking place spontaneously without any apparent conscious effort. When some are rotated, the change in

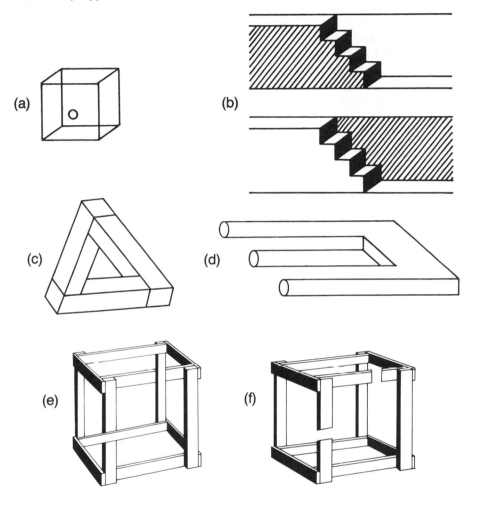

Fig. 5.9 A selection of illusions involving ambiguous perceptions of depth.

a) Necker cube; the small circle appears sometimes at the back and sometimes at the front of the cube.

b) Schroeder staircase; the steps change when the picture is inverted.

c) Penrose triangle; impossible to perceive as a real object.

d) Schuster illusion; three arms at one end becoming two at the other.

e) 'Freemish crate'; a totally false perception from a perfectly accurate picture.

f) the same construction as e) but drawn from a different perspective.

perception is sometimes quite sudden. As with geometric illusions, theories have been put forward for centuries but definitive explanations remain rare.

Other illusions involving depth and distance occur in moving displays. Phenomena called stereokinesis and kinetic depth perception can occur. An object rotating

119

on a vertical axis sometimes appears to reverse direction. An ATC antenna at an airport, rotating in the distance, may give this false impression.

Other illusions

Various other kinds of illusions have been identified. Brightness contrast sometimes creates an illusion. For example, a bright background or surround makes the centre area appear darker, and vice versa. Shadows appear at the junctions of white lattice bars on a black background. Numerous other examples of brightness contrast illusions have been cited.

Illusions of movement arising from stationary stimuli are not uncommon. An isolated stationary light in an otherwise dark visual field may appear, after observation for a while, to wander. This is called autokinesis and is typically encountered in flying.

Illusions of movement can arise from the action of the body's balancing mechanisms – the vestibular apparatus – and the visual system. There is an interaction between these two sensory systems. The somatogyral illusion refers to a false sensation of turning (vertigo) and may occur during a prolonged turn. Somatogravic illusions are those involving a false perception of attitude relative to the gravitational vertical. Neither of these are truly visual illusions as they originate from the balancing mechanisms of the ear rather than the visual system (see Fig. 5.6).

Visual illusions of movement may be oculogyral, when a false sense of turning is created, and oculogravic, which is a visual component of the altered perception following a change in the force vector. The physiological explanation for these illusions and the interaction between the vestibular apparatus and the eyes is rather too complex for further elaboration here and is well covered elsewhere (Benson *et al.*, 1973).

The ear's balancing mechanisms can be stimulated without any change in the rate of turn of a vehicle. If a person turns the head during a steady vehicle turn an illusion can arise resulting from what is called a cross-coupled (Coriolis) stimulation of the semi-circular canals of the ear.

Sometimes an illusion can arise from a moving rather than a stationary stimulus. The train moving at the next platform gives an illusion that our own train is moving. This is an illusion of induced movement.

And there are others. Illusions can arise from stimuli being presented in rapid succession. One example is the apparent slow backward movement of the wagon wheels in an old western movie. There are good specialised publications which analyse more deeply the theories behind illusion (e.g. Robinson, 1972). The message here is simply that they are complex and multifarious and that nobody is immune.

Vision and visual illusions in air transport

Flight deck geometry

The manufacturer of an aircraft, as well as the government certifying authority, assume that the pilot's eye is at all times located at a sort of cockpit keyhole, called the design eye (formerly reference eye) position (see also Chapter 12). Through this keyhole he must be able to achieve adequate visibility of the outside world and of the important displays within the flight deck.

A required visibility envelope for the flight deck of civil aircraft has been standardised by the Society of Automotive Engineers (SAE AS 580), though the design objectives are never fully met by manufacturers. The FAA no longer specifically defines the location of the design eye position; manufacturers now make use of SAE ARP 268 (as periodically updated) as guidance in the location of flight deck controls.

The effect of the pilot sitting too low is to defeat the object of the SAE standards. Not only may the pilot lose sight of some displays behind the control column and wheel but he will also lose part of the available outside visible segment in low visibility approaches (Fig. 5.10). By sitting just 2½ cm too low, 40 m of visible

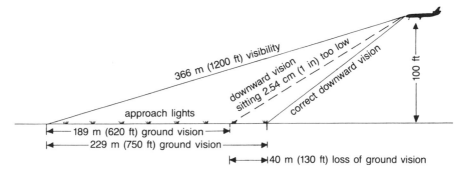

Fig. 5.10 An example of the loss of ground vision when the pilot sits 2.54 cm (1 in) lower than the manufacturer's design eye position.

segment, as well as precious time to discriminate and react upon the visual cues, will be lost in a Category 2 approach (see Appendix 1.13). It has been established that the design eye position should be located so as to allow the pilot to see a length of approach or touch-down zone lights which would be covered in three seconds at final approach speed (SAE ARP 4052/2). This represents about 200–250 m along the flight path.

A study some time ago in a DC8 flight simulator showed a very wide scatter vertically of actual eye positions, with some pilots sitting as much as 8 cm below the design eye position. As many accidents on approach have involved an apparent error of judgement, it is possible to speculate on the extent to which the pilot's seat and eye position may have been a contributing cause. Fortunately, the study also revealed that by including this item in recurrent training over an extended period, a notable improvement in the accuracy of the eye position could be achieved (Hawkins, 1977). Given the geometric downward vision angle from the design eye position and the aircraft deck angle on approach, it is possible to determine the available visible segment at different heights during low visibility approaches (Fig. 5.11).

Even with the devices installed in current jet transports to guide the pilot to the correct eye position, it is unrealistic to expect all pilots to sit at all times in this position. Pilots of some ethnic backgrounds may, in fact, have physical difficulty in doing so (see Chapter 12). It is worth noting that the FAA Special Air Safety Advisory Group (SASAG) in a 1975 report stressed the need for action to remedy

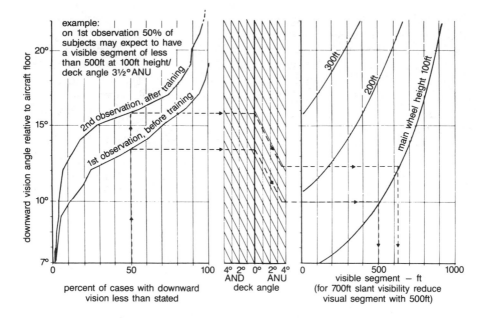

Fig. 5.11 With a knowledge of the downward vision angle of an aircraft type, together with the deck angle, aircraft nose up (ANU) or aircraft nose down (AND), it is possible to determine the ground visible segment available at different stages of the final approach. The chart also shows the improvement in pilot vision attained after training (Hawkins, 1977).

the operational deficiency which they had determined of pilots not sitting in the proper eye position (Fig. 5.12).

The upper surface of the glareshield should provide a lateral horizontal reference and should also coincide with the forward downward vision angle.

Bifocal and varifocal spectacles are in common use on the flight deck to compensate for the effect of ageing on eye accommodation. Their design is based on the usual practice of looking down to read and looking up for distance vision. On the flight deck, however, near vision may be required both above and below normal eye level and so lenses with the reading prescription only at the bottom are not optimum for working in this environment. To satisfy this particular application, varifocal spectacles have been developed which have a reading prescription at the top as well as the bottom of the lens (Aerospace, 1986).

General flight illusions

Low visibility approaches no doubt reflect a critical part of the flight. However, the predominant area of incidents and accidents arising from visual illusions seems to be in non-precision approaches, when the pilot reverts to visual control – often in quite good weather conditions. When a fatal accident occurs, the possibility of obtaining definitive evidence of a visual illusion as a contributory cause is remote indeed.

Many of the illusions occurring in flying involve the same principles as in those illusions already discussed. Some illusions are common to several stages of flight and so perhaps we should review these first.

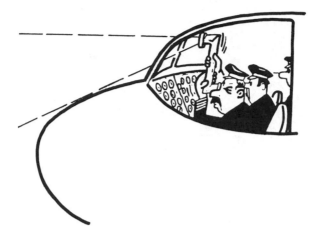

Fig. 5.12 Some pilots sit lower than the design eye position and can be expected as a result to suffer a reduction in outside vision on approach (cartoon: J H Band).

The autokinetic effect discussed earlier is typical of the night flight environment, when a stationary single light in a dark visual field appears to move. Another source of illusion may arise from flashing lights. This is called stroboscopic illumination and it can give rise to various illusory reactions. Some people react adversely to lights flashing at certain frequencies and while this is not an illusion in the usual sense, it does reflect an interaction between the eyes and parts of the brain which can influence performance.

The effect of rain on the windshield has been the subject of various studies without definitive conclusions having been reached. This may at least be partly due to the large number of variables involved. These include the exact airflow over the surface of the windshield, which may change with aircraft speed and attitude. The thickness pattern of the water over the surface may thus vary. Aircraft use various rain removal and repellant systems, such as compressed air, windshield wipers, and rain repellant fluid. In each case the pattern of water will be different and in any case is unlikely to form an even layer across the whole glass surface. It is more likely to be in the form of pools or drops of various sizes and depths. While this may create some distortion, it is likely to be unpredictable and inconsistent. Light transmission will certainly be adversely affected and this, together with the distortion of multiple lenses formed by the water on the surface, may cause some problems. Interpretation of visual cues will be more difficult and the time available for the task will be reduced.

Associated with rain on a windshield is the additional effect of the movement of the windshield wipers. This influence is similar to that of flicker, with some people being more susceptible than others, and has long been recognised by pilots of helicopters and propeller-driven aircraft. To minimise the effect, the wipers should not be operated faster than necessary and the pilot should not look continuously through them longer than is required (FSF, 1981).

Mist and fog can influence the judgement of distance (Fig. 5.13). Aerial perspec-

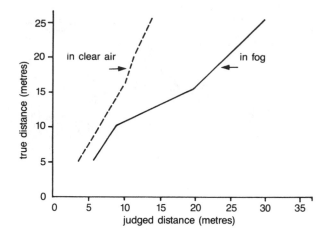

Fig. 5.13 The effect of fog on the judgement of distance (from Ross, 1975).

tive is the dimming of the image of objects with distance. This, together with the reduction of colour contrast of distant objects, acts as a perceptual clue to distance. Distances therefore tend to be overestimated in poor visibility (Ross, 1975).

The optical characteristics of the windows on the flight deck may give rise to difficulties. Electrically conductive material may be inserted between the layers of glass for ice-protection purposes. This material absorbs some of the light passing through the windshield. It is also a source of delamination which results in apparent bubbles within the windshield, effectively obstructing vision, usually at the edges. Typically, up to 7½ cm around the complete periphery (or 12½ cm in spots) of the main windshield may have deteriorated in this way before window replacement is considered by maintenance staff to be necessary.

Aircraft use different windshield shapes. The 747 uses a conical section, while the DC10 uses flat panels. Each has its own optical disadvantages. Often, there is an optical discontinuity between adjacent window panels, so a taxiway, for example, seen through two windows may appear broken or bent. This distortion may be inconvenient and disturbing on the ground, but with normal caution is unlikely to present a real hazard.

Obstruction of vision by windshield posts seems inevitable, except with the bubble-type dome on some light aircraft. These obstructions contravene the required vision envelope stipulated in SAE AS 580 and can lead to a failure to detect conflicting air traffic. There has been, nevertheless, a dramatic improvement in the forward vision provided since aircraft such as the Constellation and the Comet, if we exclude the special case of supersonic aircraft.

Illusions in taxying

Even before the aircraft leaves the gate we are exposed to illusions. As the loading bridge begins to move away an impression may be created that our own aircraft is moving, as in the train example cited earlier. At this stage of the journey it is unlikely to be hazardous. However, this illusion of induced movement (or lack of movement) may also occur when arriving at the gate at the end of a flight. There have been cases reported of inappropriate control action in these circumstances, with undesirable consequences such as setting the parking brake when still moving.

In winter, blowing snow may be sweeping across an airfield and this gives a false impression of relative movement, as any skier can testify. Quite inappropriate control action can be initiated based on this illusion. An impression can be given that the aircraft is stationary when it is still moving, brakes applied sharply, and cabin staff thrown forward and injured. Alternatively, the aircraft could slowly creep forward, colliding with an obstruction, when it was thought to be stationary.

Pilot eye height from the ground may be the source of misjudging speed. Flight instructors who are training pilots on larger aircraft than they flew previously, commonly report the phenomenon of excessive taxy speed. The 747 design eye position is 8.66 m above ground level, while that of the DC9 is only 3.48 m. Sitting higher, the pilot is taking his visual reference further away from the aircraft and this gives an illusion of reduced apparent relative motion.

Illusions in take-off

In transport flying the major accelerations longitudinally and in rotation occur during take-off and go-around. These could, if severe enough, give rise to certain illusions in which the body's balancing mechanisms and the eyes play a role (see Fig. 5.6). The somatogravic and oculogravic illusions were mentioned in principle earlier. An acceleration gives a pilot a sense of pitching up, for which the natural corrective control action would be to push the nose of the aircraft down. Deceleration gives the reverse impression. For a typical transport aircraft, the mean acceleration during the take-off run would be only about .007 g (assuming 45 sec to 140 kt). This will not cause a change in the force vector sufficient to cause an illusion. However, during rotation, a radial acceleration of about 0.5 g could occur (assuming an angular velocity of 4°/sec at 140 kt) and this could cause a momentary pitch down sensation. During initial climb, if the acceleration were high enough, a pitch-up illusion could occur, giving an impression of a steeper climb attitude than actually existed. However, these are mainly problems of military high performance aircraft with a line-of-flight acceleration of, say, 0.45 g (4.4 m/sec^2) and are not likely to be of significance in the take-off and climb of transport aircraft.

Somatogyral and oculogyral illusions are probably of little relevance during take-off. The angular rate of 3–4°/sec is close to the sensory threshold and would not be expected to produce any movement or displacement of visual targets.

Reference was made earlier to the illusion of movement resulting from blowing snow during taxying. This also has a relevance during take-off, where it can interfere with normal directional control. Sand can create a similar illusion.

Immediately after take-off, the perception of a false horizon can occur when surface lights are confused with stars or sloping cloud layers create a false horizontal. In take-off over water, lights of fishing fleets just off the coast have been known to be confused with stars, resulting in grossly inappropriate manoeuvres. A further illusion can occur when, after emerging from a shallow bank of mist or cloud just after take-off in hilly terrain, lights suddenly appear. The immediate impression may be that the lights are stars rather than on the ground and that the aircraft attitude should be adjusted to fly below them.

Sloping terrain can create an illusion at any time when flying visually at low altitude. After take-off, if the ground slopes down an illusion of excessive height may be created, and vice versa.

Illusions in cruise

Manoeuvres in cruise flight for transport, as distinct from military, aircraft are unlikely to be very significant from the point of view of illusions. However, prolonged turning as in a holding pattern, can create an illusion, particularly the cross-coupled (Coriolis) effect, if the head is moved while still turning.

Autokinesis, the apparent movement of an isolated light in a dark background, is a typical illusion associated with flying. And there have been numerous cases reported of mistaken identity of lights, such as the pilot circling round his wingtip light trying to identify it.

A common illusory problem experienced in visual cruise flight is in the evaluation of the relative altitude of approaching aircraft and the assessment of a potential collision risk. Mid-air collisions, particularly in the USA, are sufficiently common for this aspect of illusions to be taken very seriously. At a distance, an aircraft which initially appears to be at a higher level may eventually pass below the observer. Mountains at a distance which appear above the aircraft, often pass well below it. Two aircraft with a separation of only 1000 ft may appear to be approaching each other's altitude and experience has shown that both pilots can then, unnecessarily, take evasive action, possibly themselves initiating a collision. If a closing aircraft remains at the same spot on the windshield, this means that it is on a converging course and a risk of collision exists. If it appears to move on the observer's windshield, then no risk of collision exists while both aircraft maintain their current headings.

The blind spot, while of relevance at all times when flying visually, is of special significance in cruise flight where it may cause conflicting traffic to be missed.

Reference was made earlier to empty field and dark focus and this aspect of vision is important when considering searching for conflicting traffic. In such conditions the eyes tend to adjust to a focal distance on average of just less than a metre rather than to infinity, thus making detection of distant targets difficult. The Mandelbaum effect, drawing focus closer to the observer as a result of the presence of windshield posts and a windshield – particularly a dirty windshield – aggravates the problem (Mandelbaum, 1960). A powerful case can be made for keeping the windshield as clean and free from scratches, delaminations and reflections as possible.

Illusions in approach and landing

These are generally recognised as the most critical phases of flight and so visual illusions are potentially more dangerous than at other times. Crew members are most fatigued at the end of a flight, yet are then under the greatest pressure. Adverse weather conditions can have a greater influence on safety than during cruise. And time to make decisions is short and the consequences of error possibly catastrophic.

It is widely believed that visual illusion may have played a significant role in many otherwise inexplicable accidents. But, inevitably, the only tangible evidence for this is in the functioning brain of the pilot, which in fatal accidents is no longer available to the accident investigation team.

A number of the distortions discussed earlier also apply to this phase of flight but there are a few which call for special reference.

Sloping terrain can be very misleading in visual approaches. When the ground is sloping down towards the runway an impression is given of being too low, and vice versa (Fig. 5.14). A sloping runway can also give a false perception of height (Fig. 5.15).

The pilot is accustomed to seeing the runway making an angle to his approach path of about 3°. If the runway should slope down at, say, 1° then his apparent approach path would be only 2° and he would again sense that he is too low. It would not be unusual for the approach terrain and runway to slope in the same direction and so these two different illusory effects would then be cumulative. Downward sloping terrain and runway, then, for a too low illusion and the reverse for an up-slope.

Runway width may be a further source of distortion in perception during landing. When the runway is wider than normal it will appear closer and the aircraft will appear lower than they really are. And, of course, vice versa.

Lighting intensity may influence perception of distance – the effect of fog was mentioned earlier. Bright approach and runway lights may appear closer to the observer and dim lights farther away. This effect is accentuated when there are no other lights in the vicinity.

The black hole phenomenon is particularly relevant when approaching airports at night over the sea, jungle or desert. When all is dark except for the distant runway or airport lights, with a black hole intervening, there is an illusion of height. This phenomenon was cited in the report of the 1974 707 accident in Pago Pago in Samoa where 96 aircraft occupants died in the crash (NTSB-AAR-77-7).

The windshield location of an observed object, such as airport or approach light-ing, can lead to a misjudgement of height. The object will appear at the same spot on the windshield at a higher altitude with a low pitch angle (higher speed) as at a lower altitude with a high pitch angle (lower speed). A situation could arise on approach that with an inadvertent speed decay and a gradual loss of altitude, the

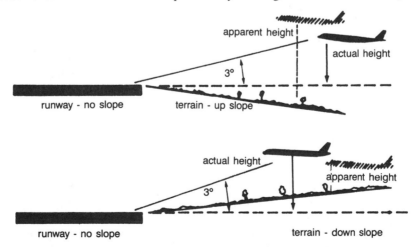

Fig. 5.14 The misjudgement of height as a result of sloping terrain on the approach to the runway (Gabriel, 1977).

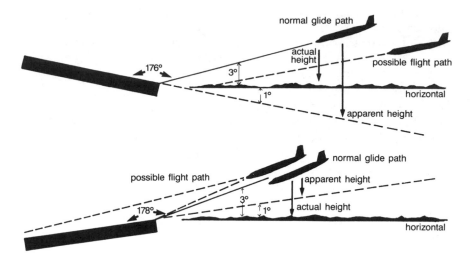

Fig. 5.15 Approach illusions arising from a sloping runway, in these examples with an inclination of 1° to the horizontal (FSF, 1965).

runway could remain for the pilot in the same position in the windshield, giving an impression of a stabilised and safe approach – until touching down some distance out. This is particularly relevant to visual flying at night around airports where the surrounding terrain is without lights.

An undulating runway can give rise to misconceived actions during landing, although this cannot strictly be classified as a visual illusion. ICAO recommends that a runway should not have a slope of more than 2%, or 1% if the runway is longer than 1200 m (Annex 14, paragraph 3.1.11). The 2920 m 34/16 runway at Cairo Airport, for example, has a slope of 1.7% and so this ICAO recommendation is not always applied. A 1% slope can give a vertical difference between the two ends of a 3 km runway of 30 m. This creates a problem where there is more than one slope on a given runway, which is also permitted by ICAO. As the eye of a 747 pilot is less than 9 m from the ground, the multiple slope runway can result in his losing sight not only of the end of the runway, but also of obstructions (including other aircraft) which may be on the runway. After landing with the end of the runway not visible, the pilot may have the impression that far less runway is remaining during his roll-out than is actually the case. He may then apply heavy braking unnecessarily, with the risk of blown tyres and resulting fire. ICAO recommends that at least half the runway should be visible to the pilot (Annex 14, paragraph 3.1.15).

Lack of runway texture can give rise to problems in depth perception. As an extreme example, the problem of landing a flying boat on a glassy water surface is such that the task simply cannot be performed safely by visual means alone. Snow covered runways, where contrast is limited, and night landings on dimly lighted surfaces give rise to similar depth perception difficulties.

The approach procedure may be followed by an overshoot or go-around and this can introduce illusions associated with accelerations, discussed earlier with those related to take-off. The investigators of the 1965 Vanguard accident at London

Airport during a low-level overshoot, hypothesised that this type of illusion may have played a role in the crash, with the acceleration giving a pitch up sensation (UK CAP 270).

Protective measures against illusions

It is sometimes said that as visual illusions result from an unconscious process within the brain, there is really nothing to be done about providing protection against them. This is not true. Very effective steps can be taken by flight crew members and operational management to reduce substantially the risks associated with visual illusions in flight operations. Perhaps these steps can be enumerated.

The first and essential stage of any such protective programme is the recognition by all those concerned that visual illusions are normal phenomena. And that they result from the way all human visual and information processing systems have been designed and so occur to all of us. This recognition is only likely to be achieved following exposure to a proper educational programme. Responsibility for this is shared primarily between operational management and the professional flying staff.

Secondly, once the vulnerability of everyone to false perceptions is recognised, the next step is to get to understand their nature and the situations in which they are likely to be encountered. This, again, involves an educational process which is largely the responsibility of operational management. However, every professional person carries an individual responsibility to ensure possession of the knowledge necessary to perform his task safely and efficiently.

Thirdly, the most general and effective protective measure against visual illusions is to supplement visual cues with information from other sources, particularly when experience and education indicate that an illusion may be expected. Prior recognition of the situation should be reflected in inclusion of the item in routine crew briefing, initiated by the captain on the flight deck. This is an area where input from other flight deck crew members can also be useful. Planning can then be made for the protective use of radio aids, radar, attitude displays, radio altimeter, distance measuring equipment (DME), inertial and other navigation systems, visual glide slope aids (e.g. VASIS) and so on, to back up basic visual information. It will be noted that the predominant area of incidents and accidents arising from illusions seems to be in non-precision approaches when the pilot replaces electronically guided flight with visual control. This often involves some relaxation of concentration which makes the pilot more susceptible to illusions. Special attention should be given to instrument flying when accelerations can be expected to induce false perceptions, such as in the case of an overshoot procedure. Head movement during visual search during cruise helps to avoid the consequences of the blind spot on the retina. Special care will be taken when the crew is suffering from fatigue, at night and in conditions of reduced visibility.

Fourthly, where geographical locations are known to be associated with visual illusions, the fact can be noted on approach and airport charts to alert the crew and encourage them to take precautionary measures.

And finally, aircraft manufacturers and certifying authorities also have a role to play in providing protection. The provision of adequate visibility from the flight deck, both in terms of vision angles and optical quality of the glass, is not directly in the hands of the operator, except in the case of replacement of defective windshields. However, he has several channels which he can use to influence policy, such as SAE and IATA. It was only after operator input that the manufacturer was required to provide means for the pilot to determine the location of the design eye position. Until that time, he was left quite in the dark as to its location and significance for safe flight operation or even its very existence. Some windshield rain and ice protection systems have been quite inadequate to maintain proper vision during approach in adverse weather. The operator must make his views known on these systems through the appropriate bodies responsible for setting technical standards.

It has been said that visual illusion elements were probably involved in ten air carrier accidents in the USA alone between 1961 and 1975 and so we are concerned with a practical issue of safety and not simply a theoretical one (Stone *et al.*, 1976). No doubt, there are some areas where further research in physiological and psychological aspects of vision and visual illusions would be useful. Nevertheless, much is already known and if this were to be fully applied, the hazard of false perceptions in flight could be notably reduced.

There is little question that visual illusions create a very real operational problem in transport aviation. One successful airline's approach can be seen in the following excerpt from a paper given by Captain J A Brown of American Airlines at the 1975 IATA 20th Technical Conference at Istanbul. Captain Brown stated, among other things that:

It is equally futile to attempt to train crewmen to cope with visual illusions by training them under all types of conditions which may produce illusions. The specific illusion that may some day cause him a serious problem may or may not be duplicated. More importantly, such experience provides no assurance that the crewman could recognize his problem much less cope with it during an actual operation. The answer lies in procedures that will assure safe flight regardless of what one crewman thinks his eyes are telling him. Training that will convince him that illusions can be expected is important, but the key to safe operations lies in compliance with protective procedures (Brown, 1975).

The visual illusions that can be encountered and many elements of the protective procedures Captain Brown referred to have been discussed in this section.

6

Motivation and Leadership

Motivation and safety

Performance and proficiency

Three out of four aircraft accidents result from a human performance deficiency of one kind or another. Yet if it had been possible to have given the pilots involved standard flight proficiency and medical checks at the time of the accident, it is probable that the majority would have passed these tests without difficulty. In most cases they would have possessed the physiological and psychological capacity, as well as the technical skill, to have conducted the flight safely. Yet they did not do so. Why not? The answer does not seem to reside in the field of competence of the flight instructor or the aviation medical staff; the pilots had regularly demonstrated that they could fly the aircraft competently and that they were medically fit. We must seek the truth elsewhere.

Traditionally, accident investigations have been primarily concerned with determining what happened and who carried legal liability. This involved confirming that all those concerned were licensed properly and that they had at the time of the accident conformed to company and state regulations. Once this had been determined, it was usually assumed that those concerned had the capacity to perform their allocated task and if they did not do so, then blame could be allocated, punishment and penalties inflicted, claims settled and the case filed.

But increasingly, intelligent people are asking, why. Why a properly qualified, highly trained, medically fit, well-paid person, failed to perform his task as expected. There is no doubt that until we are prepared to invest as much effort into determining *why* an accident occurred, as we are in determining what occurred, then only slow progress can be anticipated in reducing the percentage of accidents classified under the heading of human error.

A very few state accident investigation authorities are now coming to grips with the problem. In 1982 and 1983, the National Transportation Safety Board (NTSB) in the USA began to expand its formally qualified Human Factors staff. The Board had for long been able to call upon very highly qualified technical and engineering specialists but had previously placed less emphasis on Human Factors. A separate Human Performance Division was established in 1983. In Australia in 1984, a very significant step was taken by the Bureau of Aircraft Safety Investigation, the body

131

responsible for accident investigation. Using the Human Factors Awareness Course (KHUFAC) described in Chapter 14, the Bureau commenced providing formal Human Factors education for all its investigators. The course has units specifically dedicated to human error, motivation and attitudes, which are particularly relevant here. The CAA in Sweden was already using the same course to introduce the technology of Human Factors formally to its own staff.

But these cases are, regrettably, not typical of the approach of state aviation authorities to human aspects of flight. Most demonstrate little recognition that a technology exists which is concerned with human performance and skill and which has a direct relevance to flight safety.

Motivation and the *SHEL* model

In Chapter 1 we used the *SHEL* model to explain the interaction between the different components in the complex man-machine system which is at the heart of modern air transportation. It was said that the *Liveware*, or man, is at the centre of the system and is its most valuable and flexible component. It was also said that in order to ensure the most effective functioning of the overall system, involving both safety and efficiency, the components must be properly matched. Failure to do so, we said, could lay the foundations of future system breakdown, leading to accidents and inefficiency.

This matching cannot be accomplished without giving full recognition to the characteristics – the capabilities and limitations – of man and without understanding what makes people behave in the way they do. We need to be able to predict the way they are going to act and respond. In general, we know the kind of behaviour and performance we would like to see, but we also need to know how to encourage or enhance it.

For example, if we knew that bright colours in a workplace caused people to become irritated or aggressive, we would avoid using them in the *Hardware* design of the flight deck. If we recognised that a person's performance in a particular job notably increased when given greater responsibility, then we should try to modify procedures – the *Software* – so as to incorporate more responsibility into the system. If we found excessive errors were occurring due to high temperatures in the workplace, then we would try to control the *Environment* more effectively. If we were to determine that a more personal, informal and friendly approach of supervisors to staff would result in more conscientious work with fewer errors and greater productivity, then we would be wise to respond to this finding by modifying the *Liveware–Liveware* interface.

But all of these measures necessitate an understanding of what determines a person's behaviour. Unless we can predict this with reasonable accuracy we cannot optimise system design. What drives or induces a person to behave in a particular fashion is called motivation. This is arguably the most significant characteristic of the *Liveware* component and so commands special attention.

An approach to motivation

When two children of the same age are given some brightly coloured toys, one will eagerly investigate them, the other will simply ignore them and look elsewhere for amusement. A painter like Van Gogh could sit alone in poverty in Arles, working long hours on a painting for which he would receive a pittance. A businessman may forfeit all family life and home pleasures in the pursuit of financial success. A politician will tolerate personal abuse and insult and forgo all privacy in his quest for power. Clearly, people are different and are driven by different motivational forces.

In industry, selection, training and checking ensure that a person has the capacity to perform a given task. But it is largely motivation which determines whether he *will* do so in a given situation. In simple terms, motivation reflects the difference between what a person can do and what he will do. It is the internal force which initiates, directs, sustains and terminates all important activities. It influences the level of performance, the efficiency achieved and the time spent on an activity.

If we exclude reflex actions, hardly any human activity occurs without a motivating drive. Neither does learning take place. Within certain boundaries, the level of performance and the rate of learning may be controlled by manipulating – by influencing – this drive.

At a basic level, motivation begins with a need or a requirement for the survival of the individual or the species. For example, the hunger drive is associated with the need for food without which the organism will die. The sexual drive gives rise to behaviour which increases the probability of the propagation of the species. The relative strength of these drives determines the direction in which our energies are channelled at any one time.

The exact way we go about the process of satisfying needs depends upon our previous experience. It is possible to influence motivational drives. A child when born has certain basic physiological and biological needs which control his behaviour; hunger, thirst and so on. Other needs are manipulated by his parents – he is taught to avoid fire, not to destroy property, to try hard to achieve success – and the child gets a reward or a punishment to guide his behaviour. Such punishments and rewards can be seen as 'sticks and carrots'. We can see an influence developing between the child and the environment, such as heat and cold, and between him and his parents. This affects his motivation and thus his behaviour.

As he grows older, the motives become more complex and more subtle. It has become apparent that the development and control of his motives have become a part of his character and personality. The motivation is then described as having become internalised.

The desired result of a sequence of motivated behaviour can be seen as the goal or purpose of that behaviour. A pilot may be driven by a need for responsibility and authority and may see his goal as an appointment as Chief Pilot. An administrator may see his goal as becoming Managing Director. Such goals, and thus their associated goal-directed behaviour, can change with age, time or circumstances, though there may be resistance to such change when attempts are made to impose it from outside.

Some will set their goals at unrealistic levels. After all, there is generally only one chief in an organisation and, patently, not everyone with such a primary goal can expect to achieve it. Some who fail to reach their primary goal will turn to a secondary one. Others, who remain motivated to achieve goals which have been set unrealistically high, may become neurotic. Typical symptoms of such neuroticism are frustration, irritability, insomnia and headaches, which can be expected to be reflected at work in the *Liveware–Liveware* interface and in general levels of performance. This aspect of motivation will be raised again later as it is an important factor influencing performance at work.

Complacency, professionalism and discipline

The history of civil aircraft operations is strewn with exhortations to crew members to avoid complacency, be vigilant, display professionalism and exercise proper discipline. There have been various periods when one or other of these forms of exhortation was more or less in fashion.

They all reflect aspects of behaviour which go beyond technical skills and can be seen as indications of motivation. There can be little doubt that every crew member would see all these behavioural characteristics as relevant to safe and efficient performance in flight. Yet some are motivated towards the desired behaviour while some are not. And even those so motivated periodically experience a variation in the strength of motivation.

Attempting to modify behaviour by means of exhortation is unlikely to have any long-term effect unless the exhortation is accompanied by other measures. The work of the increasing number of confidential reporting systems is very valuable. However, the polite inferences in some of their periodic bulletins that the readers should smarten up are unlikely, by themselves, to have a significant long-term effect. A more profound enquiry into the nature of the forces which drive the activities of people is necessary in order to learn whether they can be manipulated and if so, how. It is also necessary in order to learn how to design the overall flight deck and cabin systems to live safely with the different motives which control the performance of crew members at work.

Company policy and crew behaviour

Crew behaviour cannot be properly assessed in isolation from what has been called the organisational culture in which it takes place. Increasing concern has been expressed at the apparent influence of company policy on crew behaviour and flight safety (e.g. Lautman *et al.*, 1987; Bruggink, 1988; Hackman, 1990). A well disciplined, competent, professional, conscientious crew member does not change overnight to become undisciplined, incompetent, unprofessional and complacent. A pilot who routinely follows SOPs, and FARs does not suddenly abandon that attitude towards flight safety and become indifferent to published operational requirements.

In this respect, crew motivation is no different from that found in other work places on the ground. If the boss comes in late and leaves early, he can expect his staff to adopt the same work ethic. And if a company tolerates unprofessional behaviour and is itself complacent regarding operational and training standards and their enforcement, this establishes the acceptable basis for its flying staff.

Two fatal accidents illustrate this phenomenon. In 1987 a MD-80 crashed at Detroit after attempting to take off without flaps and slats – an uncertified configuration (NTSB-AAR-88-1). Just a year later, at Dallas, a Boeing 727 crashed in remarkably similar circumstances, also taking off without flaps and slats (NTSB-AAR-89-4).

In the Detroit case, the airline had earlier experienced a series of nearly disastrous incidents resulting from poor crew discipline and a failure to follow SOPs, of which proper use of the checklist was but one example. Amid increasing public concern, the FAA stepped in to conduct a special inspection of the airline's operations. This revealed a steady erosion of professional standards, lack of cockpit leadership, violation of FARs and SOPs, inadequate check airman standards, flagrant disregard of cockpit discipline, checklists done from memory, and so on. Within about four years of this inspection, the airline's Flight 255 crashed on 16 August 1987 at Detroit, killing 156 people.

The NTSB investigation of the accident reported lack of cockpit leadership, violation of FARs and SOPs, disregard of proper cockpit discipline, improper use of the checklist, and so on. While criticising the crew's behaviour and performance, the NTSB noted that their performance had been described by peers as standard, meticulous and professional. It concluded that their manner of performance on the accident flight was one to which each pilot had been exposed and was familiar with over a lengthy period. For the crew to have become comfortable with that manner of performance, the NTSB stated, indicated that it was accepted and used by other crew members in the airline. It may be significant that the airline in its comments and recommendations to the NTSB made no suggestions for improving crew performance or behaviour, no proposals for changing training procedures and its officers considered the crew's performance to be average to above average. In the intervening years between the FAA special inspection and the accident, there had been no significant non-technical training in leadership or resource management (see pages 183, 332) and the airline did not closely follow modern training concepts applied elsewhere in the industry, such as LOFT (see page 332).

As a result of this accident, the NTSB recommended to the FAA that all inspectors emphasise the importance of disciplined application of standard operating procedures and, in particular, emphasise rigorous adherence to prescribed checklist procedures (NTSB A-88-67).

In the second example just a year later (the Boeing 727 accident at Dallas on 31 August 1988) the circumstances were almost a replay of the Detroit crash, this time with the loss of 14 lives. A series of unprecedented and serious crew-related operational incidents in the early 1980s in the airline caused public alarm in the USA, generating a six-week FAA special inspection. Their report included over 40 findings of serious deficiencies in the airline's flight operations, some of which involved violations of specific FARs. No FAA enforcement measures were taken, but 55 recommendations for corrective action were made. The airline disagreed with some, did nothing about others and in many cases defended the behaviour of the crews concerned as conforming with company policy.

A year after this much publicised inspection and criticism of the operation by the FAA, the airline's Flight 1141 crashed. The memorandum to the Board from the Chief of Aviation Accident Division of the NTSB stated that this particular crew was operating essentially in accordance with the same procedures and cockpit discipline concepts that the FAA had severely criticised a year earlier and which the airline had defended at the time as adequate.

The main report in the Dallas case cited as a contributory cause the airline's inaction in implementing improvements. A dissenting Member's report went further and cited the airline's management as being directly causal to the accident in failing to provide the required leadership and guidance to its flight crews. In other words, the organisational culture was conducive to or encouraged the inadequate crew performance which was a direct cause of the accident.

Looking at the two accidents together, several conclusions can reasonably be drawn:

- both involved a major airline with an international network
- in both cases the inadequate level of behaviour of the crew may have been normal or at least common for the particular airline's operations
- in both cases the alarm bells had been loudly sounded well ahead of the accidents with a series of incidents and severe criticism by the regulatory agency
- in both cases the public had expressed concern at the series of hazardous incidents which had occurred
- in both cases the airline tended to defend the inadequate crew performance as normal
- in both cases training was inadequate and revealed earlier to be so
- in neither case was action by the airline or regulatory enforcement adequate to correct the deficiencies and prevent the accident.

These two fatal accidents – both of which were totally avoidable – may have marked a permanent change in the manner in which the NTSB and other accident investigation bodies view crew performance. The causal chain of circumstances cited leading up to an accident now not only recognises crew behaviour and performance, but should also recognise the organisational culture in which this is allowed to flourish.

Concepts of motivation

Research and theories

Much of the research on motivation has been laboratory-based and psychology-oriented. It has frequently made use of animals as test subjects. As a result, many of the conclusions are not very relevant to the real-life working environment where many other elements – sociological and personal – interact to influence behaviour.

Nevertheless, there are many studies in industry which are available to augment the theoretical work and much can be learned from the information now available. The London-based Work Research Unit (WRU), a branch of the Department of Employment, published for the period up to 1985, reference lists of 396 papers on motivation (WRU Bibliographies Nos 37 & 37a), so there is no shortage of study material for anyone with interest or responsibilities in this area. In addition, information on motivation is to be found in many popular Human Factors books (e.g. Evans, 1975; Green *et al.*, 1991; Jensen (Ed.), 1989; Murrell, 1976; O'Hare and Roscoe, 1990; Vroom *et al.*, 1970; Warr, 1971; Wiener and Nagel (Eds), 1988).

There appears to be a complex internal organisation which drives the body towards a specific goal. We speak of a need being created when there is a deficiency in the organism. A motivational system then sets into motion a sequence of behaviour, starting with a drive and followed by goal-directed activity which continues until the goal is achieved or abandoned.

Perhaps the most fundamental drives are those that originate from the physiological needs of the individual. These include hunger, thirst, sex, sleep and even, in a negative way, pain. These may be stimulated from external sources or from hormone activity or the needs of tissues within the body. Even though these drives are basic and physiological in nature, they can be modified by learning.

Experimental studies in animals have demonstrated that some behaviour is instinctive, in that it is performed adequately the first time without a learning process, is stereotyped with regard to the particular species and progresses through an orderly sequence over a period of time. In the animal world, the migration of fish and birds is an example of such an innate behaviour pattern.

Distinct from these physiological motivations are those with a psychological or social orientation. The goal of one person may be to exercise power over others. Another may be to achieve success in a particular field of activity. Yet another may feel the need to care for the health and welfare of his less fortunate fellow men.

While we have distinguished here between physiological drives and those with a psychological or social origin, the situation may not always be quite so clearly defined. For instance, a woman's hunger drive may be modified by a conflicting motivation to remain attractively slim. This in turn may be modified if dining out as a guest, by a desire not to offend the hostess by rejecting the sumptuous meal she has prepared. The physiological sex drive is usually regulated by social and cultural factors. It is important to understand that a single behaviour may be governed by several, perhaps conflicting, motives.

Theories

A number of theories have been proposed by those concerned with the study of motivation, each one contributing to our overall understanding of the subject. One such approach is called the Learning Theory and this relates present motivation to past experience. A particular action or response which was rewarded in the past will be repeated, according to this theory, on the assumption that it will be similarly rewarded in the future. And the reverse for activities which have been penalised.

Another theory is based to a considerable extent on the studies of Freud who originally considered sex to be a primary driving force in behaviour. This Psychoanalytic Theory has been modified over the years so that less emphasis is now placed on sex and more on the ego, or a person's concept of himself.

A third basis of study of motivation can be placed under the heading of Cognitive Theories. These have moved away from the physiological and innate needs and drives and emphasise the characteristics of man as a rational being who makes choices freely as a result of perception, thought and judgement. Attitudes, beliefs, values and expectations are generated during one's life and these regulate the goal-directed behaviour.

On a different aspect of motivation, a much-quoted theory is that there may be an order of priority, or an hierarchy, in seeking the satisfaction of human needs (Maslow, 1943). Motives lower in this hierarchy are aroused first and must be satisfied first (Fig. 6.1). Once these are satisfied, the needs at the next higher level assume priority. The hierarchy rises from the level of basic physiological needs up to those related to the ego, with self-fulfilment being at the top of the scheme. Drucker, in an observation that has relevance in aviation, has pointed out that: 'As a want approaches satiety, its capacity to reward and with it its power as an incentive

137

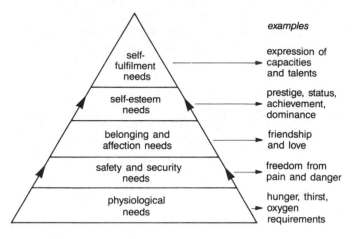

Fig. 6.1 Maslow's hierarchy of needs (from Maslow, 1943).

diminishes fast. But its capacity to deter, to create dissatisfaction, and to act as a disincentive rapidly increases' (Drucker, 1973).

While much of the research has been laboratory- and animal-based, notable studies have also been conducted with an industrial background. We shall mention here just three of these, the time studies of Taylor, the Hawthorne experiments and the Two-Factor Theory of Herzberg.

In Chapter 1, reference was made to the work of Taylor in the 1890s in the Bethlehem Steel Company in the USA. In this work, man was treated rather like a machine and money was considered to be the prime motivating force. Productivity was raised dramatically by increasing wages, while overall costs were reduced by lowering staff manning levels. This was given the title of 'scientific management' (Taylor, 1911).

This approach has since been widely criticised as it became apparent that money was not the overriding element in stimulating performance. However, it must be said that in industry today, managements frequently do not use the motivating power of money in an optimum fashion. It has been observed that charity workers, for example, often apply far greater effort in their work for no financial reward than highly paid operatives in industry or business. Some work hard in a job because they find it interesting; others, paid the same for the same job, find it less interesting and work less enthusiastically.

The Hawthorne studies at the Western Electric plant in the USA in the 1920s, confirmed that more than money is involved in motivating staff in industry. The original project was initiated to determine whether modifying the factory lighting would improve work performance. It did. However, it was then found that even when the lighting was restored to its original level or reduced to a very low level, the improvement remained. It was concluded that performance was being influenced by social or psychological factors quite independent of the work itself; in this case, possibly the demonstration by management that they cared about the working conditions of their staff (Roethlisberger *et al.*, 1939). This is still known as the Hawthorne Effect.

Since the Hawthorne experiments, a mass of literature on ways of making work more satisfying has accumulated. Periodically, new concepts have been projected and supported, almost as fashions, to claim the attention of ergonomists and occupational psychologists. Job satisfaction and the quality of working life are two such concepts and papers are rapidly accumulating under both headings. The WRU published for the period up to 1985 bibliographies listing 763 papers on these two topics (WRU Bibliographies 38, 38a, 39, 39a).

The Two-Factor Theory of Herzberg proposed in 1959 recognised what was already acknowledged by Maslow earlier, that people often appear to be motivated to do things because they are felt to be worth doing for their own sake – a fundamental departure from the concepts of Taylor. Much research has been generated by Herzberg's theory, which originated from the observation that the subjects mentioned when staff were asked what they found satisfying at work seemed to be quite different from those given when asked what they found dissatisfying. Satisfaction appeared to be derived from factors like achievement, advancement, recognition for good work, responsibility and the nature of the work itself. He called these motivating factors. On the other hand, dissatisfiers often quoted were items such as staff relations, company personnel policy, salary, security and poor working conditions. He called these hygiene factors. Herzberg did not underestimate the importance of the hygiene factors but believed it was the motivators which would induce employees to achieve better work performance (Herzberg, 1966; Herzberg *et al.*, 1959). The theory of Herzberg has notably changed traditional thinking on attitudes to work. Although it may now seem over-simplified, and so invites criticism, it has made a valuable contribution to a better understanding of motivation at work.

Murray's motives

From studies of motivation it appears, firstly, that there are separate, distinguishable drives which control behaviour, though one single behaviour may be controlled by more than one drive. Secondly, that some form of hierarchy may determine the order of priority which is put on them. And thirdly, that people may be motivated by different forces while apparently behaving in the same way.

The goals which people set for themselves are almost infinite in number and the labels we attach to them are necessarily rather arbitrary. However, a few of the more commonly held goals have been studied and labelled (Murray, 1938). Some of these are listed in Appendix 1.14. The behaviour of a person with such specific goals is likely, with respect to achieving those goals, to display some similarities with the behaviour of others with similar goals. In Human Factors we are concerned to try to predict how a person will behave in given circumstances. If there is any consistency, then, in goal-directed behaviour between individuals, we should be interested in a study of the way these more common goals can be identified and the typical behaviour which may be associated with them.

Three of the motives cited by Murray have been extensively investigated. These are concerned with achievement, power and affiliation. Of these, achievement motivation has been studied most as it has seemed particularly relevant to 20th century society and values in the Western world.

Achievement motivation has been described as the force which drives a person to achievement for its own sake rather than for the material benefits which are expected to be derived from it. Early observation suggested that someone with a high level of achievement motivation will strive to meet or surpass standards which have already been set, and if these have not been set, then he will set such standards at a high level himself. He generally works efficiently without supervision, though he may become bored with performing routine tasks. He often appears to be interested in tasks of intermediate difficulty which present a challenge but which are still attainable. He looks for good feedback on the job so as to assess how well he is doing.

A number of tests were developed to measure achievement motivation, as this seemed to be a quality which should be sought in staff selection processes for many jobs. However, attempting to estimate future performance on the basis of these measures was not very successful. As so often occurs, the situation seemed not to be as simple as first believed, and, in particular, the concept did not seem to be applicable to females.

Later studies suggested that achievement motivation may not be a single trait but may be made up of three distinct elements. These were also given certain labels. Mastery; the need to confront new challenges and surpass one's own earlier performance. Work; reflecting the satisfaction which is gained from performance. Competitiveness; the drive to surpass the performance of others. High performance in the scientist and academic may be predicted by the high level of the mastery and work elements, while the competitive element might be a predictor of lower performance. By dividing achievement motivation into these three elements it has been found possible to predict the performance of males and females equally well. It has also been possible to apply the measures to predicting the performance of airline pilots (Helmreich, 1983).

These three elements of achievement motivation appear to be already well established early in life. They can be influenced by parental training and culture, though it has been shown that even in a traditionally non-achieving culture, individuals can learn to expand somewhat their motivation for achievement, even as adults (McClelland, 1965). However, in general these traits are highly stable and modification should not be seen as an easy task.

Achievement motivation is frequently associated with leadership (Fineman *et al.*, 1971). It is often seen as being of paramount importance in jobs which require a high level of unsupervised performance, involving the use of initiative and responsible, well-disciplined behaviour.

Affiliation motivation is another motive distinguished by Murray. It is concerned with the establishment and maintenance of affectionate relationships with other people and the desire to be liked and accepted. It also involves adherence and loyalty to a friend. This motive has also been extensively investigated and a scoring system has been developed for its measurement (Atkinson *et al.*, 1954).

Power motivation reflects concern over the means of influencing the behaviour of another person. This motive, too, has been extensively investigated and techniques have been studied for measuring it (Veroff, 1957).

These are just three of the many motives which have been labelled. Most people will possess many of the drives to a greater or lesser degree. However, the strength of any particular drive will vary with each individual so that the total combination of settings of motive strength represents something of a personal signature.

What if it is impossible to satisfy these individual motivational needs by one means or another? What reaction can be expected? A typical response is one of aggressive behaviour. This may become apparent in 'fighting the system', withdrawing cooperation and sometimes developing intense dislikes for individuals or groups who are seen as being responsible. As mentioned earlier, symptoms of neurosis may sometimes appear. Good personnel management will make a point of getting to understand the individual needs of its staff and trying to encourage the satisfaction of these needs either inside or outside the working environment.

Expectancy and rewards

The Two-Factor Theory of Herzberg has been extensively applied in industry. The very simplicity of its approach to a most complex problem provided ample justification for its wide use, but also cast some doubt on its validity.

Vroom was one of the early investigators to criticise the theory as not taking into account individual differences in work motivation. He included in his own model the preferences of the individual, such as for pay, promotion or security and the valence, that is, the strength of the preference. In addition, he took into account expectancy, that is, the chance a person sees that his behaviour or performance will lead to the desired outcome (Vroom, 1964). Is the acceptance of long duty hours likely to lead to receiving a bonus, faster promotion or a commendation? Is a letter of complaint by a passenger to an airline likely to lead to an improvement in service or would it be simply a waste of time? The Vroom model was not a very practical tool for use by management as it did not provide them with guidelines on what actually motivates their staff at work. The development of personnel policy on this basis was thus rather difficult. Nevertheless, it made a contribution to an understanding of the relationship between the goals of an individual and those of the organisation.

A practical model can be used to illustrate the relationship between motivational elements in a work situation and performance (Fig. 6.2). In this model, the utility or

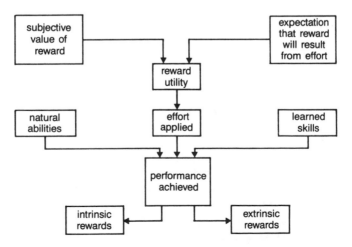

Fig. 6.2 A simplified diagram indicating the relationship between motivational factors and performance.

the usefulness of a reward can be seen as the product of the value placed upon it and the probability of its attainment. People place different values on rewards; one may value money very highly, others may favour promotion, recognition or social status. The probability of attaining the rewards is also important. If a person expects that his effort will pass unnoticed, which is not an unusual experience in working life, then he may feel that the reward has little usefulness even though he may value it highly. If he is looking for promotion, for example, then he must expect that there is a link between his effort or performance and promotion for this to influence his behaviour.

On the basis of this utility, a certain level of effort will be applied. But effort is not synonymous with performance and two other variables have an influence here, natural abilities and acquired skills. From the performance we can see the rewards emerging. These can be intrinsic, such as feelings of pride or achievement, or extrinsic, such as pay or promotion.

Other factors, of course, also play a role in this process. A wide variety of external situational and personal circumstances are likely to affect the level of performance achieved on any particular occasion. Furthermore, some rewards may be gained from the effort alone, irrespective of the actual performance. Nevertheless, this model provides a useful and practical basis for management to evaluate performance of staff and the factors which play the major role in influencing it.

Two remarks are relevant here. The most important aspect of the availability of rewards is not what management *says* is available, but what every employee sees as available. And secondly, if the high performer sees the chances of achieving a particular reward are no better than for the low performer, then management may have created a motivational problem within its staff which could have far-reaching effects. The introduction of rigid seniority systems and some trade union resistance to the application of a performance element in selection for promotion, may have originated from earlier irregular and unfair management practices. Nevertheless, such union policies clearly tend to have a demotivating influence in that increased effort and performance will not be seen as being related to desired rewards. Typically in the airline industry, there are pilots who just scrape through proficiency checks, just make the transition to a new aircraft type, take little interest in passenger relations or crew welfare and performance and fly only when they have to. They are likely to be promoted and rewarded in most airlines just the same as the highly competent, professional pilot, dedicated to safety and to his company's interests.

Influencing motivation at work

Job satisfaction

In the search to find what really motivates people to higher performance, increasing attention has been paid in recent years to what has been called job satisfaction. In literature surveys, the WRU listed 338 papers on this subject covering the period 1965–1985 (WRU Bibliography Nos 38 & 38a).

There are many factors which may influence a person's overall attitude to the job. These include, amongst others, financial rewards, management personnel policies, work colleagues, the working environment, and the nature of the task itself.

The extent to which these factors will apply an influence will depend on the person's own preferences and values. It is possible to measure the satisfaction level of the separate elements of a job and also the job itself, using specially designed questionnaires and interviews.

While job satisfaction may be a desirable end in itself it cannot be assumed that it necessarily or automatically results in improved performance (Cameron, 1973; Williams *et al.*, 1971). Eliminating proficiency checks, for instance, may raise the job satisfaction of some pilots by reducing anxiety and extra study, leaving more time for sailing or fishing. But overall performance at work could be expected to fall, not rise. There is not much evidence from industry to suggest that increasing job satisfaction will result in higher performance unless the satisfying outcome or rewards are tied to performance and seen to be tied.

In Taylor's work at the end of the 19th century, money was considered of paramount importance in creating motivation at work. Today it is known that once incomes rise appreciably above subsistence levels, dissatisfaction with *relative* incomes is a far more powerful sentiment than dissatisfaction with the *absolute* income. In other words, not that my income is not enough, but that some other comparable person is getting more.

Using material incentives to raise production or performance is not quite such a simple solution as it sounds. Expectations are constantly rising; more is needed to produce any results. And one person's higher reward may become the basis of dissatisfaction for his colleague. Incentives can become institutionalised as rights rather than rewards for good performance and their withdrawal can then be seen as a form of punishment. Annual Christmas or holiday bonuses are an example of this phenomenon. In spite of these limitations, incentives still have an important place in the overall scheme.

One airline invites all its newly appointed captains to a company dinner, attended by the company president. This certainly initially enhances motivation, but as it is a single event in the captain's career other factors gradually erode the effect. It should not, of course, be a single event and recurrent activity should be developed to sustain the high level of initial motivation apparent in many such newly appointed captains.

A rationally critical attitude of an employee towards his supervisory or management policies should not be confused with job dissatisfaction or low motivation. On the contrary, those who are greatly motivated to achieve the goals, performance, production, or product quality entrusted to them, are likely to initiate more criticisms of whatever deficiencies appear in the organisation, frustrating their efforts. Those who are uncritical and display a high level of acceptance of the status quo and apparent satisfaction, may be simply displaying apathy and complacency. The hostile response of some managements to such critical attitudes may reflect a lack of security or self-confidence rather than ignorance of the motivational factors involved, though in either case, poor management quality is indicated.

To increase job satisfaction, job enrichment and job enlargement have been used (Fig. 6.3). Job enrichment mainly involves active participation of staff in policy and decision-making concerning their work. As an illustration, airline cabin pursers could be involved in decisions on the type and timing of meals to be served on board. The check-in agent could be consulted on how to handle irate passengers

Fig. 6.3 Job enrichment and enlargement contribute to job satisfaction but are not the complete answer.

who have been over-booked by the airline. Job enlargement increases the number and variety of tasks (horizontal enlargement) or increases a person's control of the routine planning of his task (vertical enlargement). This might be seen in the delegation by the aircraft captain of certain of his tasks to the first officer, which provides horizontal enlargement of the first officer's job. Similarly, giving the flight engineer certain radio, navigation or visual lookout tasks also introduces horizontal enlargement. If crew members are allowed to become routinely involved in, say, their own duty scheduling, then this would be considered vertical enlargement. Both methods of increasing job satisfaction have shown promise in industry but are not, in themselves, the complete answer.

While the relationship between job satisfaction and performance may be variable and uncertain, other relationships are well established. Setting clear, precise and acceptable goals and targets, preferably with some degree of challenge attached to them, is very effective in enhancing performance. There should be clear targets and acceptable tolerances established for most tasks which are related to safety and efficiency. This may refer to the aircraft approach speed or the time taken to deliver a passenger's luggage to the customs hall after arrival. It can apply to an airline's schedule punctuality. Many staff feel more content if they have clear targets to meet, and providing these are realistic they can contribute to job satisfaction.

Boredom is often seen as one of the more pervasive problems facing advanced industrial societies. Surveys of job attitudes suggest that the reasons for boredom may fall into at least six categories. These are constraint, meaninglessness, lack of interest, repetitiveness, lack of a sense of completion, and finally, some people find a job boring because of its general nature, or more accurately, their perception of its nature. Where it is thought that boredom may have an influence on performance, corrective measures such as job redesign will have to be considered. In the final analysis, it is the susceptibility of the individual rather than the nature of the task that is basically responsible for boredom. What is boring to one is not necessarily boring to another.

Behaviour reinforcement

In psychological jargon, the use of 'sticks and carrots' at work to modify behaviour and performance is called behaviour reinforcement. Positive reinforcement is strengthening of desired behaviour by means of rewards; negative reinforcement is discouragement of undesired behaviour by the use of penalties or punishments. Both forms of reinforcement must remain available to management. However, positive reinforcement, though frequently neglected, can often be more effective in raising performance in industry than negative reinforcement.

Industry abounds with such attitudes as 'he is paid to do a good job, what more does he need?' This may be directed to an airline cabin attendant who manages to convert an irate, delayed, over-booked, non-smoking passenger forced to sit in the smoking section, into a satisfied customer anxious to fly on the same airline again. It may apply to an aircraft captain who skilfully handles an emergency situation or a purser effectively calming a group of drunken, aggressive passengers. And, of course, it can apply equally to administrative staff who may work extra unpaid hours to get an important operational bulletin out in good time.

Everyone reacts favourably to being told by someone he respects that he has done well. But undesirable performance cannot simply be ignored and negative reinforcement may be indicated. In such a case a number of precautions must be observed:

- If errors have been caused by lack of skill or knowledge which can be corrected by training, then negative reinforcement may have no beneficial influence whatsoever on the person concerned or any other person. It may, in fact, be directed at the wrong person altogether.
- Group punishments for the undesirable behaviour of one person in the group may be counter-productive, causing resentment and dissatisfaction, with an adverse influence on production and performance which could last a very long time.
- Petty penalties for minor indiscretions seem ineffective in stimulating the desired drives. These can also often cause resentment and resistance and, like group penalties, can be counter-productive.
- The utmost fairness must be applied, and seen to be applied, in the distribution of negative reinforcement amongst all those having a share in the undesired event. A career-long anti-company militancy can be generated from a single punishment which is deeply felt by the person concerned to be unjust. The same principle of fairness applies to positive reinforcement.

It is important to understand that individuals respond differently to different positive and negative reinforcers and it is easy, in ignorance, to generate the opposite effect from that which is intended.

For reinforcement to be most effective, it is usually preferable for it to follow closely upon the behaviour which it is intended to reinforce, so as to relate the two more directly. While we naturally think of this in dealing with animals, it is frequently forgotten with humans. Some cabin attendants may feel a greater incentive if any commission they are due from on-board tax-free sales were paid immediately after the flight rather than credited to salary at some uncertain future date.

Many of these remarks on positive and negative reinforcement may seem just common sense. Yet the skilful handling of behaviour reinforcement in industry is the exception rather than the rule, though union practices sometimes work against a more enlightened and effective approach, adding to the effects of a sometimes less than adequate management.

Fixed-crew operation
Some research in the USA suggests that a flight crew works significantly more

effectively when it is maintained as a fixed unit for a period than when the pilots are scheduled individually (Hackman, 1990). The teamwork and coordination is said to improve during this period of stability in crew complement.

This concept in civil aviation raises two fundamental issues. Firstly, such a fixed-crew operation will have less scheduling flexibility and this will require more crew members on the payroll than the variable crew system. Reserves will be needed to avoid a chain reaction developing when one pilot is sick or otherwise is unable to fill the assignment. This will cost money and it would need to be considered whether this additional money might not be more effectively used to promote flight safety in some other manner.

A second aspect of such team concept development through fixed crewing is that there is a greater risk of non-standard procedures growing during the period of rising familiarity – agreed individual techniques may become the norm during this period. This not only involves the risk of violation in any departure from SOPs but has a negative transfer effect to the following, new pairing period, when a new coordination pattern must be established.

It seems probable that the effectiveness of the fixed crew concept relevant to safety is greater when there is not a well established pattern of adherence to SOPs and where there has not been proper non-technical training in the airline. This would include leadership and crew coordination techniques such as provided in resource management programmes (see Chapters 8 and 14).

This, like so many other proposals for safety related to crew behaviour and performance, may be dependent on national and organisational cultures.

Leadership

The meaning of leadership

A powerful case has been established from extensive research on accident and incident data, notably by NASA, for a need to improve an understanding and application of the management of the human resources on the flight deck. This involves areas such as leadership, group relations, communications and motivation. We should now look at some aspects of leadership, particularly in the aircraft environment and with the situational interactions which that involves.

It was noted earlier that there appears to be a relationship between achievement motivation and the characteristics which are associated with leadership and so previous remarks should be kept in mind here.

A leader in a given situation is a person whose ideas and actions influence the thought and the behaviour of others. He is an agent of change and influence. He uses example and persuasion, combined with a personal understanding of the goals and desires of the group he happens to be leading as well as those of the company. He feels responsible for trying to implement these goals.

He should be able to contribute to solving the problems of the group and for this he should be confident that he has some influence over the decisions of his superiors. His dealings within the group must be seen to be fair. Skill in his own job helps to create respect amongst his group.

True leadership should not be confused with authority. Authority is normally assigned while leadership is acquired and suggests a voluntary following. A department head is given the authority to carry out his supervisory task and may apply dominance to do this. A dominator may assert influence primarily through threat of punishment. An optimal situation exists only when authority is combined with true leadership.

Neither is leadership just a passive status or the possession of a combination of certain personal characteristics. It involves teamwork – a working relationship of the leader within a team or group. The success of this relationship influences his quality as a leader. One of his prime tasks is encouraging the desired motives in others in his group.

There is a popular saying that a leader is born and not made. There is some truth in this; as with other skills some have more aptitude than others. But it is also true that leadership skills should be developed through proper training.

There are some occupations where leadership rests in the hands of the head of the department and the remaining staff may never expect to be called upon to adopt a leadership role. This is certainly not applicable to flying staff. A copilot may be required to demonstrate leadership whenever the captain is away from his seat. Although the cabin is under the overall leadership of a purser or senior steward or stewardess, different cabin sections may also have their chiefs who have to apply themselves as local leaders. Even a relatively junior stewardess may have a new recruit assigned to her for experience. And in the case of an emergency, any member of the crew may be called upon to adopt the role of a leader, conducting an evacuation or taking over in an accident when others are incapacitated. The Boston DC10 survivable accident in 1982 was a typical example. It occurred in the dark in wintry conditions when the aircraft overran the runway and ended in the icy water of the bay. The nose section, with the flight deck crew, two cabin attendants and two passengers, separated from the fuselage and was in deep water. The remaining seven cabin staff had the task of conducting the evacuation of 200 passengers in a situation where the fuselage was in the water and the cabin was littered with debris. Passenger opinion on the performance of the cabin staff in this situation varied but some were quite critical. Two passengers were lost (NTSB-AAR-82-15).

All members of a group contribute to the effective leadership of the group by supplying information, contributing ideas, providing support and by their general response to the leader. But within the group, it is important to know when to be a leader and when a follower; the back-seat driver does not contribute to effective leadership.

Characteristics of a leader

There are a number of characteristics which a leader often appears to possess to a greater degree than the average member of a group. He possesses more than technical skills. He is likely to have qualities related to capacity, such as judgement and intelligence. He is usually likely to have demonstrated achievement and to have shown responsibility. He will be able to participate and cooperate with others and understand group needs and objectives.

Leadership refers only to one particular situation. A leader in one situation is not necessarily a leader in another. A dynamic businessman noted for his leadership qualities may be a 'wallflower' at a social event, leaving leadership to one of his junior office followers (Fig. 6.4). In whatever situation a person assumes a leadership role, and whether he emerges or is appointed, he must acquire the respect of his group by his own behaviour.

Fig. 6.4 Leadership refers only to one particular situation. A leader in one situation is not necessarily a leader in another (cartoon: J H Band).

Tasks and problems of a leader

We have examined the nature of leadership and the characteristics which can usually be expected in a leader. We should now look at the principles which he should apply in order to carry out his task successfully.

One of his primary tasks is that of **motivating** the members of his group. This is very apparent in the case of the captain of a football team but also applies in non-sporting activities. This can be done by emphasising the objectives of the operation or activity and clarifying the targets or goals which should be achieved. For instance, on board a passenger aircraft, the purser could brief his cabin staff on the benefits of their promoting in-flight sales of tax-free goods and suggest a target for the sales on a particular route.

A second way in which leadership can be applied is by modifying habits and behaviour by **reinforcement**. The same purser could apply positive reinforcement by making a favourable written report on a stewardess who had carried out her task particularly well, resulting in some official recognition.

Another principle which the leader should apply is the demonstration of the desired goals and behaviours by **example**. A stewardess having difficulty with an irate passenger suffering from the frustrations of faulty handling by check-in staff, might call for aid from the more experienced purser. If he is a good leader he should be able to demonstrate by example the optimum behaviour in such circumstances. A common aspect of behaviour in which influence by the example of a leader is effective, is in connection with uniform or clothing standards and comportment. If the chief pilot does not wear his uniform cap, it must be expected that others will follow the demonstrated behaviour of their leader. A captain who adopts an excessively casual attitude on the flight deck sets the standard for the whole crew. The same principle applies, of course, throughout the whole management structure within an organisation.

A fourth way in which leadership is applied in practice is in concern for **maintaining the group** or team, attending to personal relationships, resolving disputes and encouraging harmony and cooperation amongst its members. This can be particularly relevant to the case of flight crews who travel around the world as a group, away from any other organisational structure or controlling authority and function almost as independent units. This task involves ensuring effective communication between members of the group. Inadequate communication between members of the flight deck crew has been demonstrated to have been associated with accidents (see Fig. 2.2). The need for good communication is emphasised in training programmes for flight deck and cabin crews (see also Chapter 14).

Finally, the leader has a **managerial role** with respect to the use of the total resources which are available to him. This involves the allocation of duties, particularly significant in high workload or emergency situations. The Everglades L1011 accident in the USA in December 1972 is a typical example of a situation in which duties wcre not properly allocated and priorities not set (NTSB-AAR-73-14). The same month at Midway Airport, Chicago, a 737 crashed and again it was concluded that the leader, the captain, had failed in this role of properly managing the resources which were available to him (NTSB-AAR-73-16). Many such cases have occurred before and since. Members of a group become increasingly dependent on a leader in high stress situations when the leader is, of course, usually under higher stress himself. They also tend to transfer some of their own responsibility to the leader. The leader must be able to perform this function of managing the resources available, and in special situations, accepting dependency of those under him and taking over their responsibilities when transferred to him.

These five principles can be applied to most leadership functions, though in some large organisations some aspects, such as group welfare, may be at least partly allocated to specialised staff or departments. A brief summary is given in Appendix 1.15 of how the principles of leadership might be applied by a captain or purser in a typical civil aviation environment.

A new element is now being introduced into leadership in the aircraft environment, that of allocating command roles to women over men on the flight deck and in the cabin. There is no reason to assume that the individual performance of female crew members is any less than that of men in both normal and high workload situations. However, it has been suggested that in some emergency situations certain male crew members may have reservations about the female ability to command and lead and that this could impair the coordinated performance of the crew. The male crew members may then wish to take over some of the responsibilities of the female leader (Helmreich, 1980). Such an assumption of responsibility has already been illustrated in analyses of the behaviour of all-male crews when a senior first officer usurps the authority and responsibility of the captain whose competence he has some reason to doubt.

It seems likely that in time the reservations entertained by some on the psychosociological aspects of women in a position of command over men in emergency and overload situations, will recede. However, in the meantime the question could possibly benefit from further study. Such study would concentrate on the emergency situation as there is no reason seriously to question the adequacy of coordination in routine operations.

Another aspect of performance and behaviour which, like leadership, comes within the scope of social psychology, relates to personality and attitude clashes within a crew. This often makes the task of a leader more difficult and can influence both safety and efficiency. The Staines Trident accident in the UK in 1972 was a case in which personality differences may have had an influence on behaviour and performance. The captain had been involved in a crew room argument before departure and graffiti directed at the captain were found on the flight deck of the crashed aircraft (AIB-AAR-4/73).

A UK survey of 249 airline pilots has confirmed the existence of a problem from this cause. The major underlying factor in flight deck cooperation problems was quoted by 54% of the pilots as unacceptable behaviour of the other pilot. A clear majority (93% of first officers and 74% of captains) said that there were other pilots with whom they preferred not to work. Personality differences were quoted by 30% of the pilots. Other factors included difficult or disruptive work practices of the other crew members, unsuitable combination of rank and experience, deviation from normal role behaviour and attitude differences (Wheale, 1983).

Maintaining motivation

In addition to the routine necessity to maintain the motivation of staff in an airline, in view of the safety implications of reduced performance, special situations have been recognised as presenting a particular challenge. One such situation results from slow promotion in periods of relative stagnation in airline growth. The career objective of almost every pilot is to be in command of his aircraft and long years playing the subordinate role of first officer is found very frustrating by many. A crewing policy which tends to aggravate this is that where the third crew member on the flight deck, sitting at the systems panel, is a pilot rather than a flight engineer. This policy has its advantages and disadvantages and discussions tend to become emotive, involving as they do union and industrial relations elements. What is recognised, however, is that long periods as a systems operator is found very demotivating by many pilots. The problem has come up in many international airline conferences. At one such meeting, British Airways which used this policy in its European Division explained that it tried to reduce the frustration by having the systems panel duty shared between the first and second officer. TWA has emphasised the importance of the selection programme in trying to employ only highly motivated pilots, but has admitted that the problem has caused concern. Other airlines, while using professional flight engineers, have also recognised the motivational problem arising from career stagnation at the first officer level. As the wartime pilot 'bulge' dissolved in the 1970s, promotion took place sometimes very rapidly, but periodic slow-down in industry growth occasionally still generates a problem in this area.

We all accept without question that the performance of materials such as gases, metals and fluids, is governed by a set of rules. We must now come to accept that human behaviour is also governed by a set of rules, though they are frequently complex and elusive. Human behaviour is to a considerable degree predictable. To the extent that it remains unpredictable, the search for knowledge must continue.

By pursuing an understanding of what motivates people and by recognising the need for skilled leadership we shall move closer to the goal of more consistent and higher quality human performance and an individually more satisfying working life.

7

Communication:
Language and Speech

Communication defined

Types of communication

If the first officer on a ship disagrees with the course which his captain is taking, he can always reserve himself a place in the lifeboat. The first officer on an aircraft has no such option. He must resolve the problem and ensure his safety by other means; he must utilise verbal communication. This has given rise to frequent difficulty, and was discussed in Chapter 2 in relation to errors resulting from system interface mismatch.

The cabin attendant checking for signs of fear or anxiety amongst passengers after a bout of clear air turbulence, will initially look for indications through non-verbal means of communication. The manufacturer wishing to guide the airline's maintenance staff in servicing an engine, will need to use written communication in the form of manuals, technical bulletins and checklists (see Chapter 10). For passengers who may be unable to understand either the spoken or printed word, communication may be by means of symbols or diagrams.

In order to enhance the effectiveness of communication, increasingly electronics is being applied. We now see electronics providing means to communicate between machine and man verbally, using voice synthesisers. And in the reverse direction, communication between man and machine, more recently using keyboards, can now also utilise the human voice directly.

Within an industrial organisation, efficient communication between departments can be seen as the lubricator of the system – without it, the system can be expected to grind to a halt.

Social, economic and technological efficiency all depend on effective communication. Its influence is such that without it, loneliness, distress and death amongst the aged can result, destructive strike action can occur in industry and aircraft can crash. The subject warrants serious attention.

In engineering we are increasingly meeting the requirements for effective communication between systems. For example, computers must receive data from elsewhere and communicate their conclusions to other systems. But communication between machines, where man is not directly involved, is outside the scope of this volume.

Even omitting the field of communication between machines, the scope of communication – the conveyance of information – is very extensive. After all, a Bachelor of Science degree in Human Communication is typically a three-year full-time course. In Chapter 8 we look at the use of communication in changing attitudes and in persuasion – the basis of all safety propaganda. In Chapter 9 some aspects of communication in training are examined and in Chapter 10 we examine in more detail communication using the printed or written word in documentation.

In this chapter we have selected for discussion the important aspects of language and speech which are fundamental to the development of most systems of communication, and look briefly at some accidents involving deficiencies in this aspect of Human Factors.

Language of communication

Verbal language

The words and symbols we use and the way they are put together and related to each other constitute a language. Language is inextricably linked with the cognitive or thinking processes as well as with communication.

We are at this time, however, concerned only with language as a form of communication. For this purpose, it can be spoken or written or even expressed by hand and finger signals, corresponding to symbols, as in the language of the deaf and dumb. It may be in the form of characters which are embossed on paper, such as the six dots of Braille which are grouped to represent letters in the uncontracted grade and words in the more advanced contracted grade. Similar in principle are the far simpler and easier to learn symbols of the Moon System, also using embossed characters and designed for reading by the blind. It could also be para-verbal in the form of specific but unstructured sounds, such as grunts or groans. It could be totally non-verbal in the form of facial expression or body movement and posture, called body language. Each application of language has its own role in communication between people and each has its own capabilities and limitations. What is suitable for one purpose may not be suitable for another.

Ambiguity is a characteristic of communication which varies with the different form of language. Sentences which may be largely meaningless or at best, ambiguous, when written, can be made fully comprehensible when spoken with appropriate stress, pitch and timing. The interpretation of identical written words with more than one meaning, such as tear or lead, can be clearly differentiated when spoken, but when written can only be identified through context. A phrase such as 'they are eating apples' is ambiguous when written, but this can be resolved when spoken by the use of stress, in this case on either the word eating or apples. On the other hand, some words, such as wait and weight, which sound the same, can only be identified through context when spoken, though no identification problem exists when they are written.

Some phrases may remain ambiguous in both the written and spoken forms. The meaning of 'time flies' depends on which word is the verb and which the noun. In languages which have a special word form for the verb and noun this particular problem does not exist. Another example of ambiguity in both written and spoken language forms is 'the steward was ordered to stop smoking in the cabin'. Here what is called the surface structure of the sentence is not sufficient to allow a proper interpretation of what is meant.

These examples illustrate the importance of structure, or the arrangement of words, in a language. The principles or rules which form the foundation for this arrangement are called syntax or grammar. Even when these rules are properly applied, as in the last example above, ambiguity can arise where only the surface or superficial structure is considered and not the deep structure, or full meaning.

The role of written and spoken language in communication is different. It is well recognised that sometimes it is preferable to write a letter and sometimes to arrange a face-to-face interview. Speech can provide a rapid exchange of messages in the form of conversation and as such also serves as a primary lubricant of social interaction between people.

Written language, in contrast to the spoken word, has historically provided the characteristic of permanence in both time and space. While the spoken words of the ancient Egyptians died with them, their hieroglyphics engraved on stone remain to communicate accurately details of their lives and times. It is easier to send information over a distance in the form of a book than in the form of speech. Advanced technology in voice recording and transmission is now giving a new dimension to these aspects of the spoken word, though for purposes of information accumulation and storage, the written word is still likely to remain the preferred method.

An important element of spoken as distinct from written language is pronunciation. This can communicate information such as the speaker's geographical origin, social class and education. Such information may also be communicated by written language, but in a different way. Pronunciation introduces problems, too, sometimes as a result of different dialects or schools and sometimes as a result of natural anomalies of the language – compare the pronunciation of rough, bough, cow and tow.

Different sensory systems are concerned with reception of the different forms of language. Speech communication uses the auditory channel, while written communication uses the visual channel. This distinction becomes somewhat blurred in the case of speech communication with the hard-of-hearing where lip-reading may contribute to reception of the message. The use of Braille or the Moon System introduces the tactile sensory channel in place of the visual one.

This chapter will be devoted to various aspects of the spoken language. In passing, however, a brief reference should be made to body language as this plays a significant role in communication and is often integrated with speech.

Body language

There is a vast range of body language with a surprising degree of universality in its interpretation. The raised eyebrow, the wink, the nodded head (though the Greeks interpret this differently), the thumbs up, the facial expression and many more. A girl in Piccadilly or Montmartre can convey her message without the use of any verbal communication at all.

On board an aircraft, the passenger may signal to the stewardess that he wants a pen, some writing paper or a drink. A pilot may indicate to his copilot by a hand movement that he wants the landing gear raised. He may use a sign to tell the ground engineer that he is ready to start the engines.

Body language is much influenced by culture. To touch and be touched is probably a basic human need, though the extent to which it is expressed or suppressed differs notably from one society to another. Spatial proximity – how close to each other we place ourselves – is also affected by culture and personality. Introverts seem to want more space around themselves than extraverts. There is thus a case when an aircraft has only a light passenger load for allowing 'free-seating' without seat number allocation. There is significance in the way people sit in relation to each other. Sitting face to face, at right angles or side by side as an optimum position, depends on the function of the communication – whether it is for normal conversation, an interview, a business confrontation, and so on. Studies have shown that in different countries people in conversation stand at different distances apart. We shall have to leave non-verbal language at this point and revert to the main theme of this chapter, the spoken language. Coverage of non-verbal language is available elsewhere (e.g. Fast, 1971; Open University, l975b).

Word intelligibility

There are a number of factors which should be considered in trying to increase the effectiveness of a message. The appropriate term here is intelligibility – the extent to which the transmitted word is understood by the listener. One such factor is the frequency with which the word is used in everyday life. Research has shown that frequently used words can be understood with a greater background of noise than infrequently used words. Not surprising, perhaps, but it is an important consideration in the design of checklists and procedures for use in and around aircraft. The word length also has an influence. Generally, the longer the word, the more readily it is identified, as hearing only parts of it may be sufficient for recognition. Both the frequency of use and the word length can vary by as much as 10–15 dB the signal–noise relationship which can be tolerated for a given level of intelligibility (Howes, 1957). These two criteria can sometimes be in conflict. For example, the longer word 'negative' in messages is more reliably understood than the more frequently used word 'no'. But the general principle remains a good one.

In noisy conditions or when there are distortions in the transmission, a word which is in a phrase or sentence is more likely to be understood than when words are on their own. The word 'fire' is more likely to be understood when it is in context such as 'there is a fire in No. 2 engine' than if it is simply spoken alone. These conclusions have been reached as a result of experimental studies (Fig. 7.1) which showed a wide gulf between the intelligibility of words from an extensive vocabulary compared with those from a small familiar vocabulary or familiar sentences (Kryter, 1972).

One other aspect of word intelligibility is relevant. Repetition usually increases the understanding of a word. 'Hurry, hurry, hurry' in an emergency evacuation of an aircraft is likely to be more readily understood than simply 'hurry'. All these conclusions fully support the use of standard phraseology in aircraft communications. However, this standard phraseology must have been developed taking into account the basic principles outlined here.

Standard word spelling alphabets have been devised to increase intelligibility when communication conditions are poor and to reduce the risk of misunderstand-

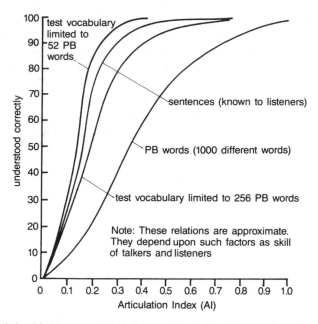

Fig. 7.1 The relationship between the Articulation Index (AI) and the intelligibility of various types of speech test materials composed of phonetically balanced (PB) words and sentences (from Kryter, 1972).

ing at all times. In aviation, the standard alphabet, finalised in 1955, is the best known (Appendix 1.16). This was specifically designed to assure high intelligibility when used by nationals of any of the partners in NATO (North Atlantic Treaty Organisation), though it has been adopted by ICAO for universal application (Moser *et al.*, 1955). It was developed with the requirement that words with Latin roots should be given preference in developing the standard phrases. Other phonetic alphabets have been developed for use by local telephone organisations. In Holland, for example, although a NATO country, the telephone operator will use quite a different alphabet, largely making use of Dutch personal first names and place names. Lodewijk and Utrecht are no doubt locally more familiar than Lima and Uniform.

Numerous standard words and phrases have been developed in aviation and their consistent use is essential for the limitation of human error in communication (ICAO PANS-RAC). It is perhaps a paradox that in spite of universal recognition of the importance of discipline in the use of standard phraseology for safe aircraft operation, the world's largest aviation country, the USA, is often accused of being one of the greatest offenders. The USA is becoming increasingly isolated in its use of local time instead of GMT and non-metric units in aviation (statute miles, degrees Fahrenheit, inches of mercury, pounds, etc.), complicating international communication. Although the US Congress authorised the use of metric measurements in 1866, the USA is the only industrially developed country which still has no commitment to conversion to metric standards. Efforts of ICAO to standardise units date from 1946. Much of the research which has led to standard international vocabulary and phraseology originated in the USA. Paradoxically, however, the international pilot flying into the USA will sometimes need to interpret a different form of radio telephony (RTF) from that in use in the rest of the developed world. The English language over Frankfurt in Germany may be closer to international standards, and so more intelligible, than that over, say, Chicago. It was very many years, and long after the rest of the world, before the USA finally applied the ICAO phonetic alphabet, published in 1955.

While these general rules for optimising intelligibility are particularly important to air/ground communications they also apply elsewhere where noise can cause interference. In a noisy hangar, on a bad telephone line or in a crowded terminal building intelligibility of a spoken message is at risk unless enlightened methods of language selection are used. In a moment we shall look at further ways in which verbal communication can be improved.

The vocal and auditory systems

Voice and speech mechanisms

The speech sounds made by man result from the interaction of several components of the vocal system. This system may be thought of as the lungs, trachea, larynx, pharynx, nose and mouth. Various other parts associated with the production of speech sounds are also shown in Fig. 7.2.

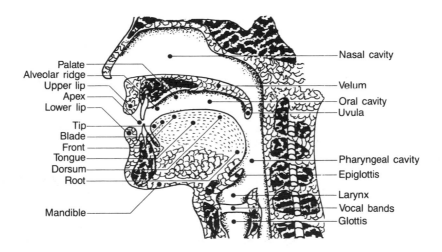

Fig. 7.2 A schematic drawing of the human speech apparatus showing the various parts involved in the production of speech sounds.

Different voices utilise different basic ranges of pitch or frequency. Regional accents will show different intensities of various frequency bands. Men employ different frequency ranges from women. However, for the same spoken words, the general visible pattern of the speech remains recognisable for all accents and all voices. This has a relevance to speech intelligibility, because although there are many ways in which speech can be deformed, so long as the pattern of frequencies is not destroyed the speech will remain intelligible.

157

While the vocal system generates speech, it is the auditory system which senses it and conveys vocal communication to the brain. As this mechanism is of vital importance in ensuring the effectivity of the communication, we should look at it more closely (Fig. 7.3).

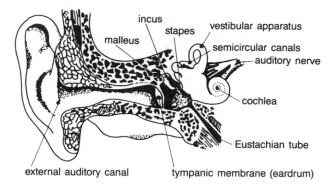

Fig. 7.3 The main components of the ear.

The external ear leads via an auditory canal to the tympanic membrane, or eardrum. The variations of pressure in the air cause this tightly stretched membrane to vibrate. Attached to this membrane, in the middle ear, are three tiny bones or ossicles called the hammer (malleus), the anvil (incus) and the stirrup (stapes). These bones are attached to the oval window of the inner ear, where a diaphragm sets in motion the fluid inside the cochlea. Within the cochlea is the Organ of Corti, a complex structure which contains the auditory nerve ends and the hair cells. The various components in this system can be seen as transformers and amplifiers.

The inner ear has another function, which was briefly mentioned in Chapter 5 when discussing visual illusions. This function is related to the maintenance of balance and the unit involved which is attached to the cochlea is called the vestibular apparatus. This consists of three semi-circular canals and the utricle which interconnects them. Like the cochlea, these canals contain fluid and delicate nerve fibres which are stimulated by the fluid. These nerve fibres transmit messages to the brain regarding movements such as turning or bending. Impairment in the functioning of this apparatus can lead to dizziness, vertigo and nausea.

This air-borne sound transmission system via the eardrum and the middle ear is not the only channel through which we hear. Hearing also takes place through bone conduction of the skull. This occurs when crunching a biscuit or when humming and speaking. Hearing by means of bone conduction by-passes the middle ear.

The middle ear is air-filled and is connected to the nose and throat by the Eustachian tube. Swallowing, yawning or sneezing, momentarily opens up this Eustachian tube, allowing pressure inside the middle ear to be equalised with that outside.

Hearing deficiencies

If hearing via the normal auditory canal of the ear is deficient, the problem may simply be due to an accumulation of wax in the outer ear, which can be easily remedied. The Eustachian tube can become blocked with a head cold, preventing normal equalisation of pressure. The air inside can then slowly become absorbed in the blood causing a pressure differential over the eardrum. When the nose is blown and the tube is opened, the ears pop as the eardrum moves outwards again as pressure is equalised. Changing altitude in flight has a similar effect in creating a pressure differential, unless the Eustachian tube can be opened up.

A common middle ear disease (otosclerosis) is characterised by the deposit of new bone or calcium material on the bony elements which are responsible for sound transmission, sometimes blocking their movement altogether. More common is deafness due to infections of the middle ear. These sometimes cause an accumulation of fluid which dampens movement of sound transmission components of the middle ear. Recurring infections of the middle ear (otitis media) can cause an increase in fibrous tissue and loss of flexibility of these components and occasionally perforations of the eardrum. Various surgical operations can be applied to correct many deficiencies of the middle ear. All such sources of interference with the transmission of sound waves through the outer and middle ears are called conduction deafness.

Because hearing by bone conduction by-passes the middle ear, this can be used as a diagnostic test to determine the source of hearing loss. If hearing by bone conduction is good, then the problem lies in the middle or outer ears. If hearing by bone conduction is also deficient, then the problem is probably due to a condition of the inner ear and is called nerve deafness.

Nerve deafness of the inner ear is irreversible. This condition can be due to damage of the sensory hair cells and the nerve fibres which are connected to the base of the hair cells. Noise is an important threat to these cells. The damage may result from exposure to a single loud noise, such as an explosion, or from long-term exposure to noise such as that from machinery or aero-engines, particularly with high frequencies involved. In both cases the cells suffer permanent damage. An audiogram of someone who has been exposed to noise at an excessive level over an extended period will typically show a dip at about 4000 Hz. Protection from exposure to noise is frequently neglected, probably because the resulting deafness usually advances gradually and insidiously. It has been suggested that hearing deficiency resulting from aircraft noise may have been an element in aircraft accidents (Hurst, 1985). The question of aircraft noise is discussed in more detail in Chapter 13.

A third cause of deafness may be called central hearing loss and arises from interference with the functioning of the region of the brain which is associated with hearing. This may be caused by injury, a brain tumour or a stroke.

Hearing can be expected to deteriorate with age, particularly in the higher frequency bands where it is notably more severe in men than women (Fig. 7.4). It also raises the threshold of hearing at the low frequency end, but eventually extends over the whole range. This natural hearing loss with age is called presbycusis.

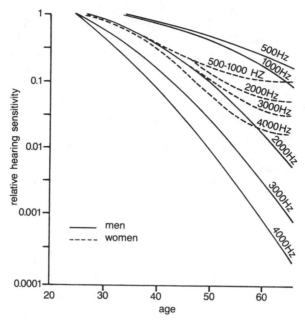

Fig. 7.4 Curves for hearing loss with ageing for men and women. The graph shows the hearing loss expressed in relation to the sensitivity at 25 years to a 4000 Hz tone. For example, the hearing sensitivity of a 50 year old man for a tone of 4000 Hz is about one hundreth of his sensitivity at 25 years (Edel, 1967).

Speech

Characteristics of sound in speech

There are four primary characteristics of sound in speech and these give rise to certain corresponding sensations. The first, intensity, is measured in decibels (dB), referenced to a standard level, and results in the sensation of loudness. The approximate meaning of decibels in everyday terms may be deduced from these examples:

quiet garden	20 dB	loud voice	86 dB
quiet voice	46 dB	discotheque	115 dB
busy street	60 dB	jet engine	140 dB

In general, we are less sensitive to frequencies below 1000 Hz than to higher frequencies. Thus a low frequency tone will not sound so loud as a high frequency one of the same intensity.

Generally, the loudness of consonants is less than that of vowels. In most languages, consonants outnumber vowels by a wide margin so contain a large part of the information in speech. Consequently, it may be said that most of the information is contained in the quietest part of speech. Because their loudness is lower than the vowels they can more easily be drowned by noise, or masking, about which more in a moment.

A second characteristic of speech is frequency, measured in Hertz (Hz) or cycles per second. Frequency gives rise to the sensation of pitch. Voice frequencies can

160

normally range from about 1000–9000 Hz. The healthy human ear can detect frequencies from 16 Hz up to about 20 000 Hz; children are sensitive to considerably higher frequencies.

A third characteristic of speech is harmonic composition and is often spoken of in terms of quality. The word 'hello' spoken by a close friend will have a different quality compared with being spoken by a telephone operator. A change in the harmonic composition can turn a sympathetic phrase into a sarcastic one. Whether or not the public address (PA) announcement from the captain to the passengers conveys an impression of authority, competence and friendliness depends upon its harmonic composition as well as its information content.

A final characteristic is related to time – the speed at which words are spoken, the length of the pauses and the time spent on the different sounds. Many aircraft passengers will have noted the occasional hurried PA announcement of the stewardess on a short flight, reflecting the time compression applied to all her numerous in-flight duties between take-off and landing.

Redundancy, masking and noise

There is much redundancy in spoken language, as in written language, and this makes it very versatile. As a result, it can be severely distorted or surrounded by noise with sufficient information remaining to convey its meaning. The elimination of certain frequency bands altogether from speech is called clipping. Even when several frequency bands are clipped, and even when total cuts in speech are made intermittently, words can usually still be understood. This is a valuable characteristic when having to communicate in unfavourable conditions or when the listener has defective hearing. However, it must be remembered that the gaps are filled in by the listener based on previous experience, learning and expectation, so there is a risk of a false hypothesis emerging (see Chapter 2). This can be very dangerous and is the source of many accidents.

Another technical aspect of speech which is highly relevant in communication is called masking. Unwanted noise from the environment often masks speech. The noise may be from an aircraft engine, road traffic, typewriters in an office, people talking in a room or machinery in a factory. It may come from electromagnetic interference within a radio-telephony (RTF) or telephone system itself.

The most effective method of protecting speech from the effects of noise interference is, of course, to control or isolate the noise at its source. Yet even when the designer has done his best within the various economic and state-of-the-art constraints with which he is faced, we are still likely to have a problem to solve. The use of standard words and phrases is helpful but, as mentioned earlier, in such conditions great caution must be used to avoid misinterpretation of partially masked signals through expectation.

Consonants, it was noted earlier, carry most information in speech but are, for the most part, weaker sounds with relatively high frequency. These are unfortunately the most vulnerable to masking. Sounds of a low frequency tend to mask not only sounds close to their own frequency but also those of a higher frequency. In noise situations, varying a signal strength by as little as 3 dB can change intelligibility from zero up to 100%. This is an important consideration in the design and operation of communications equipment.

The relationship between the loudness of the signal and that of the background noise is called the signal–noise ratio. This factor is usually more important than the absolute level of the signal or noise in determining intelligibility. Simply increasing or decreasing the volume, which changes the noise and signal together, is not very helpful. At very high levels, auditory fatigue increases and this can affect hearing performance.

Noise can be defined as sound which has no relationship to the completion of the immediate task. It is an unwanted sound. It not only interferes with speech communication but also has annoyance and health implications. Much research has been done on this subject, particularly in connection with noise in the vicinity of airports, stimulated by community protection organisations. Such activities have resulted in the development of quieter aircraft engines, operational noise-abatement flight procedures (sometimes reducing flight safety margins), sound-proofing of residential and work premises and the development of ear protectors. Control of the problem thus involves applying measures at its source and at the location of the receiver.

The degree of annoyance created by noise and its control are complex questions and they have been well studied. However, the subject is beyond the scope of this chapter and we suggest reference elsewhere for those interested (e.g. McCormick *et al.*, 1983). One point, however, is relevant here. In noisy environments, ear protection by means of ear muffs or ear plugs is essential to avoid permanent damage to the hair cells of the inner ear. In some cases, personnel avoid taking such precautions in the belief that they would be less able to hear someone speaking when there is noise. Under most noisy conditions, ear protection worn by a listener does not degrade speech intelligibility because the signal–noise ratio remains the same. In certain cases, such as with intense noise, speech intelligibility is actually improved. When both the speaker and listener are wearing ear protection some reduction in intelligibility is possible and so special attention must then be given to ensuring the clarity of speech (Howell *et al.*, 1975). When the speaker is in an intense noise field, his microphone should be protected by a noise shield.

Sudden exposure to loud noise has certain physiological effects, such as an increase in heart rate, blood pressure and breathing rate. However, they usually quickly return to normal. Noise can also adversely affect human performance under certain conditions and with certain tasks.

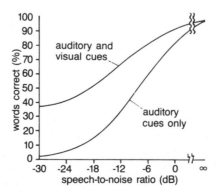

Fig. 7.5 Intelligibility of words when perceived with and without visual cues from observing the speaker (Sumby *et al.*, 1954).

162

Different speakers and different listeners vary in their ability to cope with noisy conditions. It has been shown that learning can improve the performance of both speakers and listeners (McCormick *et al.*, 1983; Orr *et al.*, 1965; Van Cott *et al.*, 1974).

Being able to see the speaker also improves the intelligibility of the spoken word, providing an input from both the auditory and visual channels (Fig. 7.5).

Expectancy and errors

In Chapter 2 we discussed the general effect of expectation on perception and in Chapter 5 its relevance to the perception of visual cues. In speech communication a similar effect can also be seen. If an answer from an air traffic controller is expected to be 'approved' then it may be so heard, even when, in fact, it was 'not approved'. The more of the speech content which is lost through clipping, distortion, noise or personal hearing loss, the greater is the risk of expectation playing a role – possibly a disastrous one – in the interpretation of an aural message. There are many examples in flying of this kind of error. In one such case, an aircraft standing at the beginning of runway 10R at London Airport was given a take-off clearance and began its roll. However, another aircraft waiting at an intersection had expected a take-off clearance, interpreted this one as being for him, and also turned on the runway for take-off. An aborted take-off from the first aircraft narrowly prevented a disaster (CHIRP Feedback No. 5).

A similar phenomenon occurs in flight. In the UK airway system there is a holding point called Eastwood. On a number of occasions a pilot, for whom English was not a natural language, has expected instructions from ATC to take up a different heading during vectoring. When instructed to turn to Eastwood, he interpreted this as an instruction to turn eastward and so altered course onto a heading of 090°. This is a combination of poorly designed nomenclature and the effect of expectation (CHIRP Feedback No. 5).

The phenomenon of expectation is particularly common and dangerous in the readback and confirmation of messages. It has even been known for two separate air traffic controllers in sequence to hear clearly a readback call from an aircraft reporting at 6000 ft and for neither to notice that he was not at 8000 ft, where he was supposed to be (ASRS Callback No. 52). The pilot who was instructed to 'runway 26, holding position', thought the instruction was to 'runway 26, hold in position'. He taxied on to the runway and a landing aircraft was forced to overshoot to avoid a collision (ASRS Callback No. 65). This was probably a combination of expectation and non-standard phraseology. Often linked with the hazards associated with expectation and anticipation are those reflecting deficiencies in short-term memory. This combination seems to be particularly dangerous and not uncommon. One example is the pilot cleared to climb to 10 000 ft but told to 'expect Flight Level 230'. The last part of the message only was retained in memory and he climbed through the 10 000 ft altitude to which he was cleared (ASRS Callback No. 51). Another aircraft was cleared to descend to Flight Level (FL) 230 and told to anticipate crossing a later fix at FL 180. He descended to FL 220, having apparently forgotten that the FL 180 was only an 'anticipation' (ASRS Callback No. 52). The inclusion of more than one altitude in a clearance introduces a hazard which must be recognised and requires a clear and careful readback and confirma-

163

tion to ensure understanding and reduce the risk of a short-term memory failure. The transmission of clearances by RTF is frequently done too fast, with the pilot noting the more critical figures in the shortest of short-hand and with the readback and confirmation by ATC too often done in a perfunctory fashion. The introduction of data-link systems, with clearances also displayed visually on a monitor or a map display, will help to resolve these problems. However, we have also learned that while data-link may well minimise some clearance misinterpretations, it can also create new difficulties (see page 166). Currently a considerable amount of research in several countries involves the optimum way to utilise this new technology, including the advantages and disadvantages of having hard-copy clearances sent direct from air traffic control to the cockpit via a cockpit printer. One such diffi-culty is that a selective hard copy transmission to the cockpit is for all intents and purposes a private message, however in many cases such a transmission contains information of vital interest to others flying in the same airspace and those others will never get that message. Therefore, compromises and trade-offs will have to be made, in many instances after considerable study and debate.

There are also other ways of protecting against the impact of expectation on the interpretation of a verbal message. For example, positive and negative messages should be worded quite differently; simply replacing 'yes' with 'no' is not enough. A repeat should always be requested when any doubt by any crew member exists as to the content of a message. Involvement of more than one crew member in important messages increases protection. Experience shows that a normal readback for confirmation is not a guarantee for detection of a misunderstanding, but perhaps it is the best insurance we currently have. It can be made more reliable if verbal stress is placed on the critical words (e.g. the cleared altitude) in the readback. In Chapter 2 we spoke of the two-pronged attack on human error and this concept is applicable here, too. Firstly, to avoid as far as possible the initial error in reception of a message through choice of language, use of standard phraseology, and so on. And secondly, to avoid the consequences of any errors which arise, primarily by means of careful readback and the involvement of more than one crew member in monitoring the communication.

Measures of speech intelligibility

There are many occasions from the factory design stage up to line flight operations when decisions have to be made on the effectiveness of communications equipment such as microphones, speakers and headsets. Practical trials and personal assess-ments are essential at the end of the evaluation process but it is helpful, and sometimes necessary, to be able to predict or calculate the level of intelligibility which can be expected. Such tests can also determine the types of speech sounds which are most, and least, subject to interference from noise.

Early speech intelligibility testing was costly, time consuming and involved problems of subjectivity. To avoid these difficulties, an Articulation Index was devised which can be calculated solely from the physical measures of the speech and noise level (McCormick *et al.*, 1983). By knowing the Articulation Index it is possible to compare the efficiency of different communication systems.

The Articulation Index can be converted into intelligibility, which is simply the percentage of spoken material of any particular type which is understood by a listener. We spoke earlier of the relative intelligibility of uncommon and common words and sentences and research has been done on this using the Articulation Index as a basis (see Fig. 7.1). An Articulation Index of about 0.3 would suggest that the speech communication system would not be adequate for use with normal speech material (Baranek, 1949). While it must be recognised that we are talking approximations, as human speaking and hearing skills vary considerably, we must also recognise that in aviation we need to have good communication at the level of the lowest common denominator of the people and the environment involved.

One other measure related to speech intelligibility is called the Speech Interference Level. It could be called the destructiveness of noise on the reception of speech. It is possible to predict approximately the voice levels required for communication, given the characteristics of the background noise and the distance between the speaker and listener. Such data have been applied to assess acceptable noise levels in conference rooms, secretarial offices and so on (Peterson *et al.*, 1978).

Automatic speech recognition (ASR)

Speech is the most obvious way for people to communicate with each other. Yet when people wish to give instructions to machines they have traditionally used knobs, levers, keyboards and CRT touch panels. It would obviously be more convenient if we could give instructions to a machine in the same way as to a human assistant, or even to the dog.

This idea has long been the subject of research. In 1952 a hard-wired piece of equipment was designed which was able to recognise ten figures spoken by its owner. But it required 20 years of development before a commercially viable device was produced and this was able to recognise 32 words. It may take a further 20 years of development beyond that stage before it is accepted for routine use in flight in civil aviation (Condom, 1983).

The technical problems are not so simple as they may at first appear. The vocal sound in human speech when analysed is virtually continuous and it is only because we know the language that we can break down this continuous flow of sound into recognisable words. Breaks which occur in this vocal sound do not always coincide with the break between words. Consequently, the ASR system has difficulty in breaking down human speech into a sequence of discrete words. This is the reason most ASR systems can only respond to isolated words.

A further problem is that people speak words differently; the broad Scottish accent may be quite incomprehensible to someone from, say, Louisiana, though both would claim to be using the English language. And furthermore, even one person will speak differently at different times; we mentioned earlier in this chapter the harmonic composition or quality of speech. This can vary for several reasons such as fatigue, sickness or emotion. For an ASR system to recognise a word reliably, then, poses considerable difficulties; for it to attribute meaning to any particular collection of words, requires many levels of complex processing and has proven more of a technical challenge than expected. Current systems have mostly been 'trained', like the dog, to recognise only the voice of the owner. These are called speaker-dependent systems and can use a much larger vocabulary than systems designed for many operators, called speaker-independent systems.

The ASR system designer is faced with having to resolve three main issues.

Firstly, the system must respond accurately to speech input from a wide range of users. Secondly, ways must be found for the system to ignore background noise and irrelevant conversation. And thirdly, words which are stored separately must be connected together in a meaningful way; this is called concatenation.

A wide potential market exists for ASR systems. Research is being done, for instance, to produce a voice-controlled typewriter and an extensive range of applications is anticipated in machine operation and control and in communicating with computers. But for use on the flight deck the essential design criterion is reliability and it is upon this that the airborne future and precise role of ASR will ultimately depend. An experimental application in flight has already been made in the USA (F-16), the UK (BAe 111) and France (Mirage) but this equipment has all been single-user-dedicated or speaker-dependent. It is probable that application to ancillary functions, such as radio and radio-navigation control will be found acceptable before direct application to the more critical control of flight guidance, which demands very high reliability and consistency.

With the development of synthesised voice and ASR systems, it is possible to produce two-way man-system voice communication, though numerous practical problems stand as obstacles to be overcome before such systems can be expected to play a significant role on the flight deck. Nevertheless, serious development work is in progress in what is sometimes called Direct Voice Input (DVI) application to avionic systems (Taylor, 1984).

Postscripts from Tenerife and other accidents

English in ATC

English is the international language of aviation and is required to be available for all air traffic control communication. However, individual countries may use their own language in air traffic control for their own national aircraft. This duplication of language in one air traffic environment undermines flight safety in an activity which, as will be seen in the following paragraphs, is already highly vulnerable to human error. It means that the English-speaking pilot will probably not be aware of instructions and clearances being given to other, possibly conflicting, traffic and will not be able to understand reports being made by that traffic. The safety contribution made by such monitoring is thus lost. In the American vernacular this is called monitoring the 'party line'. The loss of the 'party line' is one of the problems associated with future acceptance of a selective data-link between transport aircraft and air traffic control.

A possibility of this local language practice contributing to an accident has been considered in various investigations such as the 1976 Zagreb mid-air collision between a Trident and DC9 when Croatian was used with the DC9 pilot moments before the impact – both pilots had been cleared to fly at FL 330. A decision to mix English and French in Canadian ATC was only reversed after a major campaign by those, particularly pilots, interested in preserving safety rather than serving politics. The problem remains elsewhere, however, and continues to give rise to pilot warnings of its impact on safety (e.g. CHIRP Feedback No. 11).

Verbal confusion causing accidents

Inadequate communication often lies at the root not only of conflict and inefficiency but also of accidents. The history of aviation is littered with the wreckage of
aircraft which have crashed not as a result of technical failure or human incapacitation but due to ambiguous, misleading, inappropriate or poorly constructed verbal
communication.

Ambiguous voice communication between the ground and aircraft was cited as a
probable cause of the crash of the DC8 in Kaysville, Utah, in 1977 and many other
cases of such lack of communication precision have been quoted (NTSB-AAR-
78-8). An analysis in 1986 of more than 50 000 reports stored in the ASRS data
bank revealed that about 70% involved some kind of oral communication problems
related to operation of the aircraft (ASRS Callback No. 86). NASA devoted one
Research Report (No. 15 on the ASRS Publication List) to the problem of information
transfer in the aviation system. The study identified ten generic types of communication problem. It was apparent that expectation – hearing what we expect to
hear – figured very prominently amongst the many causes determined.

The ICAO Manual of Radiotelephony (Doc 9432-AN/925 paragraph 2.8.3) requires
readback of clearances in the interest of safety. It specifies that the readback shall
be done in a manner which clearly indicates that the message has been understood
and accepted. In addition, it requires that the runway in use, heading, speed and
level instructions, altimeter settings and transponder codes shall always be read
back.

Unfortunately, readback does not always provide a guarantee that the message
has been accurately received. Too often, confirmation is given of a faulty readback.
ASRS has coined the word *hearback* for this phenomenon and has cited four major
causes of error (ASRS Callback No. 127):

- similar aircraft callsigns resulting in confusion in transmission or reception
- only one pilot on board working and monitoring the ATC frequency
- numerical errors, such as confusing 'one zero thousand' with 'one one thousand'
- expectancy – hearing what one expects to hear (see page 163).

No doubt, overloaded air traffic control systems provide the basis for some
errors, but there are still practical precautions which can be taken to minimise such
errors in communication (see Appendix 1.16A)

Many cases have been reported of mis-communication on the flight deck, some
based on fact, some perhaps apocryphal. 'Back – on the power' said the captain;
but the copilot interpreted it as 'Back on – the power' (ASRS Callback No. 56).
Others are of earlier vintage. 'Take-off power' said the captain; so the copilot did.
'Feather four' was the instruction; 'All at once, Sir?' was the response. 'Feather
one', ordered the captain; 'Which one?' the careful copilot answered. 'Cheer up',
encouraged the pilot; the copilot raised the gear. But mis-communication between
air and ground may be more serious and this conclusion is supported by many
examples.

The Netherlands civil aviation authority's (RLD) report on the 1977 Tenerife
double 747 disaster made some highly relevant and perspicacious comments. It
concluded that:

● the crew took off in the absolute conviction that they were clear for take-off and nobody involved could be blamed for negligence of his own duties (see Chapter 2, the false hypothesis)

● the communication procedures and terminology employed were not perfect but were those in normal, daily use in civil aviation

● the accident resulted from a breakdown in this normal communication activity and from misinterpretations of verbal messages. Such breakdowns were known to have occurred a number of times on other occasions with resulting incidents, some of which closely resembled Tenerife

● the communication situation was more serious than generally recognised and international action was needed urgently to improve the system.

In this accident, the most critical error centred around the word 'cleared', but other verbal confusion also occurred, as illustrated in this extract from the PAA cockpit voice recorder:

PAA: Third to the left, OK
PAA CAPT: Third he said
PAA: Three
TOWER: ... ird one to your left
PAA CAPT: I think he said first
PAA FIRST OFFICER: I'll ask him again
PAA FIRST OFFICER: Must be three. I'll ask him again.

KLM meanwhile, waiting at the beginning of the runway, had reported '... now ready for take-off ... we're waiting for our ATC clearance'. The ATC controller had replied '... cleared to the Papa beacon ... right turn after take-off, proceed ...' The words 'cleared' and 'take-off' both appeared in the response from the tower. The KLM crew had taken the word 'cleared' as applicable to the airways clearance *and* the take-off clearance, as both had been requested; the tower intended it to apply only to the airways clearance. As a result, in the verbal and meteorological fog, the KLM aircraft struck the PAA aircraft which had not turned off the active runway at either the first or the third taxiway.

In another incident, a German Air Force transport aircraft was waiting for a call 'cleared for take-off'. He was instructed 'cleared into position and hold; standby for take-off'. He heard the words 'cleared' and 'take-off' in one message and, again influenced by expectancy, he started the take-off roll. Another aircraft was crossing the runway ....

The expression 'cleared for an approach' has also given rise to problems. In 1974 a 727 was approaching Dulles Airport, Washington, and was told by ATC that it was 'cleared for an approach'. The aircraft crashed into Mount Weather with a loss of 92 lives. The captain interpreted 'cleared for an approach' to mean it was safe and approved for him to descend to the final approach altitude (1800 ft), as no other altitude restriction was included in the clearance. The ATC controller interpreted it to mean 'subject to your maintaining a safe altitude yourself, you are cleared for an approach'. It appeared as a result of the subsequent enquiry that there was no standard, accepted definition for 'cleared for an approach', everyone simply made his own interpretation. Furthermore, although the aircraft had been radar vectored on to a radial coinciding with the final approach and the pilot was communicating with a radar controller, it was not technically considered by ATC as a radar arrival. As a result, no terrain clearance information was given. This was a classic case of poorly designed communication procedures and regulations leading to confusion and disaster. In fact, in its report, the NTSB cited as a contributory factor the failure of the FAA to take timely action to resolve the confusion and misinterpretation of air traffic terminology, although it had been aware of the problem for several years (NTSB-AAR-75-16).

It is of some interest that the second report received under the newly-inaugurated United Airlines (UAL) Flight Safety Awareness Program preceded this accident by only six weeks and was in effect almost an exact dress rehearsal for the accident. The UAL crew received the same clearance, made the same misinterpretation, and just barely missed the fatal hill. United's pilots and the FAA were immediately notified of the confusing clearance language but there then was no available way to efficiently and expeditiously notify the other airlines.

Depending on the intonation used when speaking, the words 'eight zero clear' can mean that FL 80 is clear or, if used with an interrogative tone, it can mean, is FL 80 clear? Just such a misinterpretation occurred near Stuttgart in 1977 and a DC9 was directed by ATC into the airspace which it was understood was reported as clear. The two aircraft passed within 400 m of each other at the same altitude. Another Airmiss, or Near Mid Air Collision (NMAC) as it is known in the USA, was recorded.

Conclusions from accident analyses

From these and many other accidents and incidents, focus was directed at several communication aspects. Firstly, the communication code of words and phrases established by ICAO in 1950 was not optimum operationally. Secondly, the pressures of air traffic resulted in abbreviated phrases being developed in practical use. And thirdly, the unofficial language of RTF that was in use was being freely interpreted and applied on an individual basis. Unfortunately, in response to time pressures and in the face of official inaction, these deviations proliferated. Even phrases used universally, such as 'ready for take-off' were not standard. This reveals a fundamental conflict. In other fields, such as law and government, messages are lengthened to ensure they are unambiguous. In aviation, phrases are being shortened, due to time pressures, but they still need to be unambiguous.

Another characteristic of aviation messages is that a critically high information element may be contained in a very small part of the message. Tenerife also gave an example of this, when the tower instructed, 'OK ... standby for take-off'. The critical word was 'standby' but apart from 'OK', which confirmed the crew's feeling that all was well, the critical part of the message was lost as a result of the other aircraft transmitting simultaneously; only a squeal was heard. If anything causes the loss of such a small but critically important part of the message – a temporary distraction, such as a cough, an irrelevant noise or voice, or radio interference, as in Tenerife – the result can be catastrophic.

Various ideas have been proposed to alleviate the problem, such as backing up aural take-off clearances with a visual traffic signal or by data-link flight deck read-out confirmation. Yet many communication errors occur at flight phases other than take-off. The two main recommendations of the Spanish government report on

Tenerife were to emphasise the importance of complying with clearances (everybody at Tenerife thought they had!) and to use standard aeronautical language. These recommendations seemed to be singularly unhelpful without having been preceded by a far more profound, realistic and enlightened study of the problem, with new methods and procedures being developed and applied. It was, after all, only three years after this accident that a 727 crashed into a mountain killing all 146 on board, also at Tenerife and again apparently as a result of communication confusion (Spanish report AAR 8/81).

Phraseology changes

In 1977, nine months after the first Tenerife accident and after a formal request from IATA, the Air Navigation Committee (ANC) of ICAO agreed to review its RTF phraseologies (PANS-RAC) and procedures (Annex 10, part 2). In 1979 the RTF Study Group was established to give effect to this decision. This Group carried out a systematic review in three phases of RTF for use, respectively, in aerodrome control, approach control and area control service. The three reports were submitted and final reviews carried out in 1981, 1982 and 1983. The amendments to PANS-RAC and Annex 10 became applicable in 1984, just over seven years after Tenerife. To support these changes, a Manual of Radiotelephony (DOC 9432-AN/925) was made available in 1985 by ICAO.

The most significant objective was to restrict the use of the words **clear/clearance** and **take-off** and so avoid some of the types of accidents and incidents quoted in this chapter. **Clear/clearance** are no longer used for start-up, push-back and taxying. The word **take-off** was replaced by **depart/departure** except for the single case of the take-off itself. The only situations in which the word **clear** may be used are the route, take-off, approach and landing clearances.

Possible confusion between negative and affirmative responses was tackled by introducing the word **affirm** to replace **affirmative**. In order to abbreviate some messages, and to bring the official procedures closer to actual practice, a number of deletions were made from standard phrases. These included the words **this is, over** and **out**, which were no longer to be required in VHF communications. Thirteen new words were added to the official list of standard terminology, though twelve of these were already in routine use for many years in the unofficial vocabulary developed by pilots and ATC controllers.

The Human Factors deficiencies in verbal communication outlined here were in no sense new and had plagued flight operations for very many years. The reader may reasonably ask why it should have taken a disaster of the scale of Tenerife to generate some corrective action. In connection with such post-accident activity, the Chairman of the NTSB has said that 'this phenomenon – a disaster response mentality rather than a disaster prevention approach – is one of the more frustrating realities we in the safety business face' (Burnett, 1984). References to the lack of awareness of the technology of Human Factors and the paucity of Human Factors expertise at influential levels in international civil aviation over many years were already made early in this volume. Communication is in the mainstream of Human Factors study.

The 1984 changes in ICAO phraseology represented perhaps the beginning of a new official recognition of the problem, though not a final solution; the ASRS reported in 1985 the case of the pilot who misinterpreted 'you are number one to approach' as 'you are clear to approach' (ASRS Callback No. 68). Further activity is required to encourage universal acceptance of English as the language of RTF, ensure consistent application of the new standards and develop techniques of communication which will reduce the risks inherent in total reliance on the spoken word. What is also needed is an on-going, specialised monitoring of Human Factors aspects of communication in civil aviation by those with the appropriate expertise and the authority to initiate change.

8

Attitudes and Persuasion

Aviation application

Personality, attitudes, beliefs and opinions

Personality traits and attitudes have a fundamental influence on the way we conduct our lives at home and at work. They determine the way we are judged by others; we become identified by them.

Personality traits may be innate and in any case are acquired very early in life. They are deep-seated characteristics which constitute the essence of a person. They are stable and very resistant to change though certain traits seem to have some tendency to alter in middle age. Attempts by psychotherapy to modify personality distortions often have very limited success, even when applied over many years (Smith *et al.*, 1980). Personality traits may be reflected in aggression, ambition, dominance or creativity, for example. Personality is often situation-related; the domineering personality at work may be submissive at home. It is far more complex than can be encapsulated in any conventional personality test, though some such tests do have validity in vocational settings.

A number of studies suggest that there are some personality characteristics which generally distinguish pilots from the rest of the population, though this does not imply that they are necessarily desirable ones (Farmer, 1984). Psychiatric disorders are second only after cardiovascular disease as the cause of permanent loss of pilot licence amongst British pilots, and anxiety state is the most common of these disorders amongst those flying professionally (Bennett, 1983). Reports from the growing number of incident reporting programmes support the view that personality, attitude and behaviour are important elements in many incidents and there can be little doubt that they play a significant role in the overall maintenance of flight safety.

It must be recognised that different kinds of flying call for an emphasis on different personality traits. The ideal personality for military missions or emergency relief operations may not be optimum for routine civil flying, and vice versa. While the evidence which is available illustrates the vitally important role of personality factors in flying, it is not yet sufficiently comprehensive to allow solutions to be applied routinely to meet the situation. It must be concluded, therefore, that more research into desirable and undesirable personality characteristics related to today's civil piloting and flight deck management task is required. In addition, more effective means of assessing personality during pilot selection and initial training are needed (Helmreich and Wilhelm, 1991).

Attitudes, in simple terms, describe likes and dislikes. More precisely, an attitude can be seen as a learned and rather enduring tendency to respond favourably or unfavourably to people, decisions, organisations, or other objects. The response becomes more or less predictable. An attitude is a predisposition to respond in a certain way; it is not a response or behaviour *per se*. We may hold a favourable attitude towards our own trade union, yet not turn up at meetings even though attendance could have been interpreted as a behaviour reflecting that attitude. While we may hold an attitude that income tax is immoral, the fear of the consequences of not paying it may cause the actual behaviour not to reflect the attitude. Many attitudes never entail any overt behavioural response other than the expression of an opinion.

A belief can be distinguished from an attitude in that it does not necessarily infer a favourable or unfavourable evaluation. It is generally an assertion about two objects or the relationship between them – skiing is a dangerous sport, or the man is a natural leader. While a belief is less strong than positive knowledge, it can definitely influence behaviour.

An opinion is a verbal expression of an attitude or belief and is one means by which others may become aware of it. Public opinion polls are often used as a means of discovering the attitudes of prospective voters to certain politicians, policies or political parties or of potential customers to certain products. The techniques used in surveys to obtain opinions are discussed in Chapter 10.

Various studies strongly suggest that many attitudes and beliefs are rooted in the personality of the individuals holding them. It has been found, for example, that there is a co-variation of many attitudes. A person who has an unfavourable attitude to one object may be expected to have a similar unfavourable attitude to certain other objects and both could be related to a single personality characteristic.

Personality, attitudes and beliefs are intangible in as much as they cannot be seen or studied directly but only inferred from what a person says or does. The precise relationship of attitude to personality will therefore depend on the theoretical approach which is adopted in defining them and constructing their hypothetical basis. In the jargon of psychology, personality, attitudes and beliefs are known as hypothetical constructs.

Relevance to aviation

It has already been pointed out that those involved in most accidents attributable to inadequate human performance probably, at the time of the accident, had the capacity to have performed effectively yet did not do so. Performance was thus influenced by factors other than the possession of technical skills or medical fitness, such as could have been determined in a routine proficiency or medical check. This conclusion is relevant to questions such as technical versus non-technical training, medical standards and compulsory retirement age. In the event, it may have been that the person concerned felt so confident that he could short-cut a standard procedure or avoid consistent use of a checklist. Or that his interpretation of leadership was dominance. Or that in difficult situations he should assume most of the tasks himself. Or that it was justifiable to by-pass some security arrangements if a flight were delayed. Or that an important radio message could be accepted without confirmation because the traffic was so heavy.

All such situations and behaviours involve personality or attitudes and in aviation have a safety as well as an efficiency dimension. In fact, attitude differences have been cited in a survey of pilots as being a cause of problems on the flight deck (Wheale, 1983). Reference was already made in Chapter 6 to the possible involvement of personality or attitude differences in the 1972 Staines Trident accident in the UK. Reports received by NASA in its ASRS programme also confirm the existence and significance of attitude clashes on the flight deck. Extracts from two reports illustrate such situations. One reads 'In the same mean voice, like that of a person very angry, the captain said ... Don't talk; do your job and I'll fly'. Another reads '... it was the hostile cockpit interruptions that I found accumulating in weight upon my mind ... I found it very difficult to keep my cockpit performance from degrading' (ASRS Callback No. 78). In a later paragraph we shall look at the extent to which it may be possible to influence attitudes generally and in the aviation environment.

The nature, function and measurement of attitudes

Origins

Attitudes are formed and changed in many ways. Their origins often lie in early life experiences and in the social environment. Some become internalised norms; others are more superficial and more easily subject to change. A new attitude often involves relinquishing an old one.

It is obvious that the establishment of attitudes is not confined to early life, though these influences, such as family political orientation, can be very powerful. We are under constant pressure from the media to adopt certain attitudes. The massive investment in this activity would not be made if it were not frequently effective in bringing about the desired change of attitude. Later in this chapter we shall look more closely at means of modifying attitudes.

Attitude components

Attitudes may be said to have three components. Firstly is a belief, knowledge or idea about the object of the attitude (cognitive). Secondly are the feelings held about it (affective). And thirdly is what is said or done about it (behavioural). If we were to relate this to the attitude of a particular pilot to, say, flying in severe turbulence, we might say that he knows what it is all about (cognitive), does not like it (affective) and tries to avoid flights into such weather conditions whenever possible (behavioural).

As mentioned earlier, there are sometimes apparent inconsistencies between these components. An alcoholic may believe drinking is wrong but still have friendly feelings towards it and his behaviour may reflect this. Similarly with a smoker. Another inconsistency also has to be considered. The verbal expression of an attitude may be different from other behavioural expressions of the attitude. Reflecting this, an employee may criticise his management verbally and display an apparently hostile attitude to them. Yet he may continue to work very hard and loyally for them. In an aircraft cabin, a purser might express a very hostile attitude towards his company because insufficient or incorrect meals have been loaded. Yet he may go to great lengths to improvise so as to obtain the best possible response of passengers towards the same company. This phenomenon was also mentioned in Chapter 6 when discussing achievement motivation. This apparent paradox arises because behaviour is determined by more than one attitude.

This leads us to question whether the cognitive and affective components of attitude – the evaluation elements – are sufficiently closely related to the behavioural aspect to consider all three as integral parts of a single construct. There is much evidence demonstrating that overt behaviour does not always reflect a verbally expressed attitude. It can occur that the behaviour associated with a particular attitude is nullified by the attitude towards the behaviour itself. For instance, the behaviour associated with a strongly-held attitude about a child molester may be one of violence towards the criminal. Yet the attitude held about violence may be such that the original behaviour is in practice nullified, either partly or totally. In the example quoted earlier, the behaviour which might be expected to follow an employee's negative attitude to his company or supervisors may be overridden by his negative attitude towards disloyalty to his company or his profession. An example in aviation is sometimes found in companies that have poor industrial relations but strong flight operations and, despite the apparent contradiction, a good safety record. Predictions about behaviour seem to be most reliable when the attitudes towards the object and the behaviour are both considered in relation to the particular individual.

The notion that behaviour is not always determined by attitude is supported elsewhere. In certain circumstances, role-playing behaviour has been shown to have an influence on attitude – the reverse of our conventional arguments. This is reflected later in this chapter when we refer to the reversal of the roles of management and union representatives in industrial negotiations. The principle is also applied in certain behaviour therapy.

It is important in industry to understand these interactions. Firstly, because incorrect interpretations can lead to undesirable consequences. And secondly, because we need to be clear as to what we are aiming to measure, assess or influence in any particular case.

Stereotypes

Many attitudes are based on what is called stereotyping. We tend to categorise or classify things and people; in fact, we do not have either the time or inclination to make a strictly individual analysis of everything and everybody before adopting attitudes about them. Any recognised characteristic of a category of people – long hair, a uniform, a smartly cut business suit – triggers various assumptions about the personality of that individual. He becomes a stereotype, with favourable or unfavourable connotations. And yet this may be unjustified and totally misleading. The red-head could, in fact, be cool and stable rather than hot-tempered and volatile. The swarthy, unshaven type could be a gentle painter rather than a Mafia agent (Fig. 8.1).

Fig. 8.1 Stereotyping involves categorising or classifying things or people. This serves a useful function in reducing time and effort expended in making individual analyses of everything and everybody. However, it is unjustified and is often based on unsound generalisations or prejudices (cartoon: J H Band).

Stereotyping is not always in a favourable direction. The confidence trickster uses stereotyping in his trade, profiting from the general association of courtesy, neat appearance and reasonable argument with honesty. The sexual responses of a pretty, seductively dressed, well-proportioned girl may not match up to that anticipated from the stereotype image.

The same risks of stereotyping apply in aviation as in other occupations. The aircraft captain who looks like the Public Relations concept of the highly responsible, healthy, intelligent and wise commander, may have a problem in scraping through his six-monthly flight proficiency or medical check.

Stereotyping is unjustified, even when it is based on certain personal experience. Frequently, the basis is not even experience, but conveyed by word-of-mouth or the media. Often stereotyping is based on totally unsound generalisations or prejudices.

In earlier discussion on human error we showed how we tend to see and hear what we expect to see and hear. As a result of a chauvinistic attitude we see our own national football team playing a cleaner game than that of foreign opposition. Or we may see our own airline providing better service than a competitor. One study of 410 industrialists showed that 92% rated their own product superior to any other while only 3% rated it as inferior (Blake *et al.*, 1962). Similarly, by stereotyping we remember the short, dirty, bearded man having the knife in his hand in a fight, rather than the well-dressed real villain.

The function of attitudes

As we have said that behaviour does not necessarily correspond to the attitude which is held, it is reasonable to ask just what purpose is served by attitudes; what is their function.

It is widely believed that attitudes serve to provide some sort of cognitive organisation of the world in which we live. They provide a structure which facilitates thinking and decision-making processes. This applies even to aspects of life about which we know very little. We may have strong attitudes about sterilisation in China, divorce in the USA or physical mutilation as punishment under Islamic law, without ever having examined the situation or arguments at close hand or in any detail. Yet by adopting attitudes about such aspects of life we can incorporate them into our routine cognitive life.

A structure of attitudes allows us to make rapid responses to people and situations – we might say, to judge them according to a pre-existing pattern – without having to make individual assessments. An attitude held by a pilot concerning the over-riding importance of safety allows rapid response to be made, such as when a conflict between commercial interest and flight safety arises.

In addition to serving as a general structure which allows us to make rapid decisions on what to do when faced with certain situations, attitudes also serve more specific or utilitarian functions. They can serve to guide a person towards rewarding outcomes, towards satisfying particular needs. A crew member may adopt an unfavourable attitude towards long flights if he has young children. A dictator will hold a hostile attitude towards free elections.

We have already noted that attitude changes can be generated from behaviour rather than the reverse. A person who adopts a particular behaviour to serve a specific purpose may not at the time hold the attitude which the behaviour appears to reflect. The attitude, we could say, is bogus. However, after some time under the influence of the behaviour, the attitude may become genuine and provide the background for future behaviour. Someone may display a behaviour relative to physical fitness suggesting his enthusiastic support, only in order to be accepted by a particular social group. However, after a time a genuine attitude favouring physical fitness and sport may emerge.

In addition to these cognitive and utilitarian functions, attitudes could serve a self-protecting or self-defensive role. A person who feels insecure socially or emotionally may obtain comfort or support by adopting hostile attitudes towards other groups – perhaps other races or social classes. An attitude may aid people in denying unpleasant truths about themselves by adopting a critical or discrediting attitude towards another person in a way which would strengthen his own ego. The need for this is often an unconscious one. Another example is reflected in a dislike of people who are sensed as superior or who make one feel inferior. This function of attitudes plays a role in many aspects of behaviour.

One of the more unpleasant applications of this self-defensive use of attitudes is in the creation of scapegoats. This arises from the need to explain undesired events and allocate blame for them. If we can put the blame for an aircraft accident on one person, this relieves everyone else from accepting responsibility and provides a tidy organisational structure in which to operate.

177

The measurement of attitudes

In science it is necessary to measure things in order to study and understand them. An attitude cannot be measured directly, but there are techniques for measuring some recognisable dimensions of it. Because attitudes are not visible and because it is generally impossible to observe all the behaviours which may reflect an attitude, enquiry is usually limited to asking a person about an attitude and then evaluating the response.

The researcher will be looking to measure different components of attitude. He may try to determine the strength of feeling which the attitude evokes and the strength of the attitude itself. He may seek to find out the degree to which the attitude is resistant to change and the extent to which the person is occupied with it. And he will want to know the extent to which the attitude is acted upon, that is, the behaviour resulting from it. It is the behavioural component of an attitude which is visible and so measurable. Yet we have already seen that an attitude may exist without necessarily any behavioural expression to reflect it at all.

Perhaps the best known way of trying to determine attitudes is the opinion poll, though there are many reasons why a polled opinion may not faithfully reflect an attitude. The poll may be conducted using a personal interview method or simply by distributing questionnaires and collecting written responses. Typically, this kind of survey is confined to narrow issues and seeks to determine the direction of an attitude rather than its strength. This is reflected in the rather fluid results of political opinion polls where a strongly held attitude is usually recorded in the same way as a weakly held one.

Various other techniques for measuring attitudes, or rather their outward expression, have been developed. We can briefly review three of them, each with its own characteristics (Oppenheim, 1966).

Probably the best known and one of the earliest of these is called the **Thurstone Scale**. A list of opinions about a certain matter is drawn up and, after careful evaluation, these opinions are placed in sequence from one extreme of opinion to the other. The person being questioned is then asked to mark the statements with which he agrees. A median value along the total scale is then calculated based on the statements which he has accepted as reflecting his opinion. This value then gives both the direction and strength of the attitude. Using this scale, described as a non-cumulative scale, acceptance of one listed opinion does not imply acceptance of the preceding ones.

Another well known scale is that devised by **Likert**. This consists of a list of opinions, each on a different aspect of the subject being studied, and the person is required to mark each one with a rating from 1 (strongly approve) to 5 (strongly disapprove). The direction and magnitude of the attitude is calculated as the sum of all the scores.

A third example, the **Guttman Scale**, is based on the assumption that a set of items measuring a single, unidirectional trait can be listed along a continuum of magnitude. This is a cumulative scale, in contrast to that of Thurstone, meaning that the person being tested would necessarily accept all items of a lesser magnitude than that which he had selected.

There are many reasons why the results of such attempts to assess attitudes may be distorted and give a misleading impression, such as a social desirability bias reflecting the wish of the person to be seen in a favourable light. This and various other sources of bias must be taken into account in the design of the technique and the evaluation of the responses. The measurement of attitudes, like its associated activity the design of questionnaires, is a skilled task calling for expertise. Questionnaire design is discussed in more detail in Chapter 10.

Influencing attitudes

Group influences

One of the issues which has been extensively investigated by social psychologists is the influence of groups on individual attitudes and behaviour. A group in this special social psychology sense is not simply a collection of people, such as those who may be assembled at an airport terminal. It implies that there are certain characteristics shared amongst those present which set them apart from others and which give them a sense of belonging. It also implies that there are shared goals, values, interests and motives amongst group members. A national sports team in the airport terminal, dressed in their national team blazers, would be clearly recognisable as a group in this special sense.

We all become members of groups during the course of our lives. For instance, the family, play groups, sports teams or school societies, social or professional associations and political parties, play a role in the lives of most of us. Some, such as flying staff, spend their whole working lives acting in groups.

This is not to suggest that an individual's attitude and behaviour cannot be influenced by others around him who do not meet this definition of a group. Experimental studies have demonstrated such an influence, particularly if the others present are capable of evaluating or judging the individual's performance or behaviour (Hood *et al.*, 1962). Other studies have shown that an individual will be influenced to make a false judgement if the majority of those around him have made the same false judgement (Asch, 1956).

A desire to join a group is normally determined by the feeling that the group members share to a significant degree one's own views. It would be surprising to see a successful businessman joining a Marxist political group dedicated to the abolition of private enterprise, or a land developer applying for membership of an environmental protection society. But in spite of the common elements which cause an individual to feel an affinity with a particular group, there is still much scope for that group having a profound and extensive effect on the attitudes, beliefs, interests, behaviour and even goals of the individual member. An extreme example of such an influence may be seen in the extraordinary sway of certain quasi religious cults over their members, even leading to mass suicide, a behaviour which few would have found acceptable upon joining the cult.

Risk-taking

One interesting conclusion of a number of studies is that when a decision is made by a group, it is likely to involve a greater element of risk than if it is made by an individual (Baron *et al.*, 1974). This conclusion may surprise some who feel that as a group decision generally involves compromises, the more risky choices should be eliminated. It might be worthwhile to examine this phenomenon a little more closely in view of the significance of risk-taking in aviation and as an illustration of the kind of experimental work conducted in connection with attitudes and the factors which influence them.

A number of reasons have been proposed for the modification of individual attitude towards risk as a result of group influence. Perhaps the first possibility which comes to mind is based on the concept of the spread or diffusion of responsibility for any adverse consequences of a decision involving risk; blame is shared amongst all members of the group.

Another hypothesis is that individuals who hold higher risk attitudes tend to be more dominant and persuasive in the group, assuming a leadership role. They thus have a disproportionate influence over their fellow members.

A further possible reason, which is quite different from the previous two, is that when an individual becomes involved in a group discussion, he becomes more familiar with all aspects of the topic upon which a choice has to be made. The increased familiarisation then provides the confidence to take more risks.

We should look at just one more of the many hypotheses which have been put forward to explain this phenomenon. In contrast to the above three, this one appears to have achieved some experimental support. It assumes that risk is assessed socially as a desirable value; many of the heroes in national life and in fiction characteristically accept high risks. The hypothesis is built on two suppositions. Firstly, that an individual normally over-estimates his personal level of risk-taking. And secondly, that once he joins a group he finds that his own level of risk-taking is not so high, relative to that of others, as he had originally thought. He then raises that level and so the risk-taking level of the whole group is shifted upwards.

While it has been well established experimentally that this shift in risk-taking in group activity occurs, the precise reasons for the phenomenon have not been established even though all the above hypotheses have been studied in laboratories. It is obviously of interest to determine how extensively these laboratory conclusions can be applied to practical working and living situations outside the laboratory. As with all laboratory findings, caution must be used when applying them due to the large number of personal and situational variables which may exist and which may not have been taken into account in the original experimental work.

Loss of inhibition
Group membership and sometimes simply the presence of other people can influence motivation, the performance of certain tasks and, as we have just seen, even attitudes towards risk. Another aspect of group influence which we should mention is the reduction of restraint and loss of inhibition which often occurs within a group. There seems to be something in the presence of others that weakens an individual's maintenance of socially acceptable norms of behaviour. This will be apparent to the airline cabin attendant on a charter flight with, say, a group of travel agents, sports supporters or package tour vacationers.

This loss of inhibition is more widely recognised in behaviour at parties or football matches where people act in a way which they would never consider proper or reasonable when acting alone. This manifestation takes on a more sinister image in situations involving mob violence, vandalism, gang rape and wartime atrocities. It may result partly from a feeling of anonymity which is generated within a large group and also partly from the diffusion of responsibility, which was mentioned earlier. In addition, the undesirable behaviour of some may act as a model for others to follow with the more extreme elements becoming leaders. There is a 'Gresham's Law' of psychology which reads that 'the behaviour of a group tends to approach the behaviour of its lowest elements.'

Conformity

We have already said that individual judgement may be influenced by the judgement of the group. If everyone else in the group estimates a 10 cm line to be only 8 cm, then one's own estimate may also be affected in the same faulty direction. This characteristic is called conformity. The larger the group holding a particular attitude, the greater is the pressure to conform.

Conformity often results in the distortion of perceptions, judgement and action. This may range from the insistence on wearing only blue jeans or long hair to the abuse of socially unacceptable drugs. But it may also sustain, for example, the proper use of uniforms and standardised, desirable behaviour within a group. The new young stewardess will be much influenced by the comportment and attitudes of her crew colleagues on her early flights. She can be expected to conform to the general pattern of behaviour of the group.

Conformity can be affected by reinforcement, that is, by rewards and penalties. The reward could simply be that others in the group approve of the individual when he conforms. In certain groups, individuality may indeed be respected and admired more than conformity. This might be the case in a debating society or amongst fashion designers. Generally, however, it seems that conformity is approved; we tend to like others who display similar behaviour and express similar beliefs and attitudes to our own. This suggests that one way of seeing conformity is as a means of ingratiating oneself with others – of achieving popularity and being liked – though to achieve this objective the conformity should appear sincere.

Group pressures to conform are felt somewhat less if an individual already has confirmation of the validity of his own personal attitudes and behaviour. For instance, if an individual had previous and personal experience of the death of a close friend as a result of the abuse of drugs, then he may be less likely to conform to the pattern of drug taking within a group with which he is associated. This also suggests that conformity is likely to be greater when an individual has not had a background of knowledge or success in the subject or attitude in question.

We have spent some time here in looking at various aspects of group influences on individual attitudes, based on the findings of extensive experimental work in the field of social psychology. While these remarks must be seen as examples only, and in no way a comprehensive review, they may provide a background against which various group activities, and not least those of flight crews, can be analysed and assessed. (Also see discussion of peer group pressure on page 36.)

In Chapter 6 we already noted the influence of leadership within a group and those remarks are also relevant to matters discussed in this chapter. This widely studied field is surveyed in numerous books on social psychology (e.g. Baron *et al.*, 1974; Open University, 1975a).

Resistance to change

One of the characteristics of an attitude, we said, was that it tends to be rather enduring. It usually takes much persuasion to change a strongly held attitude such as that towards, say, capital punishment or abortion. Attitudes towards diet, smoking and alcohol are often strongly held socially, while at work, positions adopted relative to seniority, salary and duty schedules also tend to be very resistant to change. Evidence which would appear to contradict such a strongly held attitude tends to be disputed or rejected and its source often questioned.

The ease with which people adopt an attitude is frequently quite out of proportion to the determination with which they will argue to defend it. Such an attitude can be likened to a knot; easy to pull tight, but very difficult to untie again.

Not all attitudes are so resistant to change. Those which are not central to a belief or an essential part of one's basic perception of life and living may be changed more easily.

One technique which has been applied with some success to decrease the resistance of an attitude to change, involves a voluntary role reversal, that is, stepping temporarily into the shoes of the one holding the opposing attitude. This method can be used in industrial relations and whenever a conflict of attitudes causes concern. In management/union negotiations, one side makes its case and this is followed by the other side restating the same case and arguing in favour of it. The roles are then reversed. This tends to result in the adoption of less intransigent positions and reduces the practice of seeking, considering and presenting only facts which reinforce one's own attitude and disregarding other equally relevant facts.

There are certain times when we would like to increase resistance to attitude change; we would like to pull the knot as tight as possible. We do not want a child who is brought up to reject alcohol and drugs to have its attitude changed by the influence of a group outside the family environment. Management would not want a new stewardess, who joined an airline with attitudes of loyalty and efficient service towards passengers, to have these attitudes modified by colleagues who might have become more interested in discos and daily allowances. Resistance against political propaganda and questionable advertising may be desirable, too; clearly we do not want to become puppets manipulated by the mass media. Some studies have been made on means of enhancing resistance of attitudes to change. A simple and usually effective means is to elicit some form of commitment to the desired attitude. A person who has an attitude that excessive drinking is undesirable may have the attitude reinforced if he carries out voluntary work for Alcoholics Anonymous. Even signing a petition which supports an attitude held is likely to strengthen that attitude.

Another means of increasing resistance to change has been called an inoculation method. A typhoid inoculation introduces a small quantity of the disease and this stimulates antibodies and generates immunity. Similarly, it has been found possible to expose people to small doses of certain undesirable attitudes and thus increase resistance to change (McGuire, 1969).

Yet in the working environment, all in a position of responsibility are concerned, to a greater or lesser extent, in trying to influence in a desired way the attitudes of others so as to improve safety, efficiency and the well-being of those in the industry. Good CRM programmes reinforce good attitudes.

Modification potential through training

Much airline attention has been focused in recent years on what has been called the management of flight deck resources and this is discussed in more detail in Chapter 14. It is of relevance here because personality, attitudes and behaviour play a role in such management and we need to understand the extent, if any, to which such kind of training can have an influence on them. Many airlines, with considerable encouragement from agencies such as ICAO and (in the USA) the FAA, are emphasising the management of flight deck resources through formal courses in cockpit resource management (CRM). These courses deal with both attitudes and behaviour. Basic personality, as discussed in a following paragraph, is a much more difficult question. Despite this, some believe that a good CRM programme can minimally affect personality traits in some cases.

Some programmes use a different nomenclature but the primary principles are the same. Nomenclature should not confuse anyone seriously involved. There is growing evidence that CRM programmes are successful (Helmreich, 1991; Orlady, 1992) and the trend toward broad acceptance of CRM principles (often with different nomenclature) is increasing internationally.

At the beginning of this chapter we made a distinction between personality and attitudes and this is relevant in the context of training. It is unrealistic to imagine that routine training in industry in general, or in airline captaincy or management training in particular, can have a significant influence on personality traits. These are deep-seated characteristics which are not easily amenable to change. In the aviation industry, and particularly in the flying fraternity, undesirable personality traits must be detected during the initial screening and selection process and appropriate action taken at that time.

While this may appear a rational approach, it is not without its difficulties. Even if we accept that it is possible to determine with assurance such personality traits, we still have to establish the criteria which we are going to use in deciding on their desirability and undesirability related to the particular job. In the early days of flying, and perhaps still in the military environment, a high level of courage was demanded. The early pilots had no crew to manage and no computer to program. A personality defect might have led to their own demise but is unlikely to have made the headlines as a major international disaster. The task of the pilot has been changing from an airframe driver in a relatively simple vehicle to a manager of computers and automatic flight guidance systems. He now has responsibility for the safe and efficient management of transport units reflecting a vast investment in terms of money and human lives.

It is not unreasonable to conclude that the evolving role of the pilot – and indeed, other crew members – calls for an evolution in the personality traits required. This may explain the phenomenon observed by many in the airlines, that some well-respected pre-war pilots had difficulty in adapting to the post-war civil aviation operational environment. This was similarly evident in the transition from the more leisurely piston-engine operation to the jet era. Assessment of the personality criteria to be used during selection must therefore not only take into account the job as it is, but also as it may evolve during the working lifetime of the applicant. Pilot careers can last for 35 or more years, and 'adaptability', which is an important trait for an airline pilot, is a difficult characteristic to identify during pre-employment testing.

While one must be pessimistic regarding the prospect of modifying personality through routine service training, a more sanguine approach can be adopted with respect to attitudes. The effectiveness of training in modifying an attitude will depend, of course, on the strength of the attitude. Some airlines conduct courses or workshops with the aim of developing or modifying attitudes of cabin personnel. Some attempt to evaluate cabin attendants' attitude by means of passenger questionnaires on service in flight.

Persuasive communication

Communication has several functions. The object may simply be to obtain something, like buying an airline ticket. This is called an instrumental function. It may be to find out or explain something and this is called an informative function. It can be part of a role or ceremony – 'with this ring I thee wed... '. This is a ritual function. And there are others. But one of its most significant purposes is in trying to get someone to modify an attitude or a behaviour pattern. This persuasive function of communication is met socially as well as at work.

All advertising is persuasive communication, some examples more socially desirable than others (Fig. 8.2). On social occasions, discussions on politics, religion

Fig. 8.2 All advertising is persuasive communication. It may utilise elements such as fear, humour, envy, credibility enhancement, image creation and status in order to increase the persuasive effect of the message.

and music often develop into trying to persuade someone else to change an attitude. In flight operations, many crew bulletins and other staff notices have persuasion as their primary objective. A crew briefing by a captain or a purser is likely to contain a persuasive as well as an informative element. We use it when we ask the boss to grant an extra day's vacation or try to convince the customs officer on returning home that we already had the camera when we left.

This aspect of attitudes – modifying them through persuasion – is of direct relevance to safety and efficiency and concerns everybody. Various studies have been conducted on means of optimising this form of communication, so we should briefly examine the kind of factors which influence its effectiveness.

In persuasive communication, the situation is frequently not one of an active communicator trying to modify the attitude of a passive receiver or audience. In the cabin staff workshops mentioned earlier, there is a two-way dialogue, though in that example the primary object of the exercise is still for one side to fashion the attitudes of the other in a certain direction.

A greater degree of interaction can be seen in a discussion between a smoker and a non-smoker on attitudes towards smoking in public transport. This is likely to involve each party using persuasive communication in attempts to modify the opinions expressed by the other. A confrontation between a hostile, over-booked passenger and an unyielding check-in agent will involve both sides in trying to change the attitude of the other towards the passenger's right to a seat on the flight. In both examples, the receiver as well as the communicator is active, resistant to the attitude of the other and constantly reversing roles of receiver and transmitter.

Many debates and discussions involve an active, two-way process of persuasion. This is usually maintained at a verbal level, but when this appears not to be achieving the desired attitude change, more aggressive techniques sometimes evolve.

It is convenient to divide the discussion into three basic parts and in any practical attempt to improve the effectiveness of such communication each part must be reviewed separately and relevant measures taken. The elements are firstly, the communicator, or the originator of the message. This is frequently a supervisor or leader of one kind or another. Secondly, the message or the communication itself. And thirdly, the audience or the receiver of the message – the person or group to whom it is directed. Let us look at these in turn.

The originator

The originator is the first element to be considered. A fundamental requirement is that he should have credibility. Credibility has two components – expertise and trustworthiness. When a toothpaste advertisement illustrates a doctor advising a patient to use a certain brand, it is utilising credibility as a tool to strengthen the persuasive effect of the message. The medically qualified person is believed to have expertise on matters of health and the medical profession is accepted by most as being trustworthy and not likely to misrepresent the truth. In industry, the message is more effective if it is known that the originator has done his homework properly and has acquired expertise on the subject and that he can be trusted to present an honest case and does not have an ulterior motive. This influence of credibility on the effectiveness of a message has long ago been demonstrated experimentally (e.g. Hovland *et al.*, 1951).

Studies suggest that in many cases persuasiveness will increase if the communicator has something in common with the audience. A politician would wear the colours of the local football team if he thought its supporters were in the audience.

Advertising also uses this aspect of communication in commonly portraying characters of typical or average appearance with whom the audience can identify. The washing soap powder is extolled by a typical housewife rather than a research scientist. The element of trustworthiness may again be significant here as the communicator appears to be one of us and so is not likely to deceive us. However, when special expertise is involved which it is reasonable to assume is unlikely to be possessed by a member of the audience, then an outsider may be more influential. A physician may be more persuasive in getting pilots to take action to avoid heart disease than another pilot.

Other factors can also be detected as influencing the effectiveness of the originator. The audience must respect him as believing, himself, in the validity of the message and reflecting this in his own behaviour; setting an example if you like. The effectivity of a purser's persuasive PA instructions to passengers to stop smoking is likely to be notably weakened if passengers were to see him then surreptitiously light up a cigarette in the galley. Notices to staff to persuade them to report for work on time in the morning are likely to have little effect if the manager himself is frequently seen to turn up late.

The message

The message itself is the second element to be optimised. Information is available to assist in this task. An initial consideration is whether the message should contain both sides of an argument or only one. It seems that if the audience already agrees in principle with the position being advocated, then the one-sided message may be more persuasive, simply reinforcing its existing attitudes. But if the audience does not already agree or the argument is controversial, then putting both sides of the case and arguing against the undesired one may be more effective. The more educated or intelligent audience is less likely to accept a one-sided argument.

Then there is the sequence of positive and negative arguments in the message; whether to put the arguments in favour first or last. This is called a primacy effect if the first argument is more influential and a recency effect if the second argument is more persuasive. Experiments certainly show differences, but these are not consistent. Simple rules have not yet been set to determine how to sequence arguments in any particular case.

Another question is whether a conclusion should be included in the message or whether it should contain only facts and allow members of the audience to reach their own conclusion. Usually, if the issue is very complex or the audience is not too bright, it is better to make the conclusion explicit in the message. On the other hand, if the issue is simple or the audience is either hostile or intelligent, it may be better to allow them to draw their own conclusion.

In certain cases, a high fear content in a message may have a greater influence on changing attitudes than a low fear content. But an excess of fear can be counter-productive and cause resentment and loss of credibility – a boomerang effect.

As will have been observed in the use of political posters at election time, repetition of a message usually increases its effectiveness but as in the case of fear, repetition must not be overdone.

Not only does the content of the message influence its effectiveness but so does the general manner in which it is presented. Music tends to arouse positive feelings in people and may increase their acceptance of the contents of a message and the extent to which their attitudes are influenced by it. Radio and television advertising is frequently accompanied by a few bars of music and audio-visual training programmes now usually employ music, too.

Also related to the message is the medium in which it is transmitted. Generally, a single face-to-face contact is thought to be the most effective mode of persuasion, followed by talking with small groups, showing a film or slides and, finally, using printed material.

The receiver

The audience or receiver is the third and final element in the process of persuasive communication. Several personality variables have been studied, such as the relationship between self-esteem and susceptibility to persuasion. It has been hypothesised that very high and very low self-esteem would be associated with the least amount of attitude change, though for different reasons.

It has been found that sometimes a degree of audience distraction during a message – coffee, an irrelevant anecdote or slides during a talk – seem to increase the persuasiveness of a message, though the reason for this is not clear.

It has been suggested that we may be better to think more in terms of modifying a person's perception of an object rather than modifying his attitude towards it. If it were possible, for instance, to persuade a young smoker to perceive smoking as creating an image of a weak character rather than a macho type, the attitude to smoking may change by itself. To a considerable degree, this has happened in the USA, although the negative health considerations associated with smoking have also been a major factor. This principle, of associating a negative image with an undesirable behaviour trait, can also help ameliorate a difficult problem in air transport. For instance, it makes it easier to persuade a domineering captain to modify his attitude toward other crew members if the domineering captain believes that an overbearing image creates the image of a 'poor' pilot.

When a message fails to get across effectively and when the undesired attitude persists, or even a boomerang effect occurs, then an analysis of these three elements should be undertaken.

Safety propaganda

Persuasive communication, or propaganda, related to safety is one of the most complex in the field of large scale communication. It involves a paradox which is difficult to explain or comprehend. One could imagine that it would be the easiest kind of message to get across effectively; after all, who does not want to work and travel safely. It is an exercise in getting someone to adopt an attitude to serve his own interest and ensure his own survival.

Yet safety campaigns rarely appear to have lasting effects and immediate results are usually disappointing. After nearly 20 years of expensive exhortation trying to persuade people in the UK of the personal safety involved in using car seat-belts, this finally had to be abandoned in 1983 and replaced by legal enforcement. The USA has had a similar experience. In Chapter 13, reference is made to the general reluctance of passengers to pay attention to the safety briefing given by cabin staff before every take-off. Yet it is known that those who listen to the message have a greater chance of surviving an accident than those who do not.

There are various reasons for this frequent failure to modify or strengthen an attitude which would seem to be one of the easiest to encourage, that of self-preservation. One is the lack of direct reinforcement. The listener has probably never had such an accident – they are events which are expected to occur to other people. After a passenger has had personal experience of an aircraft incident or accident, he is likely to be more responsive to the message in the cabin attendant's safety briefing on board.

Because a number of factors rather than a single one generally contribute to accidents, and because safety campaigns are not very effective, a proper overall attitude towards safety must be developed gradually. This process involves indoctrination of crews from *ab initio* training through all advanced and routine service training and checking. The attitude must be constantly fostered and reinforced, whenever possible with examples of unsafety.

Modification of the attitude of passengers towards safety presents a more difficult problem to resolve, partly due to the fact that they are not available for long-term indoctrination, as are crew members. Furthermore, the conflicting propaganda emanating from airlines and aimed at passengers does not help. On the one hand, marketing and public relations people constantly stress in advertising what good care they take of you, that they are careful and punctual and that they are very experienced, aiming to create an image of total safety. On the other hand, the operations and flight safety people would like to propagate the notion that some risks are involved in flying and that passengers should take precautionary measures to minimise these risks. The cautionary message, as currently conveyed by most airlines through PA announcements, does not appear to be effective enough.

9

Training and Training Devices

Training, education and instruction

Definitions

It might be thought that a discussion of Human Factors in training is simply of concern to the training department of a company or the full-time instructor or teacher. But that is taking a far too narrow view of teaching and learning. Almost everyone engaged in a skilled task is at some time occupied with passing on his knowledge to others. This may be a line captain passing on his experience to a junior copilot or a normal stewardess guiding the activities of a young trainee cabin attendant. It may be a check-in agent demonstrating her work to a beginner. All are concerned with training and their effectiveness in this task will be greater as a result of a better understanding of the teaching and learning processes. Furthermore, the trainee of today is the trainer of tomorrow.

It is a fundamental concept of Human Factors that the components of the *SHEL* model discussed in Chapter 1 should be matched to the characteristics, the capabilities and limitations, of the *Liveware* in order to produce an effective system. In this matching task, certain basic assumptions must be made about the nature of the *Liveware* component. Quite clearly, we would not expect to have to match the *Hardware* and *Software* – the equipment and procedures – associated with a heart transplant operation to the characteristics of a farmhand in the role of the surgeon. We expect the *Liveware* to possess the knowledge and the skills to fulfil its role in the system in the same way that we expect the other components to have the capability to fulfil their roles. The acquisition of knowledge and skills of the kind we are talking about here, usually involves a teaching process.

Education and training are two aspects of the teaching process. Education is normally used to indicate a broad-based set of knowledge, values, attitudes and skills suitable as a background upon which more specific job abilities can be acquired at a later stage.

Training is a process aimed at developing specific skills, knowledge or attitudes. Education may be seen as the precursor of training. To attempt to apply training without an adequate background of education is like building a house without first laying the foundation. This is recognised in the establishment of certain educational criteria for staff selection for all skilled and semi-skilled jobs. The application of this concept to Human Factors activities is discussed in Chapter 14. The term instruction can refer to activities associated with either education or training.

A skill is an organised and coordinated pattern of activity. It may be physical, social, linguistic or intellectual. In a technological environment, perceptual and motor skills are very important, but it will be apparent from previous chapters that social and linguistic skills play a significant and sometimes a vital role in many aspects of aircraft operation.

Most activities involve two elements each requiring some degree of skill – deciding on a course of action and then carrying out that action. A person may have skill in putting together ideas in writing but be poor in delivering them in a speech; the speech writer is not necessarily the best orator. Incidentally, the possession of a skill in a particular activity does not imply the possession of skill in teaching that activity to others; the brilliant scientist may be an appalling failure as a university lecturer. Teaching is a skill in its own right. This is important to understand in the selection of flight instructors and check pilots.

A useful distinction can be made between skill, which might be described as knowing how to perform a task, and knowledge, which is knowing what is required to perform it. A top tennis coach may have the *knowledge* but his champion student will possess the *skill*.

A glossary of training terms was published in 1981 by the IATA Flight Crew Training Sub-Committee (Doc. Gen/2728).

Alternatives to training

If any task presents a performance challenge to the extent that there is a need for a high level of skill, then there are various ways in which it can be tackled. We could, of course, take it away from man altogether and assign it to the machine. This has been done in a number of situations on board the aircraft. The manoeuvring of all higher performance aircraft has been handed over to the automatic pilot; navigation has now been largely assigned to computers and these will increasingly take over other flight management tasks. The capture of altitudes during climb and descent and altitude reporting have been to a large degree taken away from man and given to the machine, which (at least theoretically) performs the task much more reliably. (See the discussion of altitude deviations reported to the ASRS on page 18.) This has dramatically reduced flight deviations and violations. This possibility should always be kept in mind as a potential solution to problems arising from the variability of human performance.

But if it is decided to keep the task in the hands of the human operator, then there are three main channels along which we can proceed. We can, firstly, refine the staff selection process, using only those with higher education, qualifications or aptitude for the job. This may have economic penalties if the supply of suitable staff is limited.

Another possible solution is to redesign the job or the job situation so as to present less of a challenge to the operator. This objective of modifying the job to suit the man, rather than the reverse, is a central activity in the technology of ergonomics and has frequently been neglected in analyses of incidents and accidents and in the subsequent recommendations.

Finally, we can spend money to train people so that they can better cope with the job. It is with this last solution that we are concerned in this chapter. However, it is important to understand that there is usually more than one way of dealing with a task performance problem and that there is an interaction between the different ways; more of one may allow less of another.

Training principles

Training transfer

What has been learned in one situation can often be used in another. Someone who has learned to drive one type of car can usually quickly learn to drive another. This improvement in learning time as a result of earlier learning is called positive transfer. However, there may have been aspects of learning to drive the first car which interfere with learning to drive the second, such as the location of the hand-brake or light switch. This is called negative transfer. The concept can equally apply to aircraft.

Much of the training of a first officer can be directly transferred when he comes to be trained as a captain. Some of the emergency procedure training of cabin staff may involve a negative transfer and some a positive transfer when it comes to training on other aircraft types.

Many examples of negative training transfer have been demonstrated in flight simulator work. Learning to suppress a spurious warning light which routinely appeared in a simulator would certainly result in negative transfer as this would be developing a practice which had to be unlearned when transferring to the real aircraft. It is very important to be able to identify elements of training which can lead to negative transfer. The policy of standardisation of equipment and procedures tends to reduce the incidence of negative training transfer. From time to time incidents are reported where a pilot reverted to a pattern of behaviour learned and appropriate for an earlier type of aircraft (e.g. CHIRP Feedback No. 6). A risk associated with negative training transfer is that it is sometimes difficult to unlearn an earlier practice. Particularly under stress conditions, it may be expected to return and this is an insidious situation which could be dangerous when applied to critical activities such as flying.

A form of negative transfer can be a problem for pilots flying different types of aircraft, which is a relatively common practice. Negative transfer also can be a particular problem if an airline has been involved in mergers, sales, or in any other type of aircraft transfer agreements. There are many variations – e.g., the pilots of one US airline have maintained that they flew nine different versions of the same modern transport at the same time, although not all operational experts agreed with them in each case. Both within and between airlines there can be differences in instrumentation, flight configurations, and even procedures. There is far from agreement on whether 'large' or 'small' differences create the most serious problems. (See discussion under 'Deregulation', pages 28–9.)

Negative transfer is also a major consideration in the certification of aircraft with aircraft type changes. There are significant economies for the manufacturer and for the airline if a new aircraft can be considered simply a modification of an already certificated aircraft instead of as an entirely new aircraft.

191

In some cases, negative transfer problems can arise with the development of a new system or if improvements are made to an existing system. However, while strict conformance minimises negative transfer, it can be a mixed blessing for it can also virtually eliminate desirable changes and improvements.

Feedback, open- and closed-loop systems

Feedback control means simply that the output of a system can regulate or control the input. A familiar example is the central heating thermostat where the temperature sensor controls the heat output so that the system is self-regulating. This is what is called a closed-loop system. An open-loop system here would be one in which there were no feedback control – a heating system without a thermostat. The possible consequences of an open-loop system are illustrated in the behaviour of a car radiator's water temperature when the thermostat fails.

Open- and closed-loop systems are also relevant to training. When feedback is present and it is used to regulate the input, then the system can be described as closed-loop; when it is removed, it becomes open-loop (Fig. 9.1). In this model of a training system the original input, equivalent to the power supply in the central heating system, comes from the store of information which may be in the brain of the instructor, a tape, film or other source. It is then conveyed to the student. So far, this is an open-loop system which is full of potential deficiencies. The source of information may be defective, the transfer medium may be inadequate, the language of information may not be suitable for the student (see Chapter 7) or the student may not be able to absorb the information for one reason or another. The open-loop system will not adequately reveal these deficiencies.

In order to regulate the effectiveness of the system, feedback is necessary. The student's response is evaluated and this may be done by assessing the manner in which the task for which he is being trained is performed, by tests or examinations or by seeking the student's own views on the effectiveness of the training. This evaluation is then fed back to the teacher or the one controlling the training programme. He will then modify the teaching process by altering the information which is imparted or the manner in which information is transferred. The loop is thus closed.

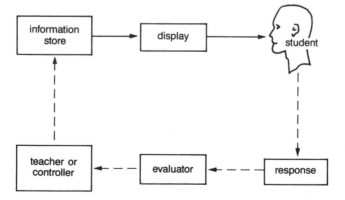

Fig. 9.1 A closed-loop training system showing the information transfer from the store to the student and the feedback loop.

While discussing feedback it is necessary to consider the direct feedback given to a student which enables him to assess the way he is performing. When driving a car, feedback comes visually from the changing picture of the world outside, aurally from engine and wind noise, tactually from the vibration or bumps transmitted through the seat and possibly via the olfactory channel (sense of smell) if the engine, brakes or tyres are overheating. In flying, the pilot gets feedback mainly through his instruments and outside visual cues. The stewardess may get feedback from burning her fingers on an oven door or from abuse from a passenger.

All these forms of feedback are called intrinsic feedback as they are available in the normal job situation. Extrinsic feedback, on the other hand, refers to additional information which is not available in the job situation. This might be in the form of information from an instructor either during or after the training exercise. If a flight simulator is 'frozen' during an exercise and the student is allowed time to evaluate the situation, he is getting extrinsic feedback. It is important to understand the difference between these two kinds of feedback. Certain kinds of extrinsic feedback may have no value in promoting learning. And in any case, the trainee must learn the task in such a way that the removal after training of any extrinsic cues – they will not be available in the real work situation – has no adverse effect on performance.

Guidance, cueing and prompting

In addition to a direct feedback of his actual performance, there are other ways of providing information to assist a learner. These have been called guidance, cueing and prompting.

Guidance may be physical, such as physically controlling the movement of a tennis student's racquet arm, or propping up a child who is learning to ride a bicycle. It might involve a flight instructor going through the control wheel movements during flare with a student pilot. In tasks where errors could have undesirable or dangerous consequences, a guidance or 'no-error' technique of training is likely to be preferable to simple feedback.

Cueing tends to be restricted to perceptual detection tasks and involves informing the trainee when a signal is about to appear or an event is about to take place. This might usefully be applied to training in a radar detection task when target signals are rare or difficult to detect. The pilot instructor may advise his student just before the glide slope indication is expected to appear on an instrument approach. The instructor is focusing the student's attention on the significant event so that he will respond correctly and in time to it.

Prompting has been found to be useful in training for certain tasks. It involves presenting the trainee with the correct response immediately after the stimulus and is most useful in the early stages of training. For example, if an altitude check must be made on passing the outer marker on approach, the instructor may himself immediately make the call on seeing the marker indication to remind the possibly stressed student of what should be done at that moment.

These three training techniques to assist a student are all in principle different, each having its own advantages. They are not equally suitable for every training situation and should be used selectively. All three, like extrinsic feedback, will not be available in the real job situation and must be removed as training progresses.

Pacing

In training, the rate at which material and experience is presented to the student is known as pacing. In a long-range navigation trainer, the time may be accelerated. In other forms of training it may be slowed to allow a student to absorb more easily new information.

Different students may need more or less time to learn a new skill or a new procedure. It may be advantageous for the student himself to control the speed at which information is presented to him. This is called self-pacing. It is obviously difficult to provide this in conventional classroom group instruction, and virtually impossible in a lecture type of teaching where student response is very difficult to assess. With programmed forms of instruction, self-pacing is more realistic and is a valuable element in the learning process. This kind of instruction is discussed later in the chapter.

Training, learning and memory

Learning processes

Learning is an internal process. Training is a control of this process. The degree of success in this process control must be judged by the changes in performance or behaviour resulting from the learning.

It is essential to recognise that learning is accomplished by the student and not the instructor. The student must therefore be an active and not a passive participant in the process. Nevertheless, it is the task of training systems and the instructors who form part of these systems, to teach in a way which enhances the learning process.

Different skills come into play in different phases of training; a student can be successful in one stage but not in another. In order to be able to analyse and interpret the problem and decide upon the appropriate corrective action, it is necessary for the instructor to be able to discriminate between these different phases. The first phase involves talking about the task and this is called the *cognitive* stage. The instructor will discuss what the student can expect, what kind of errors can be made, the level of skill to be achieved, and so on. Next comes the practical stage where proper techniques are learned and practised and where errors are progressively reduced. This is called the associative stage of learning. The final stage involves perfection of performance, speeding it up and improving accuracy and precision – a polishing-up operation. This also includes learning to increase resistance to stress and to perform in abnormal and emergency situations. This is called the *autonomous* stage.

Memory

Learning and memory are the subject of a vast field of experimental study and this is available to help optimise training methods. If you look up a friend's telephone number in the phone book, it is forgotten again within seconds. This used short-term memory (STM). If you were asked your own telephone number, you could answer correctly without any reference to a book at all. This uses long-term memory (LTM).

STM is used when information must be recalled accurately within a few seconds. In reading this text you have to remember the first part of the sentence to understand the last part. In flying, it is often necessary to read back a frequency, code or clearance. It is prudent practice to write down the figure (or set it on a selector) at once to prevent it from being lost. The capacity of STM seems to be only about six to eight items or chunks of information and the duration of retention of the material only a matter of seconds. Some processing has taken place by the time information reaches the STM store, but still at a very superficial level compared with items in LTM.

If we want to remember a phone number we have just looked up, we consciously repeat it to ourselves. This improves retention of the information. This repetition is called rehearsal. Experiments show that when there is no rehearsal, recall of information from STM decreases rapidly (Fig. 9.2). So long as the rehearsal continues, the information will continue to be recalled and in addition, it seems to help to

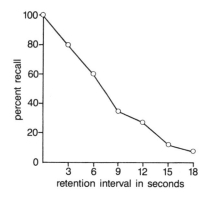

Fig. 9.2 The loss of recall from short-term memory when rehearsal does not take place (from Peterson *et al.*, 1959).

transfer information into LTM. As the number of repetitions increases so the probability of recalling the item seems to increase (Fig. 9.3).

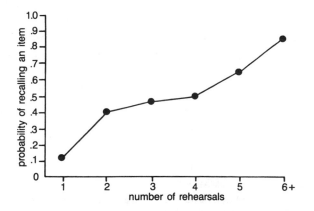

Fig. 9.3 The probability of recalling an item as a function of the number of rehearsals of that item (Rundus *et al.*, 1970).

Storage space in LTM seems not to be a problem. More significant is the way the information is prepared for storage. This might be compared with placing books in a library; the coding and organising of the books has a major influence on the ease of finding them again. Preparation for storage, however, does involve time, though such activities as summarising information and emphasising the significant items help storage and later retrieval from storage.

There are various theories as to why information appears to disappear from memory, or retrieval becomes more difficult. One is simply that the information decays with time because it is not in use. Another considers that the information storage or retrieval possibility has suffered from interference. The source of the interference may be other material which has been learned and stored before or after the required information. Another difficulty in recall is called repression and might be seen as motivated forgetting. This is a defence mechanism preventing the recall of memories which may have painful associations and which may provoke anxiety. This is not so much forgetting, as experiencing a subconscious block to memory. For example, details of an accident may be apparently lost because to recall them might involve reliving the terrifying experience of the accident (see Popplow, 1984, and Orlady *et al.*, 1988).

Memory enhancement
In spite of these obstacles to effective memory, certain methods can be used to enhance the memory process and these are relevant in optimising training. One has already been mentioned – the organisation of the material to be learned as with books in a library.

One method of organising information to aid memory is the use of mnemonics. This has proven to be effective and utilises one of three basic principles. The simplest is to organise the information into a rhyme such as 'red sky at night, shepherds' delight', and this form of mnemonic seems to be common in all languages. The Dutch say with regard to advice on drinking, '*wijn op bier geeft plezier; bier op wijn geeft venijn*' and the Spanish '*hasta el cuarenta de Mayo, no te saques el sayo*'. A second form, records of which go back to early Greek times, is to relate the items to be remembered to particular places. This involves forming a mental picture of the items and their location; the more unusual the location image, the greater the chance of recall. A third basic form of mnemonics is what might be called a key-word system. A method which in its modern form goes back to the 17th century, involves replacing the numbers which have to be remembered by consonant sounds, which are easier to organise. This method is included in many memory improvement courses and seems, with practice, to work very effectively. These and other forms of information organisation to improve memory take some time and effort but have been shown to work (Baddeley, 1976; 1982).

Overlearning is a most important concept to be built into a training programme. This not only improves the chance of recall but makes performance of the task more resistant to stress. Overlearning means simply carrying the training process beyond what is required to perform to the minimum acceptable level of performance. The effects have been demonstrated experimentally long ago using word memory tests (Krueger, 1929) but the principle also applies to skilled psychomotor tasks, that is, tasks involving muscular responses to visual, aural or other stimuli. This is a very important concept in training. Unfortunately, airlines can plan on little overlearning during training because the training of airline pilots is so very expensive (Orlady, 1992). There is an understandable tendency for airlines to do little more than the minimum training required by the regulations which govern them although there can be wide variations in the quality of such training.

One other method of enhancing memory is by chunking, or breaking down the information into chunks or familiar units. STM, we have said, is restricted to about six to eight items or chunks. It seems that although the number of chunks which can be retained remains rather constant, the chunks themselves can vary considerably in complexity and length. Although we can remember only about seven unrelated letters, if we chunk these letters into familiar words of, say, six letters each, we can remember seven words or a total of 42 letters. And if these words are grouped into small meaningful sentences, then we can remember about seven sentences, which might include several hundred letters.

Effects of sleep and age
Experiments have shown an interaction between sleep and memory. Learning right after sleep seems to be less effective than learning before sleep. This apparently damaging effect of sleep on memory has been called the prior-sleep effect. A full explanation for this phenomenon is not yet available but in research tests even half an hour's preceding sleep seemed to damage memory of the material to be learned (Barrett *et al.*, 1972). It is possible that the explanation may be associated with hormone production during the early part of a night's sleep. Memory of material learned before going to sleep might therefore be better retained than that learned after waking. Research suggests that it might be preferable not to study within at least two hours of having slept and to follow learning with at least four hours' sleep. However, research on this subject cannot yet be seen as finalised and new advice could emerge in due course. This was also discussed in Chapter 3.

The acquisition of new material, if it is to be properly stored in LTM, requires adequate preparation, adequate study time and adequate review and testing.

The effect of ageing on memory varies with the individual and it is difficult to generalise. Some experimental work shows a deterioration of certain kinds of memory as a person gets older (Fig. 9.4) but the effects usually are not very dramatic until reaching the sixties. Deterioration is also selective, with some types of memory holding up better than others (Baddeley, 1982). Memory decline with age may not be so serious as once thought and there is some evidence to suggest that actual memory of those over 50 years of age is better than they report (Khan *et al.*, 1975). Age is not the only factor associated with deteriorating memory; depression, for example, is said to be a condition damaging to memory (Mohler, 1978).

Evidence also suggests that mental functions, like physical ones, deteriorate with disuse and memory seems to be better retained amongst those who maintain a high level of intellectual stimulation. It should also be remembered that any memory deterioration which does occur, may be offset by increasing experience and knowledge.

Special attention has been given to how training programmes can be modified to suit the older trainee and resulting recommendations have been published (e.g. Newsham, 1969).

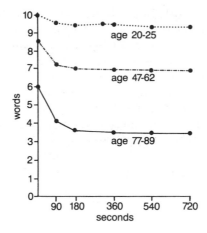

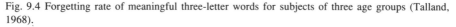

Fig. 9.4 Forgetting rate of meaningful three-letter words for subjects of three age groups (Talland, 1968).

Continuous and serial activities

Remembering what one has learned is a particular problem when the skills are not practised very often. Some forms of continuous motor activity, such as driving a car, riding a bicycle, or skating, are not easily forgotten. Other tasks which involve a series of separate responses such as putting together the carburettor of a car, checking a flight plan or making flight guidance mode selections during a departure procedure, are much more easily forgotten. Another factor which may be significant here, is that continuous psychomotor tasks require less cognitive activity, that is, less brain work and information processing, than the serial tasks calling for separate responses.

Handicaps to learning

There are a number of handicaps to learning which can interfere with the success of a training programme, apart from the more obvious ones of sickness, fatigue or discomfort.

Anxiety is common amongst students and the pre-exam period at colleges is one in which there appears to be a higher risk of student suicide. But examination and test anxiety continues for many throughout life. This can be a problem for air transport pilots, most of whom face performance checks every six months throughout their entire careers. Added to this they also are expected to successfully make a transition to several different aircraft. Stress is frequently observed building up during a training course and this can adversely affect either learning or performance or both. It is not always possible to assess the amount of learning that has taken place by means of an immediate performance test.

In flight training, a fear of flying itself may also have an adverse impact on the training programme. Reduced performance has been reported as a result of stress during flying training (e.g. Swissair, 1979). Stress may originate in the training and work situation or may have its source in social or domestic factors (see Chapter 4).

There are various psychological techniques which can be used by the anxious person to reduce a stress problem. Autogenic Training is one such technique which

198

has been used very effectively, particularly with young people (see Chapter 3). It can be used not only routinely to lower the level of general anxiety but also before and during examinations when these create an undesirable level of anxiety.

Low motivation is a handicap to learning though it is less likely to be present when career prospects are determined by the outcome of the training. Once having qualified, however, low motivation may arise. This has been reported with young pilots about a year after they join an airline from flying school. They felt that they had achieved their objective and could therefore relax. An attitude of complacency developed which was countered by putting them back into the airline school for refresher training with a particular emphasis on attitude (Holdstock, 1980).

Motivation may be lowered during training if a student feels he has been unfairly treated or if he is put under undue pressure by the instructor or the pace of the instruction. Impatience with what is felt to be a less interesting basic or preliminary part of the course can lower motivation at this stage. Poor quality instruction and instruction material are also likely to reduce motivation, as is an unsuitable instructor. Proper instructor selection and training and proper course preparation are clearly of fundamental importance in maintaining student motivation. This is discussed later in this chapter.

Reading inefficiency is a very common handicap to learning and while there are techniques to improve efficiency it is surprising how little they are applied. It is no doubt commonly assumed by training schools in industry that reading efficiency is an aspect of education which should have been optimised before exposure to training for specific skills is applied. Regrettably, this is frequently a false assumption.

The learning technique used by the student may be poor and he may have had inadequate guidance on this from the instructor. And finally, learning may be handicapped by inadequate communication. This was discussed in more detail in Chapter 7, where it was explained that the fault could lie with the instructor, the material or medium, or the student.

Training systems

Choice of training method

At an early age we find from experience that, as individuals, we prefer certain methods of learning over others. One person may learn easily from reading, another from listening to spoken instructions in the form of lectures. Another needs to use visual displays and yet another needs the actual 'hands-on' practice of doing a task. These claimed individual preferences may not necessarily always be very significant in reflecting the optimum training, though individual temperament and cognitive style should be carefully considered. The individual's basic background and level of knowledge are particularly important if major changes in technology (e.g., the change to the glass cockpit) are involved. In addition to these personal preferences and individual attributes, there are other criteria which determine the optimum training method and these must also be taken into account. In any commercial undertaking there will also be practical constraints in terms of organisation, accommodation, time, instructor staff availability, and loss of student time for productive work. All may have an influence on the training method employed.

Lectures

The objectives of training are of paramount importance in selecting the method. For example, lectures alone are inappropriate for developing job skills or changing attitudes. A lecture is useful to introduce and give a general background to a subject, to act as a briefing or to summarise a programme. And it has some notable advantages such as being virtually unlimited in audience size, and being flexible with regard to location, pacing, content and visual illustration. But it encourages passivity on the part of the student and very considerable lecturing skill is required to hold the attention of the student and to get the training message across. Feedback to the lecturer is very low unless adequate time is devoted to questions, when the technique begins to shift away from a straightforward lecture into what might be termed a lesson.

Lessons

In a lesson the material is so structured to ensure participation of the group. This permits the instructor to modify the pacing and presentation method on the basis of student response. The lesson is adaptable and permits instruction in practical skills, unlike the lecture. The lesson can be in a classroom format, where theory is often taught, or in practical sessions where skills are usually developed. It is very adaptable. However, it cannot be used with large groups and there is a tendency for the slower students to establish the pace.

Discussions

A discussion or conference method of training may be used in a number of ways. It may be a question and answer session, guided by an instructor. Or a method of learning from the experience of all participants so as to develop perhaps better procedures or attitudes. Another form of discussion is where a particular problem situation is presented by the instructor, who then encourages free and full participation. A particular training session may reflect a combination of these three types, the boundaries of which are not sharply defined. In flight simulator training in certain airlines, a video-tape is made of the actual session and a problem-solving type of discussion training follows later, attended by the instructor and crew under training and based on a play-back of the video-tape. This type of training can be very effective. Students are strongly encouraged to participate, thus enhancing learning, and interest is generally maintained at a high level. Overall knowledge tends to benefit as student experience is added to that of the instructor. There are, however, some disadvantages. The sessions cannot be highly structured and so everything depends on the skill of the instructor in remaining unobtrusive and yet adequately guiding, controlling where necessary, analysing and summarising the discussion; this is a skilled task. They are effective only with relatively small groups, allowing full individual participation, and they are generally expensive in terms of time.

Tutorial

A tutorial method of teaching is one in which an instructor works directly with one student. This is generally the case with pilot flight instruction on an aircraft and in other occupations where particularly complex skills are being taught and when certain dangers are involved with the training. It is also used for remedial training or to enhance the knowledge or performance of one individual for a job change or promotion. Clearly, the advantages of tutorial training are derived primarily from the concentration of one instructor's full attention to training a single student. Excellent feedback and participation should exist. However, the method is very expensive and the student does not benefit from observing the performance of other students or from discussions with them.

Audio-visual methods
Audio-visual teaching methods currently generally use tape-slide, video-tape or film types of presentation. While they are, by definition, non-participatory, they can be used very effectively combined with other types of instructor-guided teaching methods. They should have been prepared to present the required amount of well-researched information in an optimum fashion with audio and visual aspects refined to a degree which is hardly possible in a routine lecture or lesson. Participation can be provided by preceding or following the audio-visual with a lesson or conference type of activity in which a qualified instructor can reinforce the messages portrayed in the audio-visual, perhaps with a practical session using a training device. This type of presentation is very flexible in terms of location and audience size and the skill involved in preparing the material does not have to be available for its presentation. Disadvantages are that, used on their own, they are non-participatory and modification and updating of the programmes involve some skilled technical work. A great amount of information can be incorporated in an audio-visual package, but pacing of the presentation is totally out of the hands of the student unless question and answer sessions are interspersed in the programme. Nevertheless, repetition of the presentation is simple and does not depend on the availability of a skilled instructor.

Programmed instruction
Programmed instruction owes its origin to an awareness of the value of feedback in training and to economic pressures being applied to minimise training costs. It is usually a closed-loop system. The programme is fully structured and is based on the student signalling, by giving the correct answer, when he is prepared to move on to the next point in the programme. Incorrect answers lead to the presentation of more training material. When the system is fully automated, the student has no facility for discussion or asking questions. This kind of instruction and audio-visual method has been taken a stage further by the availability of sophisticated computers.

Computer-based training
Computer-based training (CBT), also known as computer-assisted learning (CAL), allows programmes to be highly structured, to have great flexibility and to be linked with personal records and performance. They can incorporate testing of students' knowledge and permit by-passing of instruction where knowledge appears to be already adequate. Self-pacing of instruction may be ensured by the student being required to answer questions correctly before progressing to the next stage of the programme. Participation is achieved by answering questions and by

use of touch-sensitive panel control of systems and controls displayed on the monitor. Where an organisation has staff based at different locations, these programmes can be transmitted from a central training base to any particular location by telecommunication. They can still be monitored from the central base yet minimise staff travel and lost production time. The use of touch-sensitive panel control reduces the need for training in expensive flight simulators and cockpit trainers. Software changes are rather simply achieved, though the cost of producing the original programme has been quoted by one airline as double that of a conventional tape-slide presentation (American Airlines, 1979). The kinds of applications which are and are not suitable for CBT are listed in Appendix 1.17.

A systems approach to training

A systems approach to training is one in which all the components and their inter-relations are studied before the programme is put into effect. Such an approach is essential if cost-effective training is to be achieved. A simple model might be helpful (Fig. 9.5). Once a training need has been established – as was mentioned

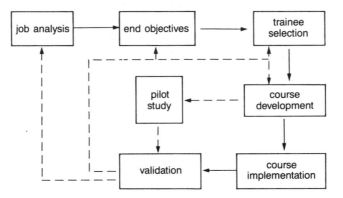

Fig. 9.5 A model illustrating the process involved in developing a training programme.

earlier, there may be other solutions – the process starts with a clear job description and job analysis. This may already give rise to problems. Debate as to precisely what is the airline pilot's job, for instance, still continues. One airline tried to resolve this by asking the pilots themselves what they felt was the scope of their job – airframe driving, staff management, commercial management, staff welfare, motivation enhancement, public relations image, passenger service, and so on. Working with the result of such an analysis, the next stage – the end objective of the training – can then be enunciated. With knowledge of the precise nature of the job and the end objectives of training, criteria can be established for the selection of trainees, their educational qualifications, age, medical standards, psychological profile, and possibly physical size and strength (see Chapter 12). The actual training course development follows and this particularly involves a selection of the training method, as just discussed, and instructor requirements. This phase will also include decisions on the type of tests which are to be utilised to determine whether students have reached the level of proficiency required. Training equipment and accommodation must also be determined at this stage.

The course implementation then follows and unless it is purely an extension of an existing and established course, a pilot study, or trial run, must first be performed. After the pilot study has been completed and modification made to the programme, full implementation can take place. Various feedback loops are shown which allow constant refining and updating of the system.

This model shows only seven blocks, but each block can be seen as a subsystem and each can be broken down into further subsections.

Training devices

Aids and equipment

The system component referred to in Fig. 9.5 as course development involves a decision on the method of instruction to be used, such as lecture, video-tape or discussion group. It also involves selection of the hardware which is to be used in support of the training.

A general distinction can be made between two major classes of training devices. Training aids may be seen as devices used by an instructor to help him present a subject. Slides and viewgraphs are commonly used training aids, as is a blackboard or a wall chart. Training equipment, on the other hand, usually refers to devices which provide for some form of active participation and practice by the trainee. The distinction becomes a little blurred when a device is used in both ways. A cathode-ray tube (CRT) monitor may be used by an instructor to support his presentation but as soon as it has a touch control facility and the student uses it in this way, it becomes training equipment. So the division refers to a distinction in the way the hardware is used rather than in the nature of the hardware itself. A flight simulator is an unambiguous example of training equipment.

It is not possible to establish specific Human Factors principles which would be equally applicable to the design of all training devices, but some generalisations can be made. In training, the hardware must be able to withstand unusual or rough use. A computer designed for home use is unlikely to withstand industrial training use for long. Special emphasis should be placed on reliability and maintainability and, whenever possible, simplicity. The device should provide for efficient learning of the specific task for which it is required. It should allow whenever possible for practice of difficult tasks and provide adequate feedback to the student. Guidelines for the Human Factors design of training devices are available (e.g. Van Cott *et al.*, 1972).

Simulation and fidelity

The development of simulators has been based on two powerful incentives. Firstly, to provide practical training in an environment equivalent to that in which the student will need to work, at a lower cost and with less risk to person and property (and therefore greater safety) than would be associated with the actual working environment. And secondly, to provide sufficient fidelity to obtain authorisation from state certifying authorities for the simulators to be used as measures of the human proficiency which can be expected in the real environment. While flight simulators are perhaps the best known and amongst the most expensive, there are many others such as the cabin or navigation simulators inside and outside the aviation industry.

One characteristic which must be determined for all simulators is the degree of fidelity which is to be incorporated. That is, the accuracy or faithfulness with which the simulator reflects the real task. It is often assumed that to achieve the best training results it is necessary to incorporate the highest degree of fidelity in the training situation. However, research has shown that sometimes different degrees of fidelity have little impact on the trainee or effectiveness of the training (AGARD, 1980; Van Cott *et al.*, 1972). In some cases fidelity could be counter-productive. We would not want to simulate faithfully the effect of a high altitude decompression, a collapsed undercarriage or a cabin fire (Fig. 9.6).

Fig. 9.6 Total fidelity in a training device may be expensive and counter productive (cartoon: J H Band).

Realistic noise could interfere with verbal training instructions. In addition, flying a new but high-fidelity simulator can require so much of the personal resources of the student that little resources are left for the learning task. In these cases, part task trainers can provide significant learning advantages. To simulate long periods of relative inaction may be a waste of time and money. On the other hand, 'freezing' the passage of time may permit an opportune intervention for relevant explanation and instruction. Distortion of time may thus be a useful tool in simulator training.

Lack of fidelity in some cases can create negative training transfer. The spurious signals which sometimes plague simulators, causing irrelevant warnings, are very undesirable. Similarly, in flight simulators, low fidelity flight control feel or stability can divert attention away from meaningful training.

Even when fidelity has been shown to be desirable, it must still be shown to be cost-effective. The six axes of motion of a large transport aircraft simulator are very expensive – each axis costs as much as a small aircraft – and it may be thought desirable to have them. But are they – all six – cost-effective? And does the cost of the simulated bump on landing pay for itself in training value? After all, we seem to get by without simulated rain beating on the windshield. Careful thought must be given to the extent of fidelity specified in a simulator.

Fidelity costs money. A hypothetical and simplified diagram may contribute to understanding the relationship between fidelity, cost and training transfer (Fig. 9.7). As fidelity increases, the cost is seen to rise increasingly steeply.

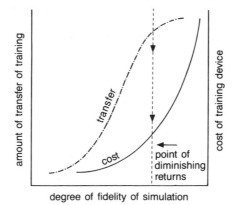

Fig. 9.7 A hypothetical relationship between fidelity, cost and training transfer or effectiveness in a training simulator (adapted from Miller, 1954).

Fidelity never reaches 100%, nor should we always wish it to do so in a training environment. But at the upper limits, even a very small increase in fidelity becomes very expensive. Looking at the relationship between fidelity and training transfer, it can be seen that although a good return in training transfer is obtained from a moderate level of fidelity, at higher levels the transfer curve flattens and the return diminishes.

From the illustration it can be seen that a point is reached where the cost and transfer lines begin to come closer together and a situation of diminishing returns obtains, shown by the dotted line. This is an important point for the specialist to determine when deciding upon the degree of fidelity to incorporate in a simulator.

Quite high levels of training transfer were achieved in a US Army study concerned with training to use a complicated control panel. Simulators varied from drawings and a cheap cardboard model up to an $11 000 high fidelity panel simulator. Training time and learning were found to be the same. In this case it seems likely that the critical aspects of the task were, in fact, adequately incorporated in all the equipment used and so high fidelity served no training purpose. This illustrates that fidelity is not an objective, *per se*. It should only be bought to the extent that it serves a definite learning purpose and is cost-effective. This also emphasises the need for a proper task analysis and an understanding of the learning processes associated with achieving a satisfactory performance of the task, before determining the nature of the training equipment to be employed (Van Cott *et al.*, 1972).

It may also be helpful to consider the need for various degrees of fidelity in the light of the *SHEL* model described in Chapter 1. In certain cases it may be cost-effective to have high fidelity in the *L–H* interface, which concerns controls and displays, but not in those parts of the *L–E* interface, which concern items such as noise, temperature and lighting. At an early stage of some training, the *L–L* interface may be less important than later. Increasing emphasis is now being given to the *L–L* and the *L–S* interfaces in advanced training of crew members in command, or as it is increasingly called, flight deck or crew resource management (see Chapter 14). To accomplish this, high fidelity is being applied to these aspects of the operational situation during training – crew members are being required to use standard procedures and cooperate, without instructor interference, as in the realistic operational situation. Another way to look at fidelity is in terms of equipment, the environment and psychological response and to determine the importance of each in the particular training situation.

In general terms, high fidelity is required in a training device when a trainee must learn to make discriminations and where the responses required are either difficult to make or are very critical to the operation. As a simple example, if discrimination had to be made between a small switch and a large lever, the exact shape of the small switch is unlikely to be important. However, if discrimination must be learned between two small switches, then high fidelity in design shape and location becomes necessary.

When only procedures are being learned, low fidelity in the equipment and the environment is not only acceptable, but in some cases, preferable. Too faithful a representation of the real situation may be confusing and even frightening to a beginner. As the training progresses through familiarisation to higher levels of skill, in principle, increasing fidelity is required.

Psychological fidelity depends upon the perception of the training device by the individual student and is not necessarily dependent on either equipment or environment fidelity.

Not only may different levels of fidelity be acceptable between these three aspects of a training situation but also within each aspect. Fidelity in control location may be more necessary than in control feel. Environmental light could be more important than environmental noise. Motion fidelity may or may not be important, depending on the task being simulated. It could be desirable when the student must make some compensatory response to the motion, but less so to achieve adequate psychological fidelity.

While these remarks can be seen as general guidelines, fidelity must be applied to each training situation taking into account the nature of the specific task being simulated. An extensive study of various aspects of fidelity in simulation was published by NATO in 1980 and this report was mentioned earlier. The report reviewed the scope and effectiveness of flight training in simulators, the status of technologies at the time and human behaviour important to the fidelity of simulators. It also identified research objectives aimed at increasing their cost-effectiveness (NATO AGARD-AR-159).

Instructors and classrooms

Instructor selection

The quality of instructors is of fundamental importance to the effectiveness of any training programme. Even when the presentation has been automated, human instructional skill must be used in its design and a human instructor is often available to support the presentation.

Airlines traditionally have taken a different approach to the selection and training of flight instructors. Some have for long considered that all instructors should have some formal education in the techniques of teaching. This may include the psychology of learning, teacher–student relationships, lecturing techniques, and so on. Others have felt it was sufficient to select good line pilots with an apparently suitable personality, who were interested for personal reasons in instructing. The technique of teaching was then acquired simply by observing other instructors, who themselves had had no specialised education in this field either.

Training management will have a greater choice of candidates if some reward is forthcoming for the special responsibilities required of a flight instructor compared with those of a line pilot. Typical characteristics sought in potential flight or simulator instructors and check pilots are listed in Appendix 1.18. They extend far beyond technical or flying competency. As the line pilot is usually unlikely to have had training in the principles of teaching, this must be included in his instructor training programme.

The annual training budget has now become so high that as in other fields of human activity in an airline, the DIY approach is no longer adequate. While the proper selection and training of instructors involves considerable expense, failing to do this effectively may turn out to be a false economy.

Lecture techniques and classrooms

When giving a lecture or instructional talk, there are a number of techniques which should and should not be used. These DOs and DON'Ts are given in Appendix 1.19. It is surprising how many instructors even after a lifetime of teaching fail to follow at least some of these basic rules.

The effectiveness of the instruction can be influenced by the teaching environment, that is, the classroom or the lecture room. Recommendations are available in connection with lighting, temperature, ventilation, acoustics, room and instructor area layout, seating and decor. Care must also be taken to minimise distraction (DOE, 1981).

Lectures and talks have an important role in training. However, they must be presented with skill if they are to be effective; poorly presented lectures can be useless and frustrating for students.

In the age of the microchip, the scope for training innovations has become unprecedented. However, innovations are only of value if they have been based upon a sound understanding of the Human Factors principles of teaching and learning.

10

Documentation

Costing documentation inadequacies

On 3 June 1966, Gemini 9, powered by its Titan rocket and crewed by Stafford and Cernan, lifted off from pad 19 at Cape Kennedy. Its most important objective was to make a rendezvous and docking manoeuvre with the target vehicle which had been launched into orbit two days earlier. This primary objective failed. From the subsequent investigation it was revealed that the failure was due to wrongly installed lanyards not pulling a protective shroud away from the docking apparatus. Wrongly installed because the written instructions used by the technicians were not adequate to ensure proper assembly of certain components in the target vehicle. The cost of this documentation failure was nearly $1 million (Welford, 1966).

There are, of course, plenty of less spectacular examples of the cost of poor documentation without going into space. One such mundane example in the UK concerns a study made of the difficulties experienced in transferring office telephone calls. It was determined that when using the operating instruction document provided by the telephone authority, only 20% of the users were able to transfer the calls correctly to another extension. As the printing was small and rather difficult to read, the researchers reprinted the same instructions larger and more clearly. This resulted in a definite improvement with 35% transferring the call correctly. But then the instructions were completely rewritten with the result that 76% were able to perform the task error-free. This illustrates how the effectiveness of instructions can be influenced by mechanical changes in the typography and clarity of the printing and even more so by the actual content and quality of the written text.

Regrettably, this example is not an isolated one. In the UK in January 1986 a popular computer magazine (*Micro Decision*) published a review of a best-selling computer. Regarding the instruction handbook, the review said that '... it is the sort of manual you really have to read all the way through ... it is hard to dip into it and find just what you want ... you are in for a lot of hacking through the manual'. The same month, the business section of a national newspaper in London (*Daily Telegraph*) went further in suggesting legal action against the computer manufacturers because their manuals do not do what they are sold to do. It said '... manuals are so appallingly written that a normal intelligent person is completely foxed ... with instructions of such comprehensive gibberish as to be unusable'. The cost here? A vast loss of office working hours, incomplete utilisation of the computer facilities which have been purchased and many frustrated operators. If the impact of such Human Factors deficiencies in documentation is considered right across industry, the penalty in lost time, inefficiency and sometimes damaged equipment, is enormous.

In aviation, of course, there is also a safety dimension. An example of this is the L1011 disaster at Riyadh in 1980 where all 301 occupants perished. The accident was in principle survivable, as the aircraft landed safely and the pilot brought it to a normal stop after turning off the runway. Survival, however, depended on time. The investigation reported that about three minutes were lost by the crew searching for the aft cargo smoke warning procedure in inadequately designed operating documentation. In the meantime, all aboard died from the effect of toxic smoke and fire (Saudi Arabian accident report summarised in Flight Safety Focus 4/85).

In addition to the large number of manuals and instruction books in daily use on board and on the ground in any airline, there may typically be some 3000 or so forms and questionnaires which require completion by staff or the travelling public. A NASA study revealed that, excluding the aircraft operating manuals, the amount of flight deck paperwork needed for a flight from Washington via New York to London had a single side area of 20 m^2 (Ruffell Smith, 1979). The quality of all this documentation has a mark on the efficiency and sometimes the safety of the airline's activities.

Technical writing is not like writing a novel or a short story where the author needs simply imagination and a skilled use of the language. The technical writer and designer of specialised documents also requires skill in the presentation of graphic material. He is responsible, too, for writing in a way to ensure accuracy and comprehension, to avoid ambiguity and to maintain motivation of the reader. He needs to be responsive to the practical needs of his reader and understand exactly how the document will be read and used. A manual, for example, is unlikely to be read in a linear fashion like a novel, and could be used primarily as a reference source or for training purposes. He will need to be able to interpret behavioural research relating to communication and information presentation. And to a far greater degree than the novelist, he will have to concern himself with the technicalities of the publication of his material.

Technical information is never likely to be so easy to read as news or other publications of general interest. Recognition that certain skills are required to present such information in an effective manner is a necessary starting point in the production of good documentation. Fortunately, there are many papers available to assist the technical and specialist writer in his task (e.g. Hartley, 1981, 1985; Wright, 1977, 1981).

General principles

We discussed at some length verbal language in Chapter 7 and compared its use in spoken and written interpersonal communication.

The advantages and disadvantages of each were reviewed so this aspect of communication methods will not be repeated here and we shall confine discussion to the written or documented medium of communication.

Any document should display consistency throughout. There should be consistency in spacing, tables and illustrations. And in the style of headings, subheadings, references, quotations, numbering, dates and foreign words. In the use of italics, underlining, abbreviations and contractions, footnotes, upper and lower case letters, decimals and money symbols. And, of course, in spelling and punctuation. Relevant to all documentation are three basic aspects which require Human Factors optimisation – the written language, printing and layout. We should now examine these in sequence. In addition, we shall consider the necessity of testing all documentation before publication.

Written language

Language is a distinctly human activity and is directly involved with the effectiveness of man in his working environment. It is thus in the mainstream of Human Factors. By language is meant not only the vocabulary and grammar applicable to different countries but also the manner in which they are used. For proper comprehension the language used must be that of the readership. For instance, a reference to a 'pilot study' can have a different meaning depending on whether the readers are airline staff or administrators of the Suez Canal; and quite different again, if they happen to be experimental researchers. A hernia to an English reader normally means a protrusion in the abdominal wall, particularly in the groin area. To a Dutchman, a hernia refers to the rupture of a spinal disc, often spoken of as a slipped disc. Even between English speaking countries the same word can have a different meaning. In the UK men used to employ suspenders to hold up their socks; in the USA they describe the straps used to hold up the trousers, which in the UK are called braces. A lift in the UK is an elevator in the USA, though an in-flight report from a Boeing 747 of a jammed elevator, referring to the galley lift, might give rise to more concern than really necessary because elevators are critical flight controls in all aircraft. This by no means suggests that galley lift problems cannot be extremely serious. An unfortunate flight attendant lost her life when a galley lift malfunctioned (see page 289).

In writing for readers in a language other than our own, care must be used and the material should be carefully checked before release. The Japanese clockmaker who published 'English' instructions with his clock, clearly had not taken proper measures in this respect:

HOW TO HANDLE THE CLOCK

This clock moves by pulling down the left edge chain, and draw up a weight once a dsy. It is right place to hang this that as you can hear balanced sound of tick of the movement when you got the right position, please fix this by against to the pillar with the nail, which there is the back of the case: If on the case of go too fast, you may pull down the weight, and go slow it, you may draw up the weight, so this clock keep the right time, If you to be correctly the time, you may turn the long hand, This hand, is free to turn left or right.

MANUFACTURER TEZUKA CLOCK CO, LTD

Language difficulty, as illustrated here, should not dissuade an author from attempting the writing task. The primary problems which must be faced in good technical writing are conceptual and procedural. Once questions of logic, rationale and principles of presentation have been successfully resolved, correction of the language such as syntax and spelling is a straightforward matter. This correction should be done, save in exceptional circumstances, by someone for whom it is the natural language.

As a general rule shorter and more familiar words are preferable but these are not the only criteria; their proper arrangement is also necessary to ensure correct understanding. This sign from a large building illustrates the point:

PLEASE

WALK UP ONE FLOOR
WALK DOWN TWO FLOORS
FOR IMPROVED ELEVATOR SERVICE

In this case, it would be reasonable for the user of the lift to assume that there would be a better, perhaps more frequent, service from another floor. But on walking to the next floor he would have found exactly the same notice displayed. What was intended was to suggest that by not using the lifts for short journeys they would be more quickly available for others who wished to travel further. This message might have been more effectively achieved this way:

TO GO UP ONE FLOOR
OR
DOWN TWO FLOORS

PLEASE WALK

And what about this notice attached to an AM/FM radio? If you do not know what it means, you may feel better to know that you are not alone ...

NOTICE

THIS RADIO USES A LONG LIFE PILOT
LAMP THAT MAY STAY ON FOR A SHORT
TIME IF RADIO IS TURNED OFF BEFORE
RADIO WARMS UP AND STARTS TO PLAY

This very poor piece of writing would better be replaced by something like this:

NOTICE

THE PILOT LAMP SOMETIMES STAYS ON
FOR A LITTLE WHILE AFTER YOU TURN
THE RADIO OFF

In giving serious warnings of one kind or another, it is even more important that the language of the message should be matched to the audience. This actual warning of the risk of death was attached to a piece of US military electrical equipment:

WARNING:

The batteries in the AN/MSQ-55
could be a lethal source of electrical
power under certain conditions

The message would have been understood far more effectively if it had been kept simple, as translated by a frustrated but enterprising technician on the job:

LOOK OUT! THIS CAN KILL YOU!

All these examples are used here through the courtesy of the Human Factors Society and have been previously brought to notice by Dr Alphonse Chapanis, a distinguished experimental psychologist. Some years ago he referred to 'this large and important area of Human Factors engineering that is almost entirely neglected' (Chapanis, 1965). The examples illustrate the importance which must be given to the language used, that is, the words and the way they are put together. What Chapanis has called 'the language and words of machines' – language associated with equipment and its use – is the province of the Human Factors specialist rather than the linguist or literary scholar.

In addressing aircraft passengers, special communication problems exist and these are discussed in a later chapter. But here again, the right starting point is to understand the audience and to apply writing skill to optimise comprehension. In trying to draw proper attention to the passenger information briefing card, the text on the left below is likely to have little effect; attention will quickly be lost. The text on the right is likely to be more effective. This is a short extract from the original and reworked texts (Hartley, 1981).

IMPORTANT INFORMATION FOR ALL OUR PASSENGERS Even though you may be an experienced air traveller, there are certain safety features of this aircraft with which you may not be entirely familiar	IMPORTANT This aircraft has special safety features Read this card carefully

Generally, shorter rather than longer sentences aid comprehension; long sentences seem to overload the memory. In 1985 the Government-sponsored British National Consumer Council reported that while British Telecom may be the biggest and most powerful UK business in the communications field, it seems to be ignorant of how to communicate with its own customers. One of its new 37 conditions of hire, running in total to 4000 words, was a totally incomprehensible single sentence of 136 words. As a general rule about sentence length it can be said (Hartley, 1981):

- less than 20 words ...   generally good
- 20–30 words ...   probably satisfactory
- 30–40 words ...   doubtful
- over 40 words ...   rewrite!

This should not be interpreted as meaning at all times the shorter the better. Keeping sentences short does not mean using telegraphese, the kind of language used in telexes and newspaper headlines where cost or space are overriding. Leaving out words such as which, that or who, can shorten a sentence but only at the cost of comprehension and perhaps the introduction of embarrassing ambiguity. 'Pilot's wife best shot' might be interpreted in quite a different way on the notice board of a rifle club at home compared with a discussion in an airline crew bar in Bangkok. Words such as soonest, which when used in a telex message means as soon as possible, cannot be used in this way to shorten a sentence elsewhere.

Other forms of abbreviation which are very familiar to the particular audience may be safely used. These might include ILS in notices to pilots and standard abbreviations such as i.e., etc., e.g. and cf. But whenever in doubt, the first time the term is used it should be spelled out in full, together with the abbreviation, and thereafter the abbreviation alone may be used. Technical writing which contains too many abbreviations can be very difficult to read, particularly when the subject is new to the reader.

Sometimes it is found desirable to use words to express approximate numerical values. The following phrases are generally understood to reflect the associated percentages (Claxton, 1980; Goodwin *et al.*, 1977):

- almost all   above 85%
- rather more than half   60–75%
- nearly half   40–50%
- a part   15–35%
- a very small part   under 10%

The language of international aviation is English and so many of those who are required to write in this language are using other than their own natural tongue. It should be remembered here that abbreviations cannot simply be translated from one language to another. For example, the Dutch for 'instead of' is *'in plaats van'* and this is abbreviated as *i.p.v.* However, if 'instead of' is abbreviated as i.s.o., as is sometimes done when writing English text in Holland, it would be understood only by another Dutchman as no such English abbreviation exists. While the error leading to the use of such abbreviations is understandable, their use is unacceptable.

Jargon seems to be an unavoidable component in technical writing. By definition, it refers to any specialised vocabulary and already in this text the reader will have met expressions such as half-life of a drug and task-dependent influences (Chapter 3), as well as the vigilance effect and the false hypothesis (Chapter 2). In this chapter we shall be referring to bias in opinion surveys. It is often not possible to avoid jargon and, in fact, it serves a useful purpose if not employed excessively. Imagine area navigation without a way-point, football without an offside, or tennis without a let. However, when jargon is used then an explanation in plain language should be readily available if there is any chance that some readers may not be familiar with the terms employed.

Ambiguity in writing must be avoided. The earlier example of 'the pilot's wife best shot' also illustrates this aspect of writing and the misinterpretation which can arise from it. In addition, unclear, ambiguous material can create annoyance and frustration amongst readers, as well as leading more directly to procedural errors. Ambiguity was discussed in more detail in Chapter 7 in the paragraph on verbal language.

Already by 1963 some 40 different techniques had been proposed for predicting the difficulty which may be expected in reading a piece of text and more have been added since (Klare, 1963). They also allow comparisons to be made between the readability of different pieces of writing and have been used in studies of a very wide range of documentation from scientific papers to insurance policies. These formulae typically use the average word length and average sentence length in a given sample of text. An example of one such simple method is given in Appendix 1.20.

This kind of formula is useful but it has limitations. Some short sentences can be difficult to understand or may open the door to ambiguity, as we have already seen. Some long words may be more familiar and easier to understand than short ones, except to a specialist; compare ergo with therefore. Longer words such as industrialisation (17 letters) and internationalisation (20 letters) are easily understood. Short words such as id (2 letters), erg (3 letters) and bice (4 letters) may be found more difficult to many readers. Neither are the effects of typography, illustrations, indexing and layout taken into account using only the simple criteria of word and sentence length.

Printing

As the typography, that is, the form of letters and printing, and the layout have a significant impact on the comprehension of the written material, the writer should have a direct interest in this aspect of his work. Typography has its own special jargon which we shall largely avoid here. Nevertheless, there are a few words which it may be useful to recognise, so as to be able to communicate more effectively with the printer and make use of further reading on the subject.

The size of print is based on a measure called the point, which is just over 0.35 mm. It is a little different in Europe from the USA and the UK. In general, printing for the main text should not be less than 10 point, that is, about 3.5 mm, measured from the top of the ascending stroke of a letter d to the bottom of the descending stroke of the letter p.

Older persons, even with normal vision, tend to prefer 11 or 12 point. There is a difference in optimum size from one person to another which is for the most part due to differences in visual acuity. Once again, it is necessary to be aware of the readership to optimise the presentation.

The legibility of small type is greatly improved by increasing the space between lines, called leading. This gives, so to speak, a little more air to the text and helps to avoid skipping of lines during reading. Excessive space between lines, like excessive letter size, usually reduces the readability. Incorporating an extra one or two points of leading often increases the ease of reading. When there is no extra leading, the print is said to be set solid. In typing, $1^1/_2$ line spacing may be preferable to single line spacing to optimise readability. Any text submitted to a printer for publication should be typed in double spacing. Measuring print and layout is simplified by using a type scale – a printer's rule graduated in points – which can be obtained from stationery shops.

The shape of the letters or type-face also affects readability. Material which is printed using typewriter-style lettering has a lower level of readability than when traditional printer's type is used (Tinker, 1963). Another disadvantage of typewritten text is that considerably more space is required for the same amount of text.

Printing type-faces come in many different styles but a general distinction can be made between those with decorative stroke ends, called serifs, and those without, called sanserif. For example:

Human Factors   Human Factors

In long passages of text, better comprehension has been shown in some experiments using serif rather than sanserif type (Poulton, 1965). But for other applications, such as questionnaires, forms and notices there is no evidence to suggest that one type is better than another. Some new sanserif type-faces have been specifically designed for good legibility and these are preferable for signs and placards.

Experiments have been made comparing the readability of upper and lower case letters. Usually, words set in capitals, or upper case letters, are less readable than those set in lower case. Some studies suggest that this is particularly true when reading text, but it is also valid for most other applications such as for maps and charts (Phillips, 1979).

It is a pity that NASA's ASRS program good though it is, still types its incidents entirely in capital letters, making them much harder to read and comprehend than if a more effective printing style was utilised.

Several reasons have been suggested for this difference in readability. As a good general rule, lower case should always be used except to begin a sentence, for the first letters of proper names and for some abbreviations such as ICAO or SOS.

Even for headings it is usually better to use heavier type lower case letters rather than capitals. For normal typing, it may be necessary to use capitals more than in printing as there is less variety of type-faces and type sizes available, though present printers in word processing packages do have considerable flexibility. However, even with normal typing it is generally wiser to use spacing and underlining instead of capitals whenever possible.

WHEN WHOLE PIECES OF TEXT ARE PRINTED IN CAPITAL OR UPPER CASE LETTERS, IT IS WIDELY BELIEVED THAT THE TEXT IS MORE DIFFICULT TO READ THAN WHEN IT IS PRINTED IN UPPER AND LOWER CASE LETTERS AS IN THIS BOOK GENERALLY. HOWEVER, CAPITALS ARE REQUIRED FOR ABBREVIATIONS SUCH AS ILS AND ICAO AND MAY BE DESIRABLE FOR PANEL NOMENCLATURE. It would look very strange if we were to write iata or ifalpa, even when the abbreviation is usually pronounced as it is printed. Even headings in text are usually better printed in lower case letters with initial capitals when necessary. They can be easily distinguished from the text by the use of **bold**, italics or type-face. *Text which is printed entirely in italics allows more characters to be inserted in each line but it is also considered to be more difficult to read than normal print. Italics are often used for foreign words, as in this book, for emphasis and sometimes in reference lists and for sub-headings.*

One last comment on typography. There are a number of typographical characteristics which reduce the clarity and effectiveness of graphical material. These are given in Appendix 1.21.

Layout

As a final general principle of documentation, the layout of the material must be considered. This has an important bearing on its readability and thus the extent to which the message it contains gets across. Document size and shape are the first decisions which must be made as these can largely determine the layout. There is an international metric standard for page sizes, A4 and A5 being most commonly used. A sheet of standard typing paper is A4 (210 mm x 297 mm) and each higher number describes a sheet which is half the size; A6 would be a quarter of A4. The USA, incidentally, has its own set of sizes and a standard typing sheet there is usually $8^1/_2$ in x 11 in. The UK and some other countries still make use of the earlier sizes of foolscap (commonly 17 in x $13^1/_2$ in) and the smaller quarto, based on three different sizes. The size and shape of the document may be influenced by the way it is to be used. Perhaps a mechanic needs to carry maintenance instructions in his pocket. Sometimes the desired functional layout determines the shape, as in a checklist. The resulting shapes may then not conform to international standards.

Much research has been done on layout. One such study looked at the way diagrams are set into the text and whether the text should be in a single column format, that is, right across the page, or in two columns. The research suggested that where ease of comprehension and reading are important, then a single column layout may be preferred for the presentation of complex instructional material. This preference for a single column layout becomes stronger when a large number of tables or diagrams are used in the document, when the broken text, in double columns, would become difficult to follow (Burnhill *et al.*, 1976).

As well as making a decision about the number of columns, there is also a choice to be made between two styles of type-setting. A text is said to be justified when the right edge of the printed page is made even. Normal typing is unjustified, but electronic printers and composing machines justify the text by varying the distance between words or letters. This makes an overall neater appearance. However, if the spacing is significantly distorted to achieve this, it is possible that it could slow down reading. It is also less flexible than unjustified text, which allows content of the sentence to determine where a line should end, rather than the fixed column width. There is no evidence that justified text aids reading (Wright, 1981).

Spacing between clauses, sentences and paragraphs, as well as between letters and words, is an important element in layout. Where paragraphs in a text are very short, it is better to use line spacing instead of indentation to identify a paragraph. Spacing is of particular relevance in manuals, reference books, checklists and questionnaires where a distinctive grouping of information aids accurate and rapid access and comprehension.

The visual impression created by a document is most important, particularly in the case of questionnaires and forms. This applies, in fact, in any case where it is desired to motivate the reader to take an interest in the document. This is illustrated in the techniques used by newspapers, which have no captive audience and must make the presentation attractive to maintain their circulation. Their presentation usually involves long columns of text, frequently broken by headings, sub-headings and sometimes pictures. There is constant variation in writing style and a wide range of different subjects so as to sustain the reader's interest. In principle, there is no reason why similar regard should not be given to sustain the interest of readers of technical documentation and so increase the effectiveness of the communication and thus the safety and efficiency of the operation.

The layout and printing standards used in a document are not always in the hands of the author. Many publishers of books and journals have their own house-style and unless strongly challenged will apply this even though it may not reflect current knowledge in Human Factors. An author who has done his homework in this area is in a better position to make such a challenge effectively, though success is not guaranteed.

Testing

One essential requirement must be applied to all technical writing and presentation of technical information. The initial version must be tested before publication. This applies not only to the text, but also to the diagrams, to ensure that the point being made is understood. Testing can be done by colleagues or by a representative sample of the eventual readership. If a reader testing the material needs to ask 'what does this mean?', in principle, he should not be told; the text should be rewritten or the diagram redrawn until it is clear without question. It is normal to expect some revision of all technical documentation after this kind of testing. If revision of some kind is not called for, then the quality of the testing must be suspect.

Checklists

Why checklists?

It is widely accepted that the proper, disciplined use of cockpit checklists is an essential element in flight safety. This reflects the view of the aircraft manufacturer, regulatory agencies, pilot bodies and airlines. It is a concept long accepted in civil aviation.

This is reflected, too, in FAR 121.315 which requires that each airline shall publish a checklist approved by the FAA and that the procedure must be such that the crew member does not have to rely on his memory. But the FAR leaves all other details, such as the form the checklist will take and its contents, to local agreement and approval.

A search from the ASRS data base reveals hundreds of cases related to checklist use (Special Report 1415).

In spite of this general agreement on the significance of the checklist to flight safety, lack of proper checklist discipline remains a major issue. It is only when an accident allows the CVR to be analysed – the CVR is the window to the cockpit – that the extent of this lack of discipline is revealed.

The 1987 Detroit MD-80 accident investigation revealed that the taxy checklist was not done at all and that none of the other three relevant checklists on the ground were done properly. The aircraft took off with no flaps or slats and crashed killing a total of 156 passengers and crew. Yet the airline had a history of lack of checklist discipline, well known to the FAA and the airline administration (NTSB-AAR-88-05). At one stage the airline itself had described the situation as intolerable.

Just one year later, in September 1988, a 727 took off at Dallas – once again, without flaps and slats as a result of the failure of the crew to do the checklist. This time 14 people were killed, 76 people were injured, and 18 survived relatively unscathed. The aircraft was in the air for only 22 seconds (NTSB-AAR-89-04). In this case, too, the poor cockpit discipline was well known to the FAA and airline as much as a year earlier.

How is it that reading of a checklist, which requires no special skill but only an appropriate, responsible attitude, seems to give so many problems within certain operational cultures? The following criteria show the purpose of a cockpit checklist even though it is acknowledged that many checklists in current use fall short of them.

- Following a checklist reduces the risk of missing items which might result from doing them from memory.
- The sequence of items may have technical significance of which all pilots may not be aware.
- Careful design of the checklist allows as many items as possible to be assigned to low workload periods.

- The checklist is carefully designed to provide convenient eye and arm movements across the various control panels.
- Optimum distribution of tasks can be assured and a high priority of checklist items can be confirmed by the involvement of both pilots. Maintaining this division of duties strictly, ensures that each pilot is aware of his own responsibility and does not assume that it will be taken over by the other pilot.
- The checklist encourages a team approach to cockpit tasks.
- The checklist should ensure compliance with the manner the aircraft manufacturer declares its aircraft should be operated.

A captain who casts aside the safety facility provided by the proper use of a checklist is taking upon himself an enormous responsibility, which otherwise is shared with his operations organisation, the aircraft manufacturer and the regulatory authority.

This discussion refers primarily to the normal cockpit checklist, because this is the use most directly associated with the safety of routine operations. However, abnormal and emergency checklists are addressed later in the section.

Some checklist history

It is paradoxical that for an operational area universally recognised as being highly significant from a safety standpoint, so few papers seem to have been written and so little serious research undertaken. It was not until 1990 that a serious review on the subject, based on a NASA contract was published (Degani *et al.*, 1990). Progress continues, for this first study has been followed by *The Use and Design of Flightcrew Checklists and Manuals* (Turner *et al.*,1991) and by the excellent and innovative *The Development and Use of a Generic Non-Normal Checklist with Applications in Ab Initio and Introductory AOP Programs* (Johnston, 1992).

Checklist development has taken place largely in-house and based on individual airline experience. The in-house study of checklists in All-Nippon Airways, reported in 1987, is typical of such developments (Maene, 1987). The generic non-normal checklist described by Johnston and referred to in the previous paragraph was created for and is in use by Air Lingus – the national Irish airline.

In view of the importance attached to checklists and an awareness of the limitations of the hand-held card variety, spasmodic efforts have been made by individual manufacturers to develop mechanical, semi-automatic or automatic models. These have not found widespread airline acceptance, though American Airlines, strongly supported by its pilots, has used a mechanical model for take-off and landing checklists for years (Turner *et al.*, 1991).

CRTs provide a facility for the display of checklists but there are several objections to their use for this purpose. Objections include: 1) it may displace another display such as radar, 2) they require more 'heads down' time than other checklists, 3) the use of a CRT for a checklist takes up valuable cockpit real estate, and 4) it can be hard to find a list or to go back to a point on the list (Turner *et al.*, 1991).

The manufacturer and checklists

'It is in the nature of man to err' – we quoted Cicero as saying in Chapter 2. Should not the aircraft manufacturer therefore design the aircraft so that erring pilots, as in the Detroit and Dallas cases noted earlier, do not destroy the aircraft and its human payload as a result of failing to perform the checklist properly?

FAR AC 25.1309/1A, published in 1988, allows the manufacturer to take into account procedures, such as checklists, when assessing the reliability of aircraft systems. In other words, the manufacturer can assume reasonable performance of checklists by the crew.

What is reasonable? What level of error should the manufacturer assume in its design concepts? What degree of protection against error can the pilot expect? Chapter 2 of this volume might provide some guidance. In tackling the problem of error control a two-pronged attack was proposed. Firstly, to reduce the incidence of errors as far as possible by selection, training, checking and by procedures. And secondly, to design the system to tolerate the residual, isolated errors by minimising their operational consequences.

However, this concept would not require the manufacturer to provide protection if reasonable action had not been taken to keep errors to a minimum. It would not cover the case of inadequate overall supervision, or poor procedures. It would certainly not cover the case of the pilot who decided not to use a checklist at all.

Manner of use

Most airlines use the hand-held-card type of checklist. It may be a 'do list', where the items are executed at the time the checklist is read. Or it may be a 'verification list' where there is simply a visual and verbal verification that the item has already been completed. A third possibility is that it is a combination of both, some items having been done previously and others at the time the item is read on the checklist.

Using the verification method requires considerable systems awareness from the crew member. Checklist items are in a particular sequence for a very good reason. For example, checking of the hydraulic pressure should possibly come after a hydraulic system has been operated. Checking the circuit breakers should not be done until after as many electrical systems as possible have been activated. Setting of flaps must not be done until clear of any ground equipment or transportation or personnel. While this method is in wide use, it would be unrealistic to assume that no sequencing errors are being made as a result.

The second method is using the checklist as a 'do list', that is, the items are not done until the checklist is read and the item is called. As this is a one-step function, as distinct from the two-step function of the verification method (do, then verify) it could be said that some degree of redundancy is lost. However, proper timing and sequencing is assured.

In both methods, a system of challenge and response is employed. The airline will establish who challenges – usually the non-flying pilot or the third crew member – and who responds. The responder may be one or both pilots depending on the location of the item in the cockpit and the importance allocated to the item. This is very dependent on airline culture and experience. An airline which has had a series of incidents resulting, for example, from deficient fuel loading, may involve both pilots in responding to the fuel quantity challenge.

The designation of who responds to each item may be given on the normal checklist itself – perhaps by means of asterisks – or may be covered in the expanded checklist. Many companies insist that the respondent should be indicated after each item on each checklist (normal, abnormal, or emergency) although in practice, as has been indicated, there is a considerable variation in both the form and the place of the designations.

The normal checklist tells the pilot what has to be done; the expanded or amplified checklist tells him how it must be done. They should therefore be seen as a single entity, one being meaningless without the other. No check pilot or flight inspector can make a proper evaluation of the performance of the pilots without a detailed knowledge of the expanded checklist. Every pilot must be fully familiar with it, without reference. Check airmen must insist on full compliance.

It is known to occur that a pilot will correctly call a checklist item and the correct response will be given, while the switch or indication does not correspond with that response. This risk can be somewhat reduced by physically pointing to the item in the cockpit when doing the check.

Content

The manufacturer publishes a checklist in its crew operating manual and each airline adapts this to its own operating culture. This is then approved in the USA by the Regional FAA inspector and outside the FAA by the national regulatory agency.

As a result of liability and litigation questions there is a tendency in world aviation to remain closer to the checklist as published by the manufacturer, even though this may not be operationally optimum for each airline.

Crew briefing

One item of the greatest importance, which is mandatory in the UK and almost universally a part of the checklist, is a before-take-off and before-landing cockpit crew briefing by the captain. This serves several other purposes. For example, a pre-take-off briefing serves:

- To crystallise the mind of the crew on the immediate critical few minutes ahead. There may have been a long hold awaiting take-off clearance when crew discussions of schedule change and other non-operational matters have taken place. Problems with despatch may have held crew attention before departure. It may be a take-off in the early hours of the morning – at the body's lowest phase of performance. It is essential for the captain, at this time to mentally prepare the crew for an emergency, with immediate and accurate response, within the following few minutes.

  To review or rehearse a possible emergency procedure which may not have been used for some time. A windshear alert requires a discussion between the pilots of how they would recognise windshear and what they should do about it (ICAO Windshear Guide, 1987). A forecast of bird activity on the airport should certainly result in a review of how bird impact is most likely to be indicated and the engine failure procedure (Airman's Information Manual). Ice, water or snow on the runway would also call for a briefing review.

- To ensure that the planned departure route – and potential deviations from it – is clearly understood between the cockpit crew members, with heading and altitudes repeated verbally. Noise abatement procedure would be included here, too.

- To discuss actions in the event of encountering forecast severe weather conditions after take-off. This would include the use of the weather radar and anti-icing equipment.
- To ensure that if any extra crew member is in the cockpit for take-off, whether for training purposes or just filling a seat, he understands the sterile cockpit rule (page 312) and his role in any emergency.
- To ensure that the F/O and any other crew member fully understands the duties which are expected of him, in accordance with SOPs. This is particularly important when the 'pairing' of crew members is new. This briefing will include insistence on the proper use of the checklist and encouragement for each crew member to speak up if he detects any form of anomaly. It will emphasise a team approach with proper distribution of duties and with competent leadership.
- To discuss any special procedure applicable to a technical deficiency, possibly a conditional despatch with a Minimum Equipment List item inoperative. (FAR 121)
- The diversion plan in the event of any engine failure or other emergency on departure; the airport of departure may be below weather limits for landing.

These, and any other relevant items, need not all be done immediately before take-off, but some will need to be and some changes may require a revision of an earlier briefing.

It is the policy of McDonnell Douglas to include the crew briefing in the normal taxy checklist, whilst Boeing has a policy of including it in the normal checklist for aircraft up to the 737 and in the expanded checklist for more recent aircraft.

While nominally the expanded checklist carries the same regulatory weight as the normal cockpit checklist, it is likely to be more effective to keep the item itself on the latter, with the detailed outline on the former.

It is rare indeed to find an airline, like that responsible for the 1987 Detroit MD-80 accident, which has no requirement whatsoever for a cockpit crew briefing by the captain. Others, such as that responsible for the Dallas 727 accident, are found to have nominally such a briefing, but applied in an ineffective manner.

In most cases, at the beginning of each separate checklist should be the call to be made by the captain, identifying the checklist to be done. Similarly, at the end of each checklist the call to be made by the F/O (or non-flying pilot) identifying the checklist is completed. A frequent modification is to specify these checklist items in pilot-flying and pilot-not-flying duties.

Terminology

It is important to insist on the use of standard terminology in the checklist, which all crew members are required to use. As soon as check airmen begin to respond *boost pumps* instead of the printed *fuel pumps*, the door has been opened to personalisation of the checklist. *Set* means something quite different from *checked* though the terms are sometimes used quite interchangeably in the cockpit. Whenever possible, the response *set* should be accompanied by the actual setting and *checked* accompanied by the status in which it has been checked. The use of such phrases as 'as required' should not be considered as an acceptable response.

Problems in performance

- Standardisation of checklist design and policy within an airline still sometimes remains a problem. This is unfortunate because pilots get promoted from one type to another, often then coming under a different chief pilot and with different publications. The situation is aggravated when an airline is made up of mergers or of components acquired from corporate take-overs. Pilots often retain an identity with their earlier culture and this can contribute to an overall undisciplined atmosphere. This became a particular relevant problem in the 1980s and 1990s when deregulation in the USA generated a desire to gobble up other companies too fast for the operational digestion. By 1986 when NWA absorbed Republic, the latter itself was already the product of three different airlines, struggling to absorb their differences. And the mergers continue.

- Distractions are one of the prime enemies of error-free checklist use. FAR 121.542 specifies the boundaries for application of the sterile cockpit as being from taxying under the aircraft's own power up to 10 000 ft on departure, and from 10 000 ft to the gate on arrival. However, a responsible captain will himself declare the cockpit sterile at any time he feels he wishes not to be disturbed. This should cover all times when the checklist is being read. A useful technique may be to list the more critical items near the beginning of the checklist before distractions have arisen, though technical factors may preclude that. Keeping checklists shorter may also contribute to assuring that all items are covered.

- Probably the greatest enemy of error-free, disciplined checklist use is attitude – a lack of motivation on the part of the captain to use the checklist in the way it should be used. No special skill – in addition to the ability to read – is needed. Simply a policy on the part of the captain to comply with the procedures as published.

- Associated with the previous item is the overall effect of the organisational culture. An organisation in which operations management, chief pilots and check airmen tolerate a lax and flexible approach to checklist discipline can expect such attitudes to thrive and bear a responsibility for them.

- The layout and typography of the checklist may be inadequate. However, there is really no excuse for such deficiencies any more as much research on these areas is now available (e.g. Wright, 1981; Hartley, 1985).

- Work pressures can sometimes build up to unusually high levels causing similar pressures on completion of the checklist. The captain – who sets the pace of all tasks on the ground – must protect performance of the checklist. This is a vital component of leadership in the flight deck environment (see also Chapter 6). Research has shown a clear relationship between the speed of conducting a task and accuracy (Wickens, 1992). A rushed checklist is thus more vulnerable to human error.

- Checklists are supposed to be conducted so frequently that the element of expectation becomes involved (see page 38). Responses may be given on such expectation rather than on what is actually seen.

Abnormal and emergency checklists

Although these checklists are not so frequently used as the normal checklist and do not appear so often in incident and accident investigations, by definition they have a direct relationship with safety.

Early in this chapter reference was made to the Riyadh 1980 accident in which all 301 occupants perished in what was essentially a survivable accident. Some three minutes were lost by the crew in searching for the aft cargo smoke warning procedure. The report criticised the documentation used by the crew (summarised in Flight Safety Focus 4/85).

An essential characteristic of this type of checklist is that reference to a procedure must be quick and easy in a highly stressful operating situation. This makes it particularly important to have well-written, well-indexed, and legible abnormal and emergency checklists, with consistency in format and location among the different types of aircraft flown by the airline (Turner *et al.*, 1991).

Nevertheless, it is important that the secondary abnormal procedures do not obscure the primary ones. This can probably best be achieved by performing such procedures, which do not require very rapid reference, by use of the crew operating manual or handbook.

The abnormal/emergency checklist should be a separate document, stowed with easy and immediate access to each crew member. Colour coding should be used to support the indexing system and the format and typography must be clear and easily readable.

Procedures which are written in a conditional format (e.g., if ..., if ..., if ...) invite errors and should be avoided if possible.

Where desirable, reference to the aircraft operating manual or handbook can be made, but it must be possible to complete the procedure from a single checklist source alone.

Before publication the checklist must be reviewed by a Human Factors specialist and thoroughly tested using line crew members in simulated abnormal/emergency conditions.

The future for checklists

Because of the concern over many years at the sometimes catastrophic effects of human failure in checklist performance, research into the application of technology to checklists will continue. NASA now has a specific research project involving checklists and in May 1990 issued a report entitled *Human Factors in Flight-Deck Checklists: The Normal Checklist* (Degani *et al.*). This is an ongoing project. One of the recommendations that the NTSB made following the investigation of the August 1987 crash of Northwest 255 was that the FAA 'determine if there is any type or method of presenting checklists that produce better performance on the part of user personnel.' The Turner and Huntley study and report are a product of that recommendation (Turner *et al.*, 1991).

But in reviewing misuse of checklists resulting in accidents it is worth recalling that behind the failure appears to have been an attitude deficiency rather than a checklist design deficiency. Authoritative comments in recent years appear to have confirmed this:

- one of the oldest and most pervasive problems in aviation is the failure of pilots to follow standard rules and procedures (Orlady 1989)
- the primary factor in crew caused accidents was pilot deviation from basic operational procedures (Lautman *et al.*, 1988). We do not seem to have progressed far enough during the last two decades.

Manuals, handbooks and technical papers

Functional design

The general principles already discussed are applicable, of course, to the design and production of manuals, handbooks and other technical papers. Good technical writing involves marrying these general principles to specific requirements. For example, is the document to be used as a reference book or is it material to be learned and then recalled from memory later? This will affect the optimum design.

Furthermore, people have different preferences for one form of presentation or another. Some have a preference for pictorial presentation, some prefer written text. Some preferences are related to cultural backgrounds; the extensive use of cartoons in the USA would not necessarily be suitable in all other countries.

Variations in the effectiveness of layout, of which the general principles were discussed earlier, already appear in the contents list or index. Page numbering on the left side of a contents list has been demonstrated to result in fewer errors from readers and is much faster to type than the more conventional one with the page numbers listed on the right. However, some confusion can then arise if the page is divided into two columns. When page numbers are listed on the right, some publications use leader lines or dots to help alignment of the page number with the relevant item. For documents such as maintenance and operating manuals, which are used primarily for reference on the job, an optimum index and contents list design is essential for efficient working and in emergency situations, for safety.

Patently, indexing of the operating procedures used in the L1011 disaster at Riyadh in 1980, was functionally inadequate and this was cited in the investigation findings (Saudi Arabian report). Abnormal procedures were distributed between Emergency, Abnormal and Additional sections and about three precious minutes were lost as a result of the crew searching to find the aft cargo compartment smoke warning procedure, possibly creating the difference between life and death.

Numbering of paragraphs is better than lettering, perhaps because people are more easily aware, for example, that 8 comes before 10 than H before J. It is also easier to remember a given number of digits than the same number of letters.

Colour coding is a useful means of distinguishing between different sections of a handbook. But as with all application of colour coding, it must be remembered that some colours can vary and even disappear under different lighting conditions and that many people have some degree of colour vision deficiency. About 5–10% of men suffer from such a disorder, though this is rare in women who, nevertheless, act as carriers of this incurable congenital defect. The most common form is red/green colour-blindness which affects about 4% of males. As red and green have traditionally been allocated certain meanings (danger v. safety, emergency v. normal), the fact that a person may only see them as shades of yellow, yellowish brown, or grey may significantly affect interpretation (see also Chapter 5).

Diagrams, charts and tables

It is often preferable to use diagrams, charts or tables instead of long descriptive text. We have a wide choice of these available and selection of the correct one is important. Research studies are generally available to help make the selection (e.g. Wright, 1977). The same data can be presented in a variety of tabular formats, some easier to use and interpret than others (Fig. 10.1). Various kinds of graphical

Range (in miles)	Payload (in kg)
100	700
200	600
	500
300	400
400	
500	300
600	200
700	
800	
900	100

Range (in miles)		Payload (in kg)
880	100	730
630	200	575
460	300	445
345	400	350
255	500	275
185	600	215
120	700	170
70	800	128
	900	97

Fig. 10.1 The same data presented in two different tabular forms (Wright, 1977).

presentations are available such as line graphs, histograms, pie diagrams and bar charts and these are useful when making comparisons between different sets of data. Each has its own special advantages and optimum applications.

One car manufacturer will publish in its owner's handbook an electrical diagram in the form of a complicated conventional flow diagram, relating the components to their actual location in the car. Another could use a simple functional diagram, which is much easier for determination of electrical current routing, but with no information on component location. So it is necessary to select not only the type of diagram but also the variation of it based on its intended use.

In aviation we have the same problem with more serious consequences if it is done wrong. There is considerable controversy over the amount of information that pilots need, and there is not always agreement among manufacturers, regulators, operators, and pilots regarding this issue. The increased flexibility of 'glass cockpit' multi-function displays complicates this problem because there is no practical limit on the amount of information that can be made available. Recently, a newly cer-tificated transport built by a manufacturer who believed in making available to pilots all of the information they might need in an emergency, provided a reasonably detailed schematic of the pneumatic system that could be called up on one of the aircraft's CRTs. Unfortunately, and apparently through an omission, the schematic was not detailed enough to permit positive identification of the source of one type of serious pneumatic fault (Armstrong, 1991).

It is interesting to note that geometrical illusions as described in Chapter 5 may also be associated with misreading of graphs. The Poggendorf illusion (see Fig. 5.8) is particularly relevant in this respect. It occurs on a graph when a point is read on a calibrated scale some distance away and the point lies on a sloping line which runs towards the scale. The error can be minimised by drawing the scales as close as possible to the point to be read (Poulton, 1985).

Perhaps the most significant development in recent years is the presentation of technical information in the form of illustrations and they serve so many different functions that it is not possible to give a simple set of guidelines about their design and use. They seem to motivate the reader, help in recall from long-term memory and aid in explanation. They are a useful way of avoiding technical jargon (Fig. 10.2).

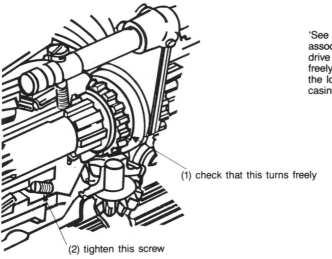

'See that the sliding dog associated with the reverse drive bevel is rotating freely before tightening the long differential casing'

(1) check that this turns freely

(2) tighten this screw

Fig. 10.2 An example of how an illustration can be used to avoid technical jargon and improve comprehension. An alternative to the drawing would be the text on the right (Wright, 1977).

In operations and maintenance manuals the use of illustrations has been extensively developed and there are numerous valuable papers available to guide those preparing such documents (e.g. Brown *et al.*, 1972; Hartley, 1985; Szlichcinski, 1979; Wright, 1977). From 50–80% of system failures have been attributed in various studies to human error (Meister, 1971). Inadequacies in manuals, handbooks and written instructions have contributed to this unreliability.

The use of colour has already been referred to in connection with manual indexing. It also has an important role in making complex information easier to understand and easier to describe. It relieves a large part of the discrimination workload. Finding one's way around London's Underground without colour coding of the different lines in the network would be far more difficult. 'Simply follow the blue line from London Airport to Piccadilly Circus ...'. Colour coding, as was mentioned previously, also has a motivational effect and encourages the reader to look at the illustration. However, apart from the caution mentioned previously in connection with colour vision deficiencies, it must be remembered that when photographing or copying coloured illustrations in black and white, some of the colours may not be reproduced and information may thus be lost.

The page and print size must take into account the working environment in which the document is to be used. In the analysis made by the Netherlands Civil Aviation Authority (RLD) of the double 747 disaster at Tenerife in 1977, reference was made to the small size of the airport chart used by the taxying pilot. It was suggested that this may have contributed to the pilot missing the correct runway turn-off and so still being on the active runway when the other aircraft took off.

Such criticisms are not new. In 1975 United Airlines had already conducted a valuable survey of pilots on this kind of chart. The airline sent out some 4000 questionnaires asking pilots to extract certain information from a number of these charts and at the same time inviting general comments on the form of data presentation. The response was enlightening with not only an unusually high return for this kind of questionnaire but with much detailed constructive criticism. From many criticisms, the following are typical: it would be difficult to extract information from the charts in the cockpit in actual flight conditions with poor light, movement etc.; the chart information is subject to too much interpretation; information on flight procedures to be followed were not clear; there was too much unnecessary clutter on the charts; the printing and pages should be larger. Many specific proposals for detailed improvement of the charts were submitted by this responsive group of pilots. It should be noted here that some airlines produce their own charts which are larger and with bigger print than the widely used charts referred to in this survey.

This survey also revealed the order of error which may be expected from the use of such charts, even in the favourable environment of the pilot's own home with good lighting, no motion and without normal operational time-pressures. Of the questions to extract certain information from the charts, most generated errors in less than 10% of the pilots. However, about one in six questions generated errors from between 20% and 50% of those responding to the survey (United Airlines, 1975). References to the difficulties experienced by pilots in reading such charts have been made elsewhere in authoritative studies (e.g. Ruffell Smith, 1979).

Technical papers

Many in aviation at one time or another find themselves called upon to prepare a technical paper on some aspect of their work. This may be for any one of the countless conferences, symposia or seminars filling the annual aviation calendar. Organisations such as IATA, ICAO, IFALPA, SAE, AEA, OAA, ISASI, ARINC and many more rely on contributions and the input of practical experience from the international aviation community. This involves those working on the job of operating aircraft and providing organisational, technical and commercial support for the operation.

The extent to which the message in the paper gets accepted and influences events (which is usually the object of the investment of time and money in preparing a paper) depends very much on the quality of the paper. The general principles discussed earlier are, of course, applicable to all technical writing but there are a number of special points which need attention.

It would be quite wrong to believe that a difficult or complex style of writing is a sign of knowledge and skill and will impress an audience. On the contrary, to write complicated new technical material in a simple fashion is the real sign of skill. This is a skill which must be learned.

Recommendations are available for the style to be used, the layout, reference listing, proof reading, typing, use of abbreviations and so on. Excellent guidance can be obtained to simplify the task and to suggest the most effective sequence for preparing the paper (e.g. O'Connor *et al.*, 1977).

There is no use in having a good message if it does not get over. Just a little homework can have a significant effect on the writer's credibility and on the extent to which the audience is influenced by the paper. All aviation papers must be produced in the English language if they are to have an international readership. This homework, then, is even more important for those who are required to write in a language other than their own.

Questionnaires and forms

Questionnaire surveys

One very familiar type of document is the questionnaire. Most forms come into this category of documentation. Surveys using questionnaires are particularly useful when we are interested in discovering people's opinions or attitudes. They may relate to the effectiveness of a new weather radar, the comfort of a passenger seat or the degree of fear of flying. The information sought may be on sensitive personal questions such as working with staff of different ethnic origins or the use of drugs by flight crews. They may relate to emotive issues such as smoking on board, over-booking or crew complement.

To obtain information on such matters it is, of course, possible simply to sound out colleagues in the office, chat to some of those concerned over a drink or simply apply one's own judgement. But these methods can be very misleading and can generate sometimes an irreversible series of actions based on false premises. When data are required which are not available directly from experimental studies then the most effective way is usually to employ a questionnaire type of survey.

The value of the survey is totally dependent upon the design of the survey programme and the questionnaire. This process is full of traps. With questionnaire forms of various kinds constituting such a substantial part of an airline's paper output some attention should be paid to seeing how these traps can be avoided and meaningful information assured. A model of the design process is helpful here (Fig. 10.3). The composition of the questionnaire itself is only one of the tasks required in designing such a survey. Using the model, it is first necessary to identify clearly the problem which it is intended to address. The information needed from the survey should be well defined and should be limited to only that which is necessary. The survey should not be used to accumulate a mass of interesting but unnecessary data.

A question like 'how many people do you know who are afraid of flying?' suggests that the designer had not clearly defined the information he needed.

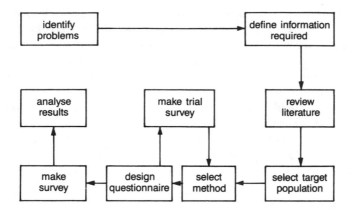

Fig. 10.3 A model illustrating the process necessary for the design of an effective questionnaire.

Is he concerned with the number of people you have actually met or those you just know about? And does he mean only those with a serious flight phobia? Is the questioner interested to know only those who will not fly because they are afraid? A question worded in such a vague way, and this is not uncommon, is worse than useless and is a waste of time and money. It can lead to false conclusions and the formulation of inappropriate policies.

Next in the preparatory sequence is a review of the relevant literature. This is needed to be certain of a full understanding of the subject under study and the technique involved in designing the particular kind of survey required. Excellent reference material is available to provide guidance on questionnaire design (e.g. Oppenheim, 1966; Sinclair, 1975). It is quite unjustifiable to repeat work already done or to design an ineffective or misleading survey simply as a result of neglecting to review the literature.

The next stage in the process is to define the target population, that is, those who are to give the answers. The survey could be to determine the quality of the public address system on board or in the departure hall of the terminal building. A decision will already have been made in the second step of the sequence on the information required. This may be to determine the technical quality of the equipment, the acoustics of the cabin or hall, or the quality of the contents of the messages broadcast. It is now necessary to establish who is to give the answers. Should it be only passengers or include crew and other staff. Should it include all nationalities, even those who do not speak the language of the broadcasts? And if on board, should it include all routes and day as well as night flights? And if crew are to be involved, should it include flight deck as well as cabin crew?

The choice of the survey method and the design of the questionnaire are the next items to be tackled. Guidance will be sought on whether it is preferable to have the survey conducted by personal interviews or whether the respondents will complete the forms themselves without assistance. The design of the questions and the selection of appropriate scales is critical to the effectiveness of the survey and these are discussed separately under the two following sub-headings.

With the questionnaire designed and the respondents selected, there is still one most important step necessary before the 'show is on the road'. A trial run, or as it is called in experimental work, a pilot study, must be carried out. This should preferably be done in three stages. Firstly, the document should be subject to criticism by one or more colleagues who have some experience in questionnaires. After revision, it should then be given to a small group, say ten, of the population to be surveyed. This should be accompanied by a personal interview with the respondents to reveal any difficulties or ambiguities encountered. After further revision, it should then be submitted to a larger sample of the intended respondents to enable an assessment to be made of the reliability and validity of the question-naire. This pilot study should be repeated until no further errors appear. For small surveys, the second and third stages may be combined. It would be most surprising if the feedback from the pilot study did not result in some modifications before the final printing of the questionnaires and initiation of the survey.

After completion of the survey, comes the analysis of results including an assess-ment of the implication of non-response. In order to do this effectively some input of expertise is needed from statistics, one of the disciplines upon which the technol-ogy of Human Factors frequently draws. This is both to assure a rational analysis of the results as well as arranging their presentation in the most meaningful way.

Design of questions

The value of a questionnaire survey can only be as good as the design of the questions (Kalton *et al.*, 1982). A most important objective here is to prevent bias in the responses. Bias is the distortion of answers through prejudice, suggestion or other extraneous influences. Using an interviewer can prevent certain biases but introduce others. Fear, social pressures to conform and a desire to please the experimenter are common sources of bias and they must be designed out of the survey.

The proper design of questions is an important element in this process. There is no possibility in such a brief introductory presentation of examining the design of questions in great detail as this is a specialised subject requiring some homework. Good guidance is available (e.g. Kalton *et al.*, 1978, 1982; Oppenheim, 1966; Sinclair, 1975; Wright *et al.*, 1975). Nevertheless, an introduction to some aspects of question design is necessary to have an awareness of the problems involved. The same question can be asked in several different ways and different answers can result from this source. As a first step, some of the different types of question should be recognised.

Factual and non-factual questions

A common, though arbitrary, distinction made is between factual and non-factual or opinion questions. 'How many hours did you fly last year?' is a factual question and typically with this kind of question, the information can generally be deter-mined or checked for validity from some other source. Research into the design of factual questions has focused on several problems:

- Definition; in this example does 'fly' mean as crew or passenger or both?
- Comprehension; if the definition is too precise as perhaps in a tax form, then comprehension may be lost.
- Memory; a perfectly honest response may be incorrect as a result of a memory fault.

● Bias; this may be related to the social desirability of responding one way or another, sometimes called prestige bias, as might result from a question like 'how many cigarettes do you smoke a day?'

Non-factual questions are more difficult and are often impossible to check or validate. 'Do you agree with the use of seat-belts in cars?' In this kind of question, research attention has been paid to the effect of the question design, the format of the questionnaire, whether a middle choice is offered, the order of presentation of choices and other ways to avoid bias. The effect of the order in which questions have been put has been studied with respect to both factual and non-factual questions (Kalton *et al.*, 1982).

Some questions involve knowledge such as 'how many lives do you believe are saved a year by the use of seat-belts in cars?' Another kind of question may involve giving a reason for some personal action; 'why did you choose to travel by train instead of by bus?' Yet another, avoided as far as possible by politicians, is the hypothetical question; 'if you were the Director of Marketing would you arrange fear reduction courses for potential air travellers who are afraid of flying?' Preference questions such as 'which of the following would you vote for if you were given the choice?' may also involve a hypothetical element.

Some questions involve a factual component which is superimposed on opinion or evaluation, like 'do you feel that your performance is significantly affected at the end of a long flight?' This technique of trying to ascertain facts by using individual opinion is very common but is fraught with risks; low correlations between perceptions and facts are often experienced.

Open-ended and closed questions

Questions can be open-ended or closed. 'How do you compare flying with travel by rail' is an open-ended question. This type of question usually provides wide information with little bias. But it is time-consuming to answer and processing is difficult. A similar question in a closed form might say 'compared with rail travel, do you find flying ...', followed by a series of alternatives such as more/less comfortable or faster/slower. The closed question is quicker to answer and easier to process but important information may be lost. In this question the element of fear which may be present in each kind of travel will not be revealed. There is also a greater risk of bias.

Rating scales

There are a number of variations of the closed question. Checklists are one such variation. These are quick to handle and are versatile, but as with other closed questions there is a risk of bias and interpretation may be difficult. Rating scales are extensively used as they are versatile and result in numerical scores which are easy to process. However, there is a high risk of bias and the apparently simple results may be difficult to interpret. These scales can appear in different forms.

A graphic form of rating scale is a simple 10 cm line with the two extremes marked at each end, for example, poor and excellent. A numerical value can easily be extracted by measuring the number of millimetres along the line where the respondent placed his mark. A common bias with this type of scale is that people are often reluctant to select extremes and tend towards the middle. This is called an error of central tendency. Clear instructions must be attached to this kind of scale, otherwise some respondents will simply assume that there are only two choices and opt for one end or the other.

Sometimes rating scales have descriptive phrases attached to them. One study compared the result using this kind of 10 cm scale with an identical one without phrases. The respondent in each case was asked to mark a cross on the line to reflect his opinion about the noise on a helicopter flight. The result was quite different in each case (Oborne *et al.*, 1975). This illustrates the care needed in designing the questionnaire and interpreting the results (Fig. 10.4).

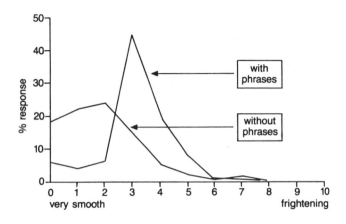

Fig. 10.4 An example of the influence of rating scale design on the response to the same question, which asked about the noise on a helicopter flight. Two different patterns of response emerged (Oborne *et al.*, 1975).

Middle and neutral options

When designing questions a frequent consideration is whether the respondent should be given a middle or neutral option. If an even number of options is offered there is no possibility of selecting a neutral position. This forces a reluctant respondent to declare for one side or the other. On the other hand, using an odd number, say, five or seven choices, allows him to take possibly the easy or middle way out when he has an opinion but prefers not to express it. The middle option may be a 'don't know' or a 'do not wish to commit myself'. Or it may be a true expression of a middle or no option opinion. Research has been done in this area and guidance is available (e.g. Kalton *et al.*, 1980; Presser *et al.*, 1980).

Order effect

Another consideration is the order effect resulting from the sequence in which the alternatives are listed. There is some evidence that in written questionnaires the first-mentioned alternative may be favoured, but the effect is small (Kalton *et al.*, 1978). The order in which questions are placed relative to other questions may also have an influence on the response. This makes comparisons between surveys difficult and risky.

233

Wording

Sensitive questions require special attention if meaningful responses are to be achieved. Such questions may involve professional or social matters which might have an impact on personal respect, prestige or even legality and generate disapproval. 'Have you ever used alcohol within the prohibited period before flying?' is a sensitive question and in this simple form little reliability can be placed on the responses. 'Do you ever use sleeping tablets when away from home on a long intercontinental flight schedule?' can also be expected to produce responses of doubtful validity. In such cases, it may be necessary to utilise a so-called leading question such as 'many crew members find it necessary to use sleeping tablets away from home on long flights. Do you?' However, great caution must be applied if a leading question is used and this is often considered an unacceptable source of bias.

This leads to some other aspects of question design. Only one question should be asked at a time. What would an affirmative answer mean to this question: 'do you suffer from headaches and stomach pains?' Words themselves may be loaded, that is, they may have an emotional appeal so that the respondent reacts to the word rather than to the question. 'Do you agree with the way union bosses handle industrial disputes?' might generate a different response if the word bosses were replaced by leaders, possibly then causing a loading in the other direction. When people are asked to agree or disagree with a given statement, those who are uncertain tend to agree or acquiesce.

Some sources of response bias and unreliability originating in the wording of questions
These questions call for simple YES/NO or AGREE/DISAGREE answers

Loaded words
'Do you think that the union bosses are properly reflecting the views of their members?'

Leading questions
'Are you against giving too much power to the police?'

Pleasing the questioner
'Do you find our newly redesigned seat comfortable?'

Social expectation
'When you are smoking do you always try to avoid causing discomfort to others?'

Professional expectation
'Do you always avoid the use of alcohol within the prescribed prohibited period before flying?'

Multiple questions
'Do you enjoy travelling by road and rail?'

Acquiescence
'Do you AGREE/DISAGREE that individuals rather than social conditions are more to blame for crime in society?'

Imprecision
'Do you watch television in the evenings?'

Complexity (and use of negatives)
'Are you not unhappy with the decreasing negative response of the minority to drug avoidance?'

Ambiguity
'Did the cabin attendant stop smoking in the cabin before landing?'

Layout

The general layout of the questionnaire or form is important and must be optimised. Good headings, good type-face and possibly the use of colour coding facilitate comprehension and enhance motivation in the respondent to answer the questions conscientiously (Gray, 1975). In the UK the National Consumer Council (NCC) sponsored a booklet entitled 'Gobbledegook' as a challenge to bureaucrats to show good reason why they should not drastically improve the quality of the forms and questionnaires they generate (Vernon, 1980). It claimed that the 100 000 official forms in the UK were mostly incomprehensible. It argued that even 1% improvement in efficiency of forms would save the taxpayer £1 million a year. The NCC encouraged use of a sticker which they supplied and which said 'This is gobbledegook. Please use plain English'.

This would be attached to incomprehensible or badly designed forms which were then returned to their originators. This may not have produced a more sympathetic response from the income-tax inspector but it made the one struggling to read the form feel better. ...

The questionnaire is not a precise tool even when produced well. And when produced on a DIY, unskilled basis it can be grossly misleading and costly. We have seen that the result can be influenced by many factors such as the selection of respondents, the survey technique, the layout of the questionnaire and the design of the questions. And finally, the way a person responds may be conditioned by the society in which he lives and his own social background.

The design of questionnaires and forms requires a skill which must be learned. In general, there is vast scope for improvement in the design of such documents.

Forms

Not all forms are questionnaires – some simply give instructions or tender advice. Deficiencies here may not be potentially so costly for the originator as in the case of questionnaires, though they might result in additional personal interviews being required to convey the message or correct misinterpretations. However, they will certainly involve the recipient in frustration, loss of time and perhaps financial loss.

Considerable work has been done on Human Factors aspects of the design of forms and there is really no longer any excuse for such deficiencies. Tax and social security documentation seems to be particularly vulnerable, but similar problems also exist throughout industry. Quite dramatic improvements can be made by the application of expertise to this design task (Fig. 10.5). Numerous papers and books are available to provide guidance (e.g. Hartley, 1985; Wright *et al.*, 1975).

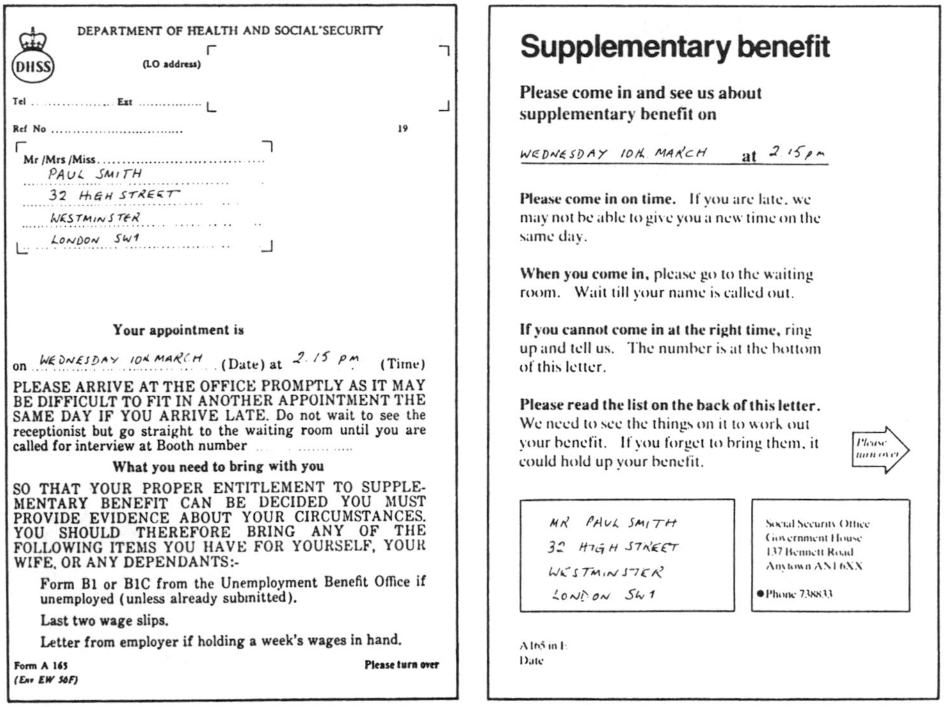

DHSS

DEPARTMENT OF HEALTH AND SOCIAL SECURITY

(LO address)

Tel .... Ext ....

Ref No 19

Mr /Mrs /Miss ....
PAUL SMITH
32 HIGH STREET
WESTMINSTER
LONDON SW1

Your appointment is

on WEDNESDAY 10th MARCH (Date) at 2.15 pm (Time)

PLEASE ARRIVE AT THE OFFICE PROMPTLY AS IT MAY BE DIFFICULT TO FIT IN ANOTHER APPOINTMENT THE SAME DAY IF YOU ARRIVE LATE. Do not wait to see the receptionist but go straight to the waiting room until you are called for interview at Booth number ....

What you need to bring with you

SO THAT YOUR PROPER ENTITLEMENT TO SUPPLEMENTARY BENEFIT CAN BE DECIDED YOU MUST PROVIDE EVIDENCE ABOUT YOUR CIRCUMSTANCES. YOU SHOULD THEREFORE BRING ANY OF THE FOLLOWING ITEMS YOU HAVE FOR YOURSELF, YOUR WIFE, OR ANY DEPENDANTS:-

Form B1 or B1C from the Unemployment Benefit Office if unemployed (unless already submitted).

Last two wage slips.

Letter from employer if holding a week's wages in hand.

Form A 165 (Eav EW 56F) — Please turn over

Supplementary benefit

Please come in and see us about supplementary benefit on

WEDNESDAY 10th MARCH at 2.15 pm

Please come in on time. If you are late, we may not be able to give you a new time on the same day.

When you come in, please go to the waiting room. Wait till your name is called out.

If you cannot come in at the right time, ring up and tell us. The number is at the bottom of this letter.

Please read the list on the back of this letter. We need to see the things on it to work out your benefit. If you forget to bring them, it could hold up your benefit.

Please turn over

MR PAUL SMITH
32 HIGH STREET
WESTMINSTER
LONDON SW1

Social Security Office
Government House
137 Bennett Road
Anytown AN1 6XX

● Phone 738833

A165 in E
Date

Fig. 10.5 An example of the improvement which can be achieved by the application of expertise to the design of forms. The form originally used by the government department is on the left and the revised version on the right (Cutts *et al.*, 1984).

From research on language and comprehension a number of rather simple rules can be applied to optimise design. We might call these the three Ls of form design – layout, logic and language. While we have already discussed these in general terms related to all documentation, it may be worthwhile to review them here briefly in the context of forms.

A good layout will generate more response and less misunderstanding from the recipient. Good layout enhances motivation in the reader. Headings should be clear and prominent. Spacing and typography (including size and type-face) must be optimised.

The information in the form should be set out logically. This is particularly necessary when the reader must carry out a number of steps in a given sequence. The use of flow charts, decision trees or algorithms may assist in assuring logic in the presentation.

Language refinement involves the use of short sentences, preferably with only one clause. The active (open the door) rather than the passive voice (the door should be opened) is more direct and more clearly understood – politicians and scientists are said to be particularly fond of the passive verb form. We have already seen the comprehension difficulty which negatives – particularly more than one – can give in the list of questions inviting bias and misinterpretation. However, in certain cases, use of the negative may be justified (e.g. 'do NOT touch the live wire'). Familiar rather than unfamiliar words should be used. 'Pecuniary advantage' should not be written when we simply mean 'money' and 'remuneration' can usually be replaced by 'pay'. Sometimes prose can be beneficially replaced by symbols or a flow chart.

With proper attention to layout, logic and language, the losses and frustrations associated with badly designed forms can be eliminated.

Maps and charts

Historical background

As commonly understood, certain distinctions have developed between maps and charts. Maps are generally topographical and are applied for many purposes. Charts, on the other hand, are nautical or aeronautical and are used only for navigation of one kind or another. Maps tend to have a more permanent nature, while charts may be annotated and then discarded or be subject to periodic revision and replacement as the data which they display change.

The technology used in the design and production of maps and charts is called cartography. It is about 100 years since suggestions were first made on how maps could be annotated to assist air navigation, though at that time based only on the experience of ballooning. The first aviation chart is thought to have been produced early this century in Germany but a systematic and professional application of Human Factors to this field of documentation has since then been notably lacking. In general, it can be said that the needs of the user have not had sufficient influence on either the content or the appearance of most aeronautical maps and charts though there have been a few exceptions.

There seem to be two basic reasons for this neglect. Firstly, the design of maps and charts is an activity involving long tradition, with responsibility in the hands of specialised cartographers and with inadequate input from psychologists, engineers or the users of the material. Many of the colour conventions of 19th century land mapping were followed in aviation cartography without much serious consideration of their merits (Taylor, 1985). And secondly, research into the Human Factors of maps and charts is both complicated and expensive. When attempts to apply

Human Factors data have been made, these have often not gone beyond the application of general principles of displays, an unsafe and frequently inappropriate practice. The 1975 survey of pilots of United Airlines, quoted earlier in this chapter, confirmed the existence of a very wide range of deficiencies in the type of charts widely used for approach and landing. Not totally unexpected was the remark made in the RLD report of the 1977 Tenerife double Boeing 747 disaster, that the aerodrome surface chart used by one of the crews who missed the taxyway turn-off in bad visibility may not have been adequate.

Task and environment analysis

Human Factors expertise is required in several activities associated with the design of maps and charts and we should briefly review these here. As with the design of tools in other work situations, the initial step is a detailed task analysis, in this case, the navigation task. This will be used to determine items such as the type of projection to be employed and how geographical orientation is to be presented – latitude and longitude, grid coordinates or bearing and distance. A detailed task analysis will reveal how the map or chart is used and what information is required to be displayed. One of the most common complaints of air pilots and navigators is that either excessive information clutters the sheet, obscuring vital data, or that valuable information is omitted.

Different types of operation demand different requirements for the information to be displayed. High or low speed flight, high or low altitude, military or civil transport, helicopter operations, or light general aviation aircraft, all have different navigational needs. They are all represented on an SAE G-10 Charting Subcommittee, which is reinforced by charting organisations and by other interdisciplinary experts. At the present time this very active subcommittee is considering nearly 50 technical issues involving aviation maps and charts, which range from the 'timeliness and accuracy of preferred route requirements' to the 'portrayal of active volcanoes', and includes electronic charting display and symbology.

A task analysis leads logically to an environmental analysis. The maps and charts must be optimised for use on the aircraft flight deck, though secondary requirements may be set for use in an operations room, meteorological office or air traffic control centre. This analysis will reveal constraints such as working space, which will affect the size and format of the document as well as viewing distance and angle. It will indicate the kind of stowage to be expected and the degree of rough handling in routine use which must be anticipated. It will provide information on the level of vibration and turbulence to which the document and its user will be subjected and this will influence a decision on printed character size. Vibration may have an influence on design when it is in the range of 1–30 Hz (Hopkin *et al.*, 1979). Various studies have determined the frequency bands in which visual acuity is affected; this is, of course, also relevant to other flight deck documentation. Vibration is more of a problem in helicopters than in fixed-wing aircraft, emphasising caution when using the same documents in different operational environments. In contrast, turbulence is more of a problem in fixed-wing aircraft than in helicopters, though there is a limit to the extent to which map design will be able to accommodate the kind of turbulence encountered in high speed flight.

The type of lighting, including colour and brightness, in which the documents are to be used will also affect design. This represents a particular difficulty as the flight deck illumination can routinely vary from the extreme intensity of direct sunlight at high altitude to a dimly lit environment at night, with possibly marginally effective map reading lights. Ultra-violet lighting was used in USAF aircraft and in the Luftwaffe during the Second World War. Red lighting was used by the US Navy, the RAF and in the 1950s, the USAF. This also became a civil aviation standard for some years before the general adoption of low colour-temperature white in the 1960s. Some flight decks still use red lighting. In the UK in the early 1920s red lighting was first recommended, but it was not long before the adverse impact of this on colour coding, particularly the use of red, was realised and special maps were developed. The most widely used airway charts, Jepperson, use blue on white with some green. Their approach and landing charts are published in black on white, thus compromising for use on all flight decks. However, the compromise deprives the user of the benefits of colour contrast, reducing the effectiveness of map use. Glare resulting from reflectance of a white map surface can affect visual acuity; this aspect of vision was discussed in Chapter 5. The environmental analysis will determine, in addition to whether the document will need to withstand rough handling, whether it is likely to be exposed to heat, moisture or direct sunlight.

Presentation optimisation

The task and environmental analyses will provide much fundamental information for the designer. However, before actual drawing and printing can start, some profound questions related to the optimum presentation and interpretation of complex visual information need to be answered. It has been argued that insufficient research has been directed to this aspect of design. An understanding of the sensory capabilities and limitations of man, visual acuity and the process called visual search, that is, eye movements to extract relevant details, is needed in order to optimise the presentation of information. An understanding is also required of factors related to interpretation such as attention, memory, expectancy and set, which are discussed in other contexts elsewhere in this book.

In civil air transport most charts are used for the primary purpose of displaying aeronautical information such as airports, radio facilities, airways and airspace boundaries. For visual flying, however, this aeronautical information must be superimposed upon topographical information. Many of the symbols used were standardised in the 1940s by ICAO (Annex 4) mainly on the grounds that they were in common use rather than as a result of Human Factors research and definition.

The paper basis of conventional maps and charts not only provides flexibility in that they can be folded to almost any size or shape but it allows them to be easily annotated, makes them very portable and permits convenient stowage. As a material it is inexpensive and easily replaceable. However, it has disadvantages in the operational environment. Consulting a conventional chart while at the same time instrument flying in a complex terminal area poses difficulties due to the location of the chart, its illumination (at night), the lack of selectivity in the information displayed, the fixed orientation and the lack of any continuous indication of the aircraft position on the chart. Progressively, new technology has been applied to overcome some of these difficulties as well as to provide entirely new facilities unavailable on conventional charts. The new 'map modes' on the ND CRTs of many 'glass cockpit' transports are a good example. They are very popular with pilots.

Perhaps electromechanical instruments such as the HSI (Horizontal Situation Indicator) may be seen as a primitive form of map display. But in a more truly cartographic form, development of automatic chart displays has tended to follow three routes. Direct view map displays such as the Decca Navigator take the form of a strip map, moved on rollers, with a cursor indicating the aircraft position. Optically projected map displays consist of a microfilm transparency of an actual map projected from behind onto a screen, with a symbol to represent the aircraft. Electronic map displays have replaced these two earlier forms of moving maps and are more reliable, more accurate and more flexible in display content.

Many of the Human Factors aspects of conventional maps and charts discussed here are also applicable to these newer automatic devices, but each new display technology brings with it new problems which require resolution.

Reference sources

The AGARD (Advisory Group on Aerospace Research and Development) branch of NATO has published a comprehensive document in which the actual and potential contribution of Human Factors to the design and evaluation of aviation maps and charts are discussed (Hopkin *et al.*, 1979). This impressive document provides a reference list of 863 sources of relevant information. A later review of colour design in aviation cartography discusses the problems of airborne map displays with respect to colour coding and proposes a set of principles and guidelines for the best use of colour in these displays (Taylor, 1985).

Conclusions

In industry today, poorly designed documentation is the source of very substantial financial loss; quite as expensive though perhaps not so spectacular as the Gemini 9 docking fiasco. This loss appears in the field of training, in operations, maintenance and in every area where an airline uses its vast armoury of manuals, bulletins, forms and questionnaires. Time is wasted, learning is retarded, bulletins aimed at raising the levels of safety and efficiency are less effective, incorrect policies are established on the basis of misleading survey responses or interpretations and frustrations are generated.

The Human Factors design of documentation is a skilled activity; university departments and other specialised institutions have been established to study its various aspects. NASA research has revealed errors in airline operation induced by deficiencies in documents and charts and has called for increased effort to be devoted to their redesign (Ruffell Smith, 1979). An extensive store of Human Factors information related to paperwork is now available and this is sufficient to eliminate current deficiencies in aviation documentation.

11

Displays and Controls

Historical background

The advanced electronic displays and controls in modern transport aircraft reflect over a century of development. The evolution has taken place not in order to make the flight deck more comfortable or convenient as a workplace but as a result of two major pressures – safety and economics – between which there is an inescapable interaction.

As air transport has become a more popular means of travelling and as aircraft under economic pressures have become larger, with more victims per accident, the public and the litigation industry have focused greater attention on air disasters. References were often made to wrongly operated controls such as flap instead of landing gear, and faulty or failed instruments. At the same time aircraft performance was increasing, again under economic pressures, and this required more information input to the pilot, more sophisticated controls and more complex displays to permit satisfactory operation.

The earliest form of flight instrument was the barometer, used in balloons from the 18th century, and this remains the basis of today's pressure altimeters. It was not until the advent, around 1930, of the need to fly without visual references – blindflying as it was then called – that serious attention began to be given to the development of displays. Earlier displays ranged from the piece of string used by the Wright brothers as a slip indicator to conventional engine RPM indicators.

The breakthrough which permitted this new form of flying in both military and civil aviation, came from the development of a usable gyroscope which could be applied in the form of an artificial horizon. While today's pressure altimeters owe their origin to the barometers used in balloons, so the sophisticated gyro-stabilised platforms which form the basis of modern flight guidance systems can trace their origins to the primitive gyroscopes in the artificial horizons of the inter-war years.

With the development of electronics, servo-driven instruments became possible in the 1950s and this permitted a very notable improvement in instrument design as the sensor could be remotely located away from the instrument.

One other technical development that has had a major impact on displays is the CRT. This was already in service in some military aircraft in the early 1940s but it took another quarter of a century before it became available for display of primary

241

flight information. Early airborne approach and navigation systems such as BABS and LORAN also used the CRT.

Development of controls shows an equally spectacular evolution since the wires and levers of early flying machines. Direct mechanical controls gave way to servo-assisted controls and, with the facilities provided by electronics, to advanced semi-automatic and automatic flight control using fly-by-wire techniques.

Avionics and ergonomics

Aviation electronics, or avionics, has been one of the necessary pillars upon which progress has been built. Another, regrettably frequently neglected, has been ergonomics. The history of ergonomics or Human Factors was briefly traced in Chapter 1, where several milestones were noted. Yet it was too often considered adequate to have information available on the flight deck without considering adequately how this was to be sensed and processed by the crew member. Aviation, of course, is not alone in being guilty of such neglect; it is apparent throughout industry, including nuclear power stations where inappropriate human control can be no less catastrophic than in air transport.

Man is adaptable as an operator and this adaptability often masks display and control deficiencies which nevertheless remain to trap the unfortunate or unwary. However, several important Human Factors studies have aided instrument development though the findings of such work have not always been properly applied. Studies such as those of Fitts, Jones and Grether at the Aero-Medical Laboratory in Dayton in the USA made a valuable contribution to the adoption of a more scientific approach to display design. The study of reading errors in three-pointer altimeters was a notable example (Grether, 1948). At about the same time in the USA studies were being made on the effect of dial shape on legibility (e.g. Sleight, 1948). These were followed shortly afterwards by research on subjects such as scale reading accuracy, scale graduation intervals and pointer design. By the 1960s numerous studies were being conducted on electronic displays. Research on the Human Factors of CRTs expanded rapidly so that by the 1980s it became possible to publish extensive reviews of current knowledge of ergonomic aspects of visual display terminals (VDTs). These are discussed later in this chapter.

Operational input

A third pillar upon which progress has been built is operational experience. The means by which this has found its way into the design process has been largely through the input of airline development pilots. They have had the task of translating the experience accumulated in training and line operations into a language which can be used by manufacturers and equipment designers. These few dedicated individuals in the major airlines, with names well recognised in the industry, formed the vital bridge between air transport operations and the manufacturing industry for the first quarter of a century of post-war civil aviation. Their task was to try to ensure that new technology was applied to meet the rapidly developing operational requirements. They strove for years, often against the forces of prejudice and obsolete attitudes, to obtain recognition of the need for a more enlightened approach to Human Factors in flight deck design.

In recent years such input has been increasingly organised through bodies like

SAE and IATA. Higher development costs have reduced the scope for individual initiative and enterprise and it has become increasingly necessary to develop only to a standard already established by a consensus of airlines. The avionics package on a passenger jet aircraft can now amount to as much as 15% of the total aircraft cost (Flight International, 1984). It is worth remembering, however, that standardisation can be the enemy of progress and that some diversification is an essential prerequisite to take advantage of new technology and new knowledge.

Although a minimum instrument list for flight was specified as early as 1917 in the USA, it was not until the 1930s that serious attention began to be given to the layout of these instruments on board. Further reference is made to this aspect of displays in Chapter 12. Good short histories of flight instrument development have been published (e.g. Chorley, 1976).

Displays and controls are in the mainstream of Human Factors. If we refer back to the *SHEL* model in Chapter 1 it will be apparent that we are concerned here with the *Liveware–Hardware* interface. In the case of displays, the transfer of information is from the *Hardware* to the *Liveware*. Controls are used to transfer information and energy in the other direction, that is, from the *Liveware* or operator to the *Hardware* or the system. Usually there is an information loop involved and engineers have the responsibility of optimising the flow around this loop. We should now examine displays and controls separately to determine some elements which should be considered in this optimisation process.

Displays

Visual, aural and tactual

A display is any means of presenting information directly and it usually makes use of the visual, aural or tactual senses. A stall warning using a stick-shaker is using the tactual sense as well as aural (sound) and visual (warning light) senses. A household gas system which had a smell added to odourless gas so as to provide a warning in the event of a gas leak, could be said to use an olfactory display.

The purpose of a display in an aircraft is to transfer information about some aspect of the flight accurately and rapidly from its source to the brain of the crew member, where it is then processed. This introduces the first major Human Factors problem area. The human sensory capacity is enormous but the human information transmitting rate is very limited, as is man's short-term memory capacity. This imbalance results in a bottleneck arising when the information which is being fed to the brain is being filtered, stored and processed. This is of fundamental importance in the design of flight deck displays. The display must not only present information, but present it in such a way as to help the brain in its processing task. Furthermore, it is of little use in the overall flight deck system to display more information unless the system is so designed that the crew member will be able to utilise it. Not only under normal circumstances, but also when his performance is affected by stress or fatigue. There are a number of ways in which display design can facilitate this and as an introduction to the subject we shall review some of them here.

Classification and description of visual displays

Displays may be described in many different ways and the choice of the dimension to be used as a basis for the description is rather arbitrary. In general terms, a display may be classified as dynamic or static. Dynamic displays are those subject to change through time and include instruments such as altimeters and attitude indicators. Static displays, in contrast, remain unchanging over a period and these would include placards, signs and graphs.

Another method of describing displays is in terms of the type of information which they present. This could be quantitative such as altitude or heading, or qualitative such as instruments which provide information on rates or direction of changing values. The type of information could be simply one which advised on the status of a system such as a landing gear indicator. Displays may be used for warning or cautionary purposes such as an engine fire warning light and this provides yet another description within this particular taxonomy. There are several more types of information which can be displayed and which can thus form a basis for classifying the display. Some displays may, in fact, be either static or dynamic and may contain more than one type of information.

Flight displays, which are necessarily dynamic, may be described in different ways. We can talk in terms of command displays such as the flight director, predictor displays like certain landing monitors, and situation displays such as the HSI. The earliest situation displays were artificial horizons and these raised the issue of whether such displays should be based on an inside-out or outside-in concept. That is, whether the display reflects what the situation would look like from inside the aircraft with a fixed aircraft symbol and a moving background, or from outside the aircraft with a moving symbol and a fixed background. This conceptual dilemma also largely applies to map displays. The aircraft symbol may be fixed over a moving map, or can move as a cursor over a roll-map or as a rotating symbol on a north-up moving map. In certain ground-based applications such as in ATC or at the instructor's station in a flight simulator, the map may be totally fixed with the aircraft symbol tracking over it, thus fully reflecting the outside-in concept.

Some years ago the industry proclaimed the arrival of so-called integrated instrument systems. Such systems, which incorporated the HSI and the ADI (Attitude Director Indicator), were not truly integrated displays except to the extent that various different displays were incorporated in a single instrument case. The operational benefit of doing so was a reduction in scanning area for essential information and a saving of panel space.

These objectives have been achieved more effectively since, through use of the CRT. This permits time-sharing of high priority panel space and allows information which is not relevant in the particular phase of flight to be removed from view altogether.

Some general aspects of visual display design

A few general remarks on display design should be made here, reflecting deficiencies which have often been found in practice to influence operating efficiency. It will become apparent that it is necessary at the design stage to understand how and in what circumstances and by whom the display is to be used.

A fundamental difference between aural and visual displays is that aural signals are generally omnidirectional while visual signals normally are not. If it is desired, for example, that both pilots should be able to see the colour of the landing gear warning lights, then the lamp may not be monodirectional, as is the case on some aircraft. The gear warning horn, of course, provides an omnidirectional display when it is activated.

Another characteristic of visual displays is that, by definition, they need light and visibility to be effective. In a cabin emergency at night, with no electrical power available and in dense smoke, static visual displays are of little value.

Most flight displays originated in an analogue form but increasingly digital alternatives have become available. In some instances, such as mach indication, the information may be presented in both analogue and digital form in a single instrument. Most altimeters now display altitude in both forms. In other cases it may be necessary to make a decision on which concept to apply. Although a digital presentation may be expected to occupy less space, the decision should depend primarily on the manner in which the information is to be used. In an aircraft clock, GMT is used mainly for log keeping and position reporting and so, when displayed, a digital presentation is appropriate. However, a sweep second-hand analogue display may be preferred for short-term elapsed time and time-to-go during timed procedures.

Where precise values are required for recording or system monitoring then digital displays usually provide greater accuracy. On the other hand, they demand more time to be read. When the numerical values are changing frequently or when it is necessary to observe the direction or rate of change, then analogue scales – usually a fixed scale and a moving pointer – are preferable. More detailed recommendations are available elsewhere (e.g. Heglin, 1973).

The availability of newer display techniques such as LEDs (light-emitting diodes), LCDs (liquid-crystal displays) and CRTs increases the scope for the use of digital displays as they avoid some of the deficiencies of earlier mechanical and electro-mechanical equipment. These techniques are discussed under the next heading in this chapter.

These are all fundamental questions which must be answered before a display is designed. But there are many more items of significance which are sometimes neglected at an early stage of design – perhaps altogether – and which have an influence on the effectiveness of the display. It used to be the practice of the manufacturer to emblazon his name across the face of the display, though this had no interest whatsoever to the pilot. The maintenance department found the face of the instrument a convenient place to display the part number of the equipment so as to facilitate replacement though again, this was redundant information for the pilot struggling to distinguish the important information from the clutter. While over the last 30 years or so much progress has been made in getting extraneous information removed from displays, success has not been total and it remains necessary to include this requirement in the display specification.

The angle at which a visual display is to be viewed has an important influence on design and must be clearly stated in the specification. The optical characteristics of a lamp, as we have already mentioned, may limit its usable viewing area but the same is also true of certain mode annunciators. A somewhat different source of loss of display visibility is the instrument bezel or display case, when viewed from an angle other than normal to the face of the display. This was reported, amongst numerous other display and layout deficiencies, in the NASA study on operational workload already mentioned (Ruffell Smith, 1979).

Another problem related to viewing angle is parallax. This occurs when the pointer or index is in a different plane from the scale against which it moves. A display can be designed to minimise this effect and, of course, with a CRT it does not exist.

Viewing distance is also of relevance to visual display design. This is often assumed to be about 71–78 cm, which is about the distance to the main instrument panel from the design eye position on most large transport aircraft. But an instrument installed on the lateral or systems panel could be some 2 m from the pilot's eye position (see Chapter 12). The viewing distance should be reflected in the size of characters used in the display.

A display which is in a standby mode and which is not currently active should itself clearly annunciate that fact; it must be recognisable from the display itself and not from a nearby switch position or annunciator. Some radio control panels have a dual selector head allowing a frequency to be pre-tuned. If the inactive frequency remains exposed and with the same appearance as the active frequency, then errors resulting from the ambiguity can be predicted.

Ambiguity increases the cognitive load on the crew member as well as inducing errors, so must be avoided. The to/from indication on some VOR displays is an example where this deficiency was not properly anticipated.

It is a basic principle of display design that information which is suspect should not continue to be presented to the crew. Reference to this is made in a later paragraph when reliability is discussed.

Alphanumerics

The display of letters and numbers has been the subject of many Human Factors research studies. As in documentation, the type-face, stroke width, width-height ratio, size, background, etc. can all influence legibility of the characters used. This was covered in Chapter 10. Here we are concerned more with Human Factors associated with different display methods.

In earlier flight decks, mechanical techniques were mainly used and these were based on wheels or drums. A question long discussed with this type of display is whether they should rotate up or down as the numbers increase. Magnetic wheels, which were later introduced on grounds of cost, space and their digital drive, were not faced with this problem though they brought with them their own problems such as instability after movement. There now seems to be some agreement that a drum should rotate downwards when displaying an increasing value. However, this is not yet fully standardised and equipment is still in service where in one panel display read-out, some digits move up when increasing and some down.

This Human Factors issue has now completed a full circle as a simulated rolling counter indication is incorporated into certain solid-state displays such as the LED digital read-outs for engine instruments for the A310.

Mechanical and electromechanical alphanumeric displays are rapidly being replaced by electronic displays of various kinds. Each of these has its own Human Factors problems which require attention. In general, apart from the characteristics of the alphanumerics, it is necessary to examine:

- brightness – whether it can be seen in direct sunlight, or in shadow when the pilot is in direct sunlight
- colour – some have limited colour capability
- flicker – a problem with certain kinds of display
- ambiguity – what is displayed if there is a partial failure.

It might be worthwhile to review some of the main display techniques currently in use and aspects of them which may require special consideration:

- mechanical and electromechanical – direction and rate of movement, stability, ambiguity, lighting
- electroluminescence (EL) – brightness, colour, contrast ratio, maintainability
- incandescent lamps – uniformity, life, maintainability, fragility
- liquid-crystal displays (LCD) – colour, lighting, viewing angle, contrast, brightness
- light-emitting diodes (LED) – colour, contrast ratio, brightness
- gas-discharge plasma – colour, flicker, contrast, brightness
- cathode-ray tube (CRT) – bulk, contrast ratio, brightness, flicker, resolution (raster display).

Applications of these technologies differ. Some can be used in seven segment displays, though a complete alphanumeric set cannot be made with this type of display. In dot matrix displays, the dots or cells must be sufficiently packed to assure adequate legibility. Each kind of display must be evaluated in the light of the technology used, the specific application and the environment.

In addition to legibility, that is, the ability to discriminate one character from another, we are increasingly concerned with readability or the ability to recognise total words or groups of letters and numerals in combination. CRTs now provide the facility to display text and expanded data-link systems involve the reading of more extensive messages. Readability is also influenced by spacing between characters and lines and the use of capitals and lower case letters (see Chapter 10).

Dial markings and shapes

Perhaps the most familiar aspect of display Human Factors is dial shapes and markings. There are three basic kinds of dynamic displays for use in presenting quantitative information. These are fixed scales with moving pointers, moving scales with fixed pointers and digital read-outs. The difference in application of analogue, the first two, and digital displays was discussed earlier.

A wide variety of shapes and scales is in use (Fig. 11.1). These examples all show fixed increments in the scale. However, scale progressions are sometimes variable with several different progressions with irregular major graduation markings. In principle, varying progressions should be avoided whenever possible. The easiest progression system is in single units. Steps of 10 are also good and to a lesser extent steps of 5. Steps of 2 are generally reasonably acceptable. Decimal points should preferably not be used and in any case the 0 ahead of the decimal point should be omitted.

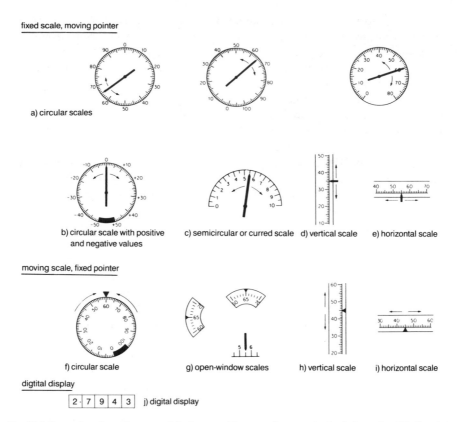

Fig. 11.1 Examples of certain types of displays used in presenting quantitative information (McCormick *et al.*, 1983).

The graduation base is usually on the outside with the major markers extending inwards. However, the reverse is also possible; this is sometimes called a 'sunburst' design.

A few studies have looked at the design of pointers. The design is of particular importance when the instrument also contains a digital read-out which can be partly obscured by the pointer. With scale graduations as shown in Fig. 11.1, it is usually recommended that the pointer tip should just touch the end of the small graduations but should not overlap them. The whole pointer should be visible to the centre of the dial unless it will obscure a read-out on the face of the instrument.

A common problem with all such displays is parallax resulting from the distance between the pointer and surface of the scale. Some instruments have avoided this by using the sunburst scale format and raising the level of the graduations around the periphery to that of the pointer tip. This problem disappears when the dial is portrayed on a CRT, as in current electronic primary flight displays. However, this characteristic of a CRT brings its own disadvantage as considerable complication arises when trying to ensure that one symbol appears to be in front of another, such as the flight director in front of the aircraft symbol.

The size of the display, in particular graduation spacing and alphanumeric characters, must be related to the viewing distance (see also Chapter 12). Various figures have been quoted for minimum character size but when referring to laboratory research, care must be taken to ensure that an environmental correction factor is added. This must allow for poor lighting conditions, vibration, sometimes non-optimum viewing angles, human fatigue and the full age range of the users. A figure often used for 100% legibility is 0.007 radians or approximately 7 mm at one metre distance (Smith, 1979). While many displays use character sizes less than this, it must be recognised that in less favourable conditions some people may not be able to read them with complete reliability.

Another basic division of indicator types is between round and longitudinal (vertical or horizontal) scales. Here different criteria are relevant to dynamic flight displays such as airspeed and altimeter, and system status displays such as engine instruments. The earliest United States Air Force aircraft to use vertical scale flight instruments was the C141, introduced into service in 1964. This aircraft also used vertical scales for the display of engine parameters. The C141 was followed five years later by the C5. These were, of course, all electromechanical displays. Civil aviation chose not to follow this lead. However, vertical scales are now often an integral part of CRT primary flight displays, notably for altitude, airspeed and vertical speed.

With respect to engine instruments, a number of airlines installed vertical scale indicators following a study conducted at the Netherlands Aerospace Laboratories (Lamers, 1968). This study demonstrated their superiority over conventional round dials for detection of an engine fault as well as speed and accuracy of power setting. In CRT displays, engine parameters can appear in either round or vertical formats.

Cathode-ray tube (CRT) displays

The introduction of CRT technology for the display of flight systems information represented a milestone in the evolution of the flight deck. The so-called glass cockpit provided a release from the many constraints of earlier electromechanical displays, permitted integration of displays, a more effective utilisation of high priority panel space and greater flexibility.

The CRT was first used as a laboratory oscilloscope more than half a century ago and it was in such a form that it was first installed in aircraft. Conception as an airborne pictorial display of information came much later.

Valuable development was carried out in the 1970s by British Aerospace and Smiths Industries at Weybridge, England, on what was known as the AFD or Advanced Flight Deck (Wilson *et al.*, 1979). Many Human Factors problems were resolved while at the same time the technology of producing a satisfactory colour CRT was progressing, particularly in Japan. The KSSU (KLM/SAS/SWR/UTA) group of airlines was already studying layouts for a CRT flight instrument panel during their DC10/L1011 evaluation studies in the 1960s, though this was finally abandoned due to development time-scale problems.

Applications

It was in the early 1980s that the all-digital A310 and Boeing 757/767 introduced CRT flight displays in civil aviation and this marked the watershed in the evolution of the glass cockpit. While the technology employed in the displays was not significantly different, conceptually the A310 and A300-600 displays were more advanced than those of the Boeing aircraft. Boeing used an Electronic Attitude Director Indicator (EADI), the display details of which were similar to those of the electromechanical ADI which it replaced. On the other hand, Airbus took advantage of the research on the Weybridge AFD and elected to introduce a Primary Flight Display (PFD) which incorporated the main airspeed indication, selected altitude and deviation, full flight mode annunciation and various other information. The A310 flight director was conventional but could, on selection, be changed to a flight path vector display.

There was little difference between the CRT Navigation Display (ND) on these aircraft, providing a map mode, a reproduction of the conventional HSI and super-imposition of weather radar. However, the concept of system display was quite different. The Airbus Electronic Centralised Aircraft Monitoring (ECAM) system employed two CRTs, one for warning displays and the other for systems. The systems display automatically related to phase of flight but had to be manually selected. Airbus continued to use conventional round engine indicators. The Engine Indicating and Crew Alerting System (EICAS) of Boeing used the upper of two CRTs for primary engine parameters with secondary information on the lower screen. No system diagrams or guidance on corrective actions were displayed on these CRTs, as were available on the Airbus ECAM.

The use of CRTs for flight instrument displays is just one of three flight deck functions for the application of this technology. A second is for the display of systems information. This involves engine data as well as other aircraft systems. The flexibility of this time-sharing form of display enables systems information to be presented only when required, either because of the phase of operation, such as engine starting, or when a system deviates from its normal operating range. This function includes use as part of the warning system.

The third use of CRTs on the flight deck is for the flight management system (FMS). These systems are increasingly being installed, particularly to optimise operating efficiency with a primary aim of reducing fuel consumption. The FMS interfaces with the navigation system. Introduction of the digital FMS is easier on the new all digital aircraft; retrofit on earlier analogue aircraft such as the Boeing 747 is expensive but in certain cases it could prove cost-effective. In principle, a distinction should be made between a FMS retrofitted as a fuel-saving device in aircraft already in service, and the more complex models designed into new aircraft. The latter can typically ensure reduced workload, compile complicated lateral and vertical profiles and supply data for the electronic flight guidance system.

Problems

As with all new display technologies the CRT posed its own set of Human Factors problems. Brightness and brightness contrast have been the subject of study as have the relative merits of monochrome and colour displays. Much has been said about the possible fatiguing effect of long periods of CRT monitoring, particularly in the ATC environment. Much attention has been given to symbology, though in early installations this simply mirrored the old electromechanical displays. Software aspects of switching and time-sharing have perhaps involved the greatest amount of Human Factors development time – what should appear where and when on the screens.

Research has shown that reading or checking text from a CRT is slower than from printed paper but the reasons for this are not clear (Gould *et al.*, 1984; Wright *et al.*, 1983).

The CRT is relatively cheap, is versatile and can display fully shaded pictures. However, it is heavy, bulky, and has a high power consumption. Alternatives to the CRT are LCDs, electroluminescence and plasma panels. Of these, LCDs are proving most promising but are more expensive and have their own set of Human Factors problems (see discussion on flat panel displays on page 252).

Office use of VDTs increased rapidly as microelectronics opened up new vistas in data storage and presentation. This expansion was accompanied by various complaints from operators of the equipment, including visual discomfort and fatigue, blurred vision, headaches, nausea, cataracts, and muscular disorders. Glare, lighting and flicker were often cited as being sources of discomfort or annoyance. Although the physiological symptoms involved in these complaints remained largely unexplained, several countries and international bodies have taken them seriously and have tried to establish certain standards to provide protection for the operators.

The International Organisation for Standardisation (ISO), the International Labour Organisation (ILO) and the World Health Organisation (WHO) all became involved in setting standards and input was also made by organisations such as the Human Factors Society. Germany took an early lead in Europe in establishing extensive VDU (Visual Display Unit) requirements. In France VDU work was classified as hazardous, which entitled workers to special benefits and exposure time limitations. In the USA several states introduced legislation to provide protection.

Certain visual stimuli, such as particular striped patterns, have long been known to be unpleasant to look at and sometimes to create visual illusions though, as mentioned in Chapter 5, the physiological mechanisms underlying such illusions are speculative. For an individual suffering from light-sensitive epilepsy, the CRT display of a television set can cause a seizure. It is thought that frequencies even above the CFF may have an influence on viewing comfort and readability of text.

Lighting of VDUs may give rise to difficulties; the task of using a CRT has been compared with simultaneously reading a book and looking at a film. CRTs have typically used light characters on a dark background because the reverse provided inadequate resolution for continuous use. Designers have generally elected to use more restful colours such as green or phosphor bronze on a black or grey screen. Increasing contrast on the screen improves the readability of the text but may also increase visual discomfort.

The research and concern over VDTs (Visual Display Terminals) has been related primarily to the office type of environment; that of the aircraft flight deck is quite different but warrants no less concern and study. VDUs have been used for many years for the airborne display of alphanumeric text of flight plans in navigation systems and have an essential role in data-link systems. They are also used for the text of checklists, though as with other flight deck applications their use is more intermittent than often in the office. On the other hand, the environment is sometimes more hostile with extremes of brightness, an unstable platform, vibrations and probably a less than optimum location.

The general concern over health and efficiency aspects of VDTs is reflected in the literature (e.g. Grandjean *et al.*, 1983; Koffler Group, 1985; Sandelin *et al.*, 1984;

Wilkins, 1985). Advanced techniques for the control of CRTs are mentioned later in the chapter.

Flat panel displays

One of the newest technologies available is the LCD 'flat panel' display. Flat panel displays will be inaugurated in the Boeing 777. They are used in place of CRTs and offer considerable economies in weight and space, require considerably less cooling, and are thought to significantly increase reliability, although the latter test will come when flat panels are used routinely in line service.

Flat panels utilise an entirely new technology and are not simply an improved CRT. Not surprisingly engineers and Human Factors experts are still working out some of its problems. For example, off-angle viewing is still a difficult issue and there are others.

Head-up displays (HUD)

The development of the HUD provides a useful illustration of the interaction between technological progress, economics and safety and also provides a basis for discussion of fail-passive and fail-operational concepts in flight guidance systems. We shall devote more space, then, to the subject than might otherwise have been justified.

The HUD in civil aviation was conceived with the objective of easing the transition from head-down flying on an instrument approach to head-up flying for the visual landing. A study made in 1978 by Questek for the FAA covering flights into US airports from 1964 to 1975, revealed that although only 2% of the landings were made in weather conditions worse than 200 m (600 ft) cloudbase and 2.4 km (1½ miles) visibility, these landings produced more than half of all the fatal accidents (Graham, 1978). In another study of accidents during approach and landing for the period 1970–1975, the NTSB concluded that 'almost every mishap occurred after the flight crew had seen either the ground, the airport, or the runway environment and was trying to transition from instrument to visual flight procedures' (NTSB-AAS-76-5).

The Air Line Pilots Association (ALPA) in the USA, supported by many other pilots, felt that the solution lay not in simply improving the quality of head-down displays. What was needed was an increase in the information available when finally searching for visual landing cues, head-up (DeCelles *et al.*, 1979).

The HUD programme has been conducted within a continuous philosophical debate on its place within low visibility operations. Much early British work on automatic landings was based on the concept that in very low visibility, intervention by a human pilot into the control loop during landing could only increase the hazard; the automatics must therefore be fail-operational. It was also said in support that if the pilot were to intervene, the information given to him to do so must be at least as reliable as that being supplied to the automatic systems; not always an easy thing to do.

The cost of such automatic systems is high and their utilisation is relatively low. In addition, there were many pilots who, conditioned by personal experience of questionable avionic reliability, did not feel that the pilot should be taken out of the loop. This view was particularly strongly held by pilots in the USA, by operators of smaller transport aircraft and by those whose network was not frequently exposed to low visibility weather conditions.

The HUD protagonists further developed their case. Not only would it have an application for low visibility instrument/visual transition, it was stated, but it could be used in most other phases of flight to enhance safety. Take-off, climb, cruise, good visibility approaches, automatic landing monitoring, roll-out guidance and windshear protection were all cited as applications for the HUD.

Certification

Stimulus was given to the programme in France by Air Inter which had been operating in Category III weather (see Appendix 1.13) since 1967, though with an airborne equipment package which was not universally accepted for such operations. Air Inter specified a HUD in its Mercure aircraft, which received certification in 1974. Further impetus was given by the FAA in 1976 in initiating an experimental programme together with NASA-Ames Research Center. Already by 1978 it was possible for the FAA to publish a bibliography of nearly 300 reports and studies on HUD (FAA-RD-78-31).

Douglas completed certification for the DC9-80 Sundstrand HUD in 1981 and this was initially introduced into airline use by Swissair. The A300 of Air Inter had already used a partial HUD system for a number of years. This was called a Windshield Guidance Display (WGD) and was produced in France by SFENA. It also allowed a reduction in take-off visibility limit to 100 m by providing a simple roll-out guidance display. This system requires the windshield to be coated. The coating absorbs light transmission through the windscreen and is continuously visible, an aspect of the system which some pilots find disturbing. Sundstrand developed a Visual Approach Monitor (VAM) which went into service in 1976. These developments were all seen as having limited objectives and not meeting all the requirements of the HUD advocates. Recognised standards had not yet established the role which the HUD could play in Category III certification.

In 1984 the FAA published an Advisory Circular (AC-120-28C) which detailed means for obtaining approval for Category III operations. This was based on the concept that the primary mode of Category III operations was by automatic landing without pilot intervention. However, it required that pilot intervention should be anticipated if the pilot suspected inadequate aircraft performance or if the automatic landing could not be accomplished within the touch-down zone. The concept did not preclude Category III operations with the pilot actively in the control loop if the systems installed allowed him to do this safely. This Advisory Circular officially opened the door to FAA-approved manual Category III operation, based on the use of a HUD system. The first system to benefit from this new policy was the Flight Dynamics holographic system certified on a Boeing 727 in 1984.

The system certified consisted of a CRT mounted over the pilot's head, the display being projected onto a combining glass located about 34 cm ahead of the pilot's eyes. The combiner consisted of a gelatin sheet sandwiched between two sheets of glass and allowed about 90% of outside light to pass through. As a CRT is an integral part of the Flight Dynamics system, all the Human Factors considerations related to CRT displays are also relevant for this form of HUD – symbology, brightness contrast and so on.

In addition, psychophysiological Human Factors questions are generated as a direct result of superimposing a synthetic display on to the real-world picture. What field of view, horizontally and vertically, is required? Should the display provide simply raw data such as displacement from flight path, or should it display command information? And what selection possibilities should be given to the pilot? How is the threshold of unacceptable deviation displayed and how reliable is manual take-over close to the ground? How is the hardware located ahead of the pilot's eyes to be delethalised to provide impact protection? How much light transmission loss through the combiner or windshield coating can be accepted? Perhaps most work has been done on the symbology to be used with less accomplished in other areas (Fig. 11.2).

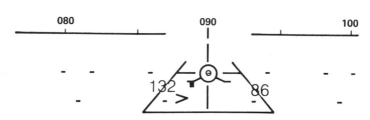

Fig. 11.2 An example of symbology for a HUD developed by NASA, showing the flight path symbol array with an indicated airspeed of 132 kt and a height of 86 ft. The aircraft symbol is at the top. The aircraft is on the flight path about 300 m from the runway threshold (NASA Technical Memo 81199).

Fail-passive and fail-operational

Reference should be made here to what are called fail-passive and fail-operational flight control systems, as these concepts have significant Human Factors connotations. A fail-passive system is one where in the event of a failure, no significant aircraft deviation occurs, no out-of-trim condition remains and the pilot is presented with no control problems which are not readily apparent. In other words, the automatic pilot simply hands over the aircraft in a steady condition to the human pilot. A fail-operational landing system, on the other hand, provides sufficient redundancy that in the event of a failure an operational capability still remains to touchdown and, if applicable, through roll-out. As already noted, the 1984 FAA Advisory Circular permits this redundancy to be automatic, manual or a combination of the two.

A fail-operational landing system was formerly considered to exist where one automatic system took over from another in the event of a failure. The extent to which a human pilot can handle this take-over safely and reliably has been a controversial issue and remains so. In any case, it is widely believed that a fail-passive aircraft landing in conditions below Category II should not be authorised without at least a HUD to assist the pilot in his role as the standby, or take-over system.

The original objective of ALPA in pursuing the development of the HUD was primarily to raise the level of safety when operating within current landing weather limits. The primary objective of airlines in installing a HUD is to improve schedule regularity and reduce the cost of diversions by permitting landing in lower visibility than otherwise would be permitted. These objectives are obviously quite different.

There is little doubt that most pilots using a HUD installation feel that it contributes to making a successful transition from instrument to visual flight on the final part of the approach. And that making this transition, in given visibility conditions, the operation is safer with than without a HUD. However, the original safety objective will only be achieved if the contribution to safety resulting from the HUD is greater than the contribution to risk resulting from any HUD-related lowering of operating weather limits. This cannot be simply demonstrated by 1000 simulator or 100 aircraft landings as required for certification, or by an Advisory Circular from a regulatory authority.

Warning, alerting and advisory systems

Perhaps the flight deck display system which has been expanding and developing faster than any other and involving Human Factors in depth is the warning, advisory and alerting system. From little more than a fire bell and a few lights this has now grown rapidly. Even within the jet era, warnings increased from 172 on the DC8 to 418 on the DC10. And from 188 on the Boeing 707 to 455 on the 747. In addition to the visual signals, a cacophony of aural signals were incorporated including a bell, clacker, buzzer, wailer, tone, horn, intermittent horn, chime, intermittent chimes and, later, a synthesised voice. These were to draw attention to situations such as fire, take-off configuration, landing configuration, stabiliser trim, overspeed, altitude, autopilot disconnect, evacuation, ground proximity, decision height and cabin call.

Regardless of what can be learned in a classroom, it is quite unrealistic to expect the warning sound to be identified at once in the rare event of an operational abnormality. It will be identified perhaps more from the phase of flight in which it occurs than from the nature of the sound, *per se*. This conclusion is recognised in the DC10 design philosophy where the same sound is used for take-off and cabin pressure warnings. The proliferation of warnings was based on the notion that the problem was mainly a perceptual one – that more information was needed to be made available. However, the difficulty is now seen as more cognitive in origin, involving attention, learning, memory and understanding. Greater emphasis is thus being given to electronic processing of information and guidance upon the appropriate action.

The Boeing 757 and 767 aircraft warning systems evolved from that of the 737, which also incorporated a Master Warning and Caution feature. However, the use of CRT displays enabled data from a particular engine parameter to be highlighted on the screen automatically if it became out of tolerance.

SAE published in 1980 a recommendation for an integrated warning system (ARP 450D) which addressed the problem of the excessive number of discrete sounds. It introduced the general concept of an attention-getting sound (*attenson*) supplemented by voice alert messages and discrete visual supplementary information located on a centralised panel. It strongly recommended against the use of several discrete aural alerts. The attenson was intended for use with emergency, abnormal and advisory conditions of alert.

The 757 and 767 both adopted as a basic configuration this general concept in principle, though retaining the traditional bell for fires. A two-tone siren was used for unsafe situations and a new sound altogether to cover all cautions. The conventional stick-shaker was retained for stall warning and the new standard voice warning used for the ground proximity warning system (GPWS). The A310 also made use of the CRT in its warning system.

In the design of all flight deck warning systems three fundamental principles apply. They should alert the crew and call for their attention, report the nature of the condition and guide them in the appropriate corrective action (Fig. 11.3).

Fig. 11.3 The three basic objectives of a warning system (cartoon: J H Band).

Many systems are particularly deficient in the last of these areas. There is a long history of aircraft crashing when, for example, the crew have closed down the wrong engine after an engine failure. They were properly alerted, probably well-informed, but were given no direct guidance to the control which needed actuation to manage the situation. Many aircraft now provide the guidance function in this particular situation by having the fire warning light also appear in the appropriate fuel lever which must be closed.

There are a number of Human Factors considerations which the system designer will need to consider. The first is that system reliability is paramount because credibility will be lost if a crew member has come to expect false warnings. This has been confirmed in findings of investigations into aircraft accidents and incidents. It has been estimated by ALPA that some 65% of the GPWS warnings before 1982 were unnecessary and greatly reduced the credibility of the system. Even when actuation of the caution or warning is technically (though not operationally) justified, excessive appearance of an alerting signal will reduce response to it and also create a nuisance. In other words, it should not appear in normal operation. The interpretation of multiple aural warnings requires learning and it cannot be expected, as already noted, that this learning will be retained adequately to ensure immediate response, without voice addition. Such a multiplicity of discrete sounds attracts attention but may generate an error or delay in corrective action. The temptation in design to avoid additional system complexity by introducing a warning, calling for manual intervention, must be avoided whenever possible.

Classification

Alerts and warnings on the flight deck can broadly be divided into four functional classes. First there are those related to performance or departures from safe flight profiles. This group would include stall, overspeed and ground proximity. These are usually of high urgency. A second class includes those related to aircraft configuration such as landing gear or flap position warnings. Third there are those associated with the status of aircraft systems, and this group should include limiting bands and flags on instruments as well as lights and aural alerts. The fourth and least urgent category is that related to communications, such as SELCAL and the interphone system.

In grouping these signals in degrees of urgency, the highest priority reflects an emergency condition, the term 'warning' is used and the colour coding is red. This implies that immediate action is required for the safety of the aircraft. A second level of priority implies an abnormal condition which could become an emergency if neglected. Usually the term 'caution' is used here and the colour code is amber. A third level of urgency involves a condition where advice is needed and where some crew action may become necessary. A final and low urgency level is associated with alerts which provide information but where no special crew action is demanded. The colour used in these last two groups is not very specific but is often blue, white or green (SAE ARP 450).

All the colours used have been established by tradition rather than research. For example, in Chapter 5 it was noted that the eye is most sensitive to light in the middle of the spectrum. Yet we traditionally use green which is in this area for a safe condition, and red in the least sensitive area for warning and distress signals.

Synthesised voice

Considerable thought has been given to the use of synthesised voice in flight deck warning systems. The B58 Hustler supersonic bomber of the late 1950s used recorded voice warnings and attracted much attention. During the 1960s much effort was made to develop synthesised voice to an acceptable quality and one of its early applications was in the GPWS, which became mandatory in civil aircraft by the mid-1970s.

The advantages of voice communication warning systems are, firstly, that like other aural signals, they are omnidirectional and so attract attention wherever the crew member's interest happens to be focused. Secondly, the message also contains the information content formerly only available in visual displays. The disadvantages are that the message can easily be interfered with, as any other voice message, and can itself interfere with other communication activity. And, as with other aural warning systems, a programme of priorities must be established so that only one message is presented at a time. Certain technical problems such as the quality of the synthesised voice have now been largely overcome.

Reliability

In Chapter 2 we discussed human reliability and the human failure rates which can be expected. In this chapter we should mention system reliability and the consequence of technical failure of displayed information on human performance and behaviour.

Calculation

It is sometimes necessary to make an estimate of the reliability of a system. This calculation depends on whether the components of the system are arranged in series or parallel. For example, we might say in simple terms that a particular system consists of three components, the pilot, the airborne display and the ground station transmitting the displayed information. The total reliability of this system, which is arranged in series, would then be the product of the three individual reliabilities. The actual figures and components used in the following example are purely hypothetical and are not intended to be realistic or representative. We could then say that the overall system reliability would be:

0.90 (pilot) x 0.92 (display) x 0.95 (ground transmitter) = 0.7866 or nearly 79%.

On the other hand, if the components are arranged in parallel as in systems incorporating redundancy or back-up components, the overall reliability would be calculated quite differently. Let us say that there are two pilots who could be seen (unrealistically) as quite independent components performing the same function and each with a reliability of 0.90, then this combined reliability could be calculated as $1 - (0.10 \times 0.10) = 0.99$. We can also assume that there are three ground transmissions each functioning independently and each with a reliability of 0.95, as before. The reliability of the ground transmission is then $1 - (0.05 \times 0.05 \times 0.05) = 0.999875$.

We can now recalculate the reliability of the original system which was arranged in series but which now has two pilots in parallel, and three ground transmitters, also in parallel, and a single display. The calculation would then look like this:

0.99 (pilots) x 0.92 (display) x 0.999875 (ground transmitters) = 0.910686 or about 91% instead of the original 79%.

Monitoring of the task of one person by another or one piece of equipment by another introduces a parallel element into the system which increases its overall reliability.

Consequences

The direct consequence of a technical failure is that the pilot may be presented with unreliable information or possibly no information at all. The source of the failure could be in either the airborne or the ground-based equipment and this can often be determined easily by checking with other airborne displays using the same information source.

Whether the origin of the fault is in the air or on the ground, it is essential that the user of the display is not presented with unreliable information in a manner that it can be used. It is not sufficient for a remotely or even adjacently located annunciator to advise that the display is unreliable. The display itself must annunciate this by, for example, removing the unreliable information by biasing or shuttering it out of sight. So long as the information can be seen, then it can be predicted that sooner or later it will be used. The 1961 Comet crash at Ankara and the 1978 Boeing 747 disaster at Bombay both resulted from the pilot following invalid guidance information which remained on display.

The most insidious situation is when the information is still available and on display but is unreliable. This represents a challenge to the designer, using techniques such as electronic self-monitoring and comparators, to ensure that the user is not misled. In certain cases a technical failure can cause a change in the information presented by a display which otherwise appears to be normal. If one segment of a seven-segment numeric display of the figure 8 were to fail, the result could appear as a 0, 6 or 9 (Fig. 11.4).

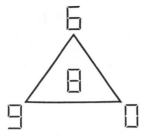

Fig.11.4 If one segment fails in a 7 segment display, a figure 8 can appear as a 0, 6 or 9.

Apart from the direct impact of display unreliability on performance, it can condition the user to behave and react in an undesirable manner, with an adverse influence on flight safety. A display which is known to be unreliable will generate a loss of confidence on the part of the user who will then employ other, perhaps even less reliable, information sources. The majority of GPWS alerts with early installations were false or unnecessary alarms. It was predictable in such circumstances that pilots would be reluctant to take the drastic emergency action necessary when the warning display actually spoke the truth.

Another situation which generates an undesired response, or rather no response at all, is where habituation occurs as a result of a warning display being frequently activated in conditions which require no corrective action. This may result merely from an inappropriate setting of the warning signal threshold or another design characteristic. This reflects not a technical failure but rather a system design fault. The unreliability is in the design of the display rather than in its technical functioning.

Unreliability in displays in flight simulators and other training devices introduces a further phenomenon, negative training transfer. This was discussed in Chapter 9 and so no further reference need be made to it here.

One additional consequence of inadequate display or system reliability is that it sometimes becomes necessary to devise procedures to provide protection. One such case relates to airspeed indication, where verbal cross-checks have been used by many operators in the absence of electronic monitoring, to detect any discrepancy between the two airspeed indicators. This procedure has been applied particularly in the take-off phase of flight. Many years ago, a special procedure was introduced on aircraft using a certain type of dual fire detection system to obtain confirmation of the validity of the warning before taking remedial action. This resulted from the frequency of false warnings which were experienced. The development of procedures to avoid the consequences of inadequate technical reliability introduces another reliability dimension, that of the human operator carrying out the procedure.

We have looked at different consequences of unreliability of displays. The most obvious one is the loss or degradation of information, but there are others. These include the changed human behaviour which can arise from being required to utilise displays with inadequate reliability and the unfortunate need to create additional procedures, vulnerable to human reliability limitations, and the need to provide operational protection. The argument for high display reliability is well founded.

Controls

Functions and types

A control is a means of transmitting a message from the operator to some device or system. The purpose may be to transmit discrete information, such as in selection of a transponder code or an on/off command to a no-smoking sign. Alternatively, it could be to transmit continuous information as in a cabin temperature selector or in the setting of the command bars of a flight director.

The purpose may be simply to control a display directly as with the altimeter sub-scale knob. Or it may be to give a signal to a system, such as with the throttles, which in turn will subsequently be reflected in changes in various displayed parameters.

These functional requirements will influence the type of control device which is used. In addition, selection of the device will be influenced by the manipulation force required. Generally, electrical systems require a low control force, while mechanical linkages using rods and cables require a higher force. Hydraulic systems may require something in between but this will vary with system design. A landing gear actuator controlled by a mechanical linkage would call for quite a different device from one where the means of control were electrical.

Using the two criteria of function and force, we could say that for discrete functions when the forces are low, push buttons, toggle switches and rotary switches are suitable. When manipulation forces are high, then detented levers (e.g. flap lever) or large hand- or foot-operated push buttons would be suitable. For continuous function controls with low forces, rotary knobs, thumb wheels (e.g. autopilot pitch control) and small levers or cranks are suitable. For high forces, handwheels and large levers (e.g. the control column and wheel) as well as large cranks and foot pedals are most effective.

Sight and reach

In the following chapter further discussion is devoted to the layout of controls. One aspect which should be repeated here has notably influenced the basic design of displays. Early instruments were always supplied together with their controls in a single package. This was perhaps acceptable when performance on board was not seen as being very critical. However, the optimum location for installation of a display for sight, is frequently not the optimum location for its control, for reach. The development of the glareshield for flight guidance control, on the initiative of KLM in The Netherlands, provided a near optimum location for many controls but required manufacturers to provide displays with the servo-mechanisms for remote control (Hawkins, 1966).

Design principles

A primary Human Factors requirement with regard to controls is their location but there are several others which must be applied and there are good publications available to guide the designer (e.g. McCormick *et al.*, 1983; Van Cott *et al.*, 1972). We should look at five of these requirements.

Control-display ratio is the distance of movement of the control related to that of the moving element of the display such as the read-out or the pointer. It can be seen as the sensitivity of the control. It is often a critical design factor affecting operator performance. Outside aviation, poor control-display ratio in appliances such as radios is very common. The concept of control-display ratio may not always be applicable in the case of complex and indirect relationships between a control and a displayed parameter.

Direction of movement of the control relative to the display is another aspect requiring Human Factors attention. Human performance can be improved by assuring correct design in accordance with human expectation and physiology. The matter is not always as simple as it may seem. Normal expectation is that a control knob should rotate clockwise to increase the value and vice versa. However, it has also been demonstrated with vertical scales that when a knob is adjacent to the display which it controls, then the indication is expected to move in the same direction as the side of the knob closest to it (Fig. 11.5). This is known as Warrick's Principle. With horizontal scales, different stereotypes emerge. A good designer will try to avoid conflicting expectations of this kind.

Control resistance is another important element in design. There are four kinds of resistance which can affect the speed and precision of control operation, the 'feel' of the control, smoothness of control movement and the susceptibility of the control to inadvertent operation. These are called elastic, static and sliding friction, viscous damping and inertia.

The purpose of control coding is to improve identification and so reduce errors and time taken in selection. Coding may be by means of shape, size, colour, labelling and location. Much work has been done on shape coding of controls and recommendations are available (e.g. McCormick *et al.*, 1983). With white cockpit lighting, colour coding is now a practical method; a good example is the red fire control or emergency door control handle.

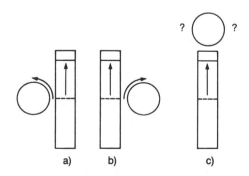

a)   b)   c)

Fig. 11.5 An illustration of Warrick's Principle. To move the indicator in an upward direction this principle rules:
a) anti-clockwise rotation of the control
b) clockwise rotation
c) no clear application of the principle is possible
(Brebner *et al.*, 1976).

The last of the five primary principles of control design which we should mention involves protection against inadvertent actuation. It has already been mentioned that proper control resistance will help to avoid accidental operation but there are also other methods. Gating is used frequently in the design of the flap handle. The landing gear lever has a locking device which may be both electrical and mechanical. Switches can be lever-locked or guarded or they can be recessed or placed out of the way. Controls can be interconnected to guarantee an operational sequence – the reverse thrust levers may not be operable until the power levers are in the idle position. In some cases, inadvertent operation may not be blocked but advice of the incompatible action will be given by means of an aural or visual warning. This applies to the closing of power levers when the landing gear is still retracted. This kind of protection applies not only to inadvertent operation such as knocking off a switch or catching the flap lever with the jacket sleeve. It also provides some protection against inappropriate action, that is, operating a control deliberately but at the wrong time. The gating of a lever or the guarding of a switch calls for additional movement and thought which provide some time for reflection and retreat from the intended action.

Keyboards

The use of keyboards on the flight deck has been steadily increasing and reflects the requirement to give instructions to computerised systems. The introduction of inertial and area navigation systems accelerated the use of more extensive keyboards, and the installation of flight management and data-link systems expand their use further.

Push-button controls such as on audio-selector or autopilot panels have been around for a long time. But their operation involved generally one or two discrete actions and not a 'conversation' with the equipment as is now required. The CRT touch screen and automatic speech recognition systems permit other means of inserting data into a machine, but the keyboard is likely to remain the main channel of communication for some time.

An experienced operator in a day's work may press as many as 80 000 keys on a keyboard with a layout similar to that of a typewriter, with an uncorrected error occurring once in about 2000–4000 keyings. The occasional, inexperienced operator will not do so well. Not only will his keying rate be far less, but undetected errors may average one in 200 keyings in a favourable environment but with a very wide variation between operators (Klemmer, 1971). In flight deck application we should classify the crew member as an unskilled or inexperienced keyboard operator when assessing the kind of error rate which can be expected. The information which is inserted into flight systems may be of critical importance, as illustrated by the potentially catastrophic consequences of navigation errors in certain parts of the world. The Korean civil aircraft destroyed by the USSR in 1983 was one such example. Application of the two-pronged attack on errors described in Chapter 2 is essential in keyboard design and use, to minimise the occurrence of errors and reduce their consequences.

The most familiar keyboard is that of the typewriter. This has what is known as a QWERTY layout, after the first six letters of the top letter row. It was developed by Christopher Sholes in 1873. Because of the relatively primitive design of the machines at that time, which jammed if operated too fast, the design objective was certainly not to enhance the speed of typing. It is an inefficient layout with individual finger loading varying from 1% up to 22% and with nearly 60% of the work being done by the left hand.

By the 1930s, machines had evolved so much that far higher typing speeds could be used but this was prevented by the ergonomically poor keyboard layout. August Dvorak, a psychologist at the University of Washington, then applied ergonomics to produce what has become known as the DSK or Dvorak Simplified Keyboard (Fig. 11.6). This was specifically designed for speed and efficiency of operation, putting all five vowels and the most common consonants in the centre row. This centre row, directly under the fingers, can produce about 3000 English words on the Dvorak layout compared with only about 100 on the QWERTY version in the same amount of time. The balance of work has been shifted to the right hand and individual finger loading better distributed from 8% up to 18% of the total work (Dvorak, 1943).

Yet in spite of the claims for improved efficiency of this new keyboard for something like half a century, there has been a reluctance to change. However, the advent of the microchip now allows rapid conversion of one layout to the other.

Dvorak keyboard

QWERTY keyboard

Fig. 11.6 The traditional QWERTY keyboard layout compared with the more efficient Dvorak version.

Furthermore, the increasing use of keyboards for data entry rather than only typing has meant that many operators are not trained typists and so can start from scratch using the more efficient layout. While Dvorak, who died in 1975, did not live to see the acceptance of his application of Human Factors to keyboards, there is now increasing interest being shown and some computer manufacturers are incorporating a Dvorak conversion capability in their products or offering low-cost conversion facilities. A good survey of comparative studies of the QWERTY and DSK versions, some contradictory and some of doubtful experimental design, as well as of other aspects of keyboard design, is available elsewhere (Koffler Group, 1986). There have been more than twenty serious proposals to rearrange the QWERTY keyboard during the last 100 years; the DSK is the best-known of these.

Of course, for on-board use, speed of entry is not the most important criterion in design; accuracy is far more significant. And other factors also play an important role, such as use in turbulent conditions and currently a less than optimum panel location relative to the operator. Furthermore, keying operation on board may utilise one rather than two hands on systems currently visualised. But the QWERTY/ Dvorak controversy illustrates how standardisation can delay progress in the application of Human Factors to improve the efficiency of man-machine systems, even when the advantages have been well demonstrated.

Where numeric keys are to be used frequently these are better placed in a separate block than typewriter-style across the top of the keyboard. And when in a block, the layout standardised by SAE with 1, 2, 3 at the top can be operated faster and with fewer errors than the calculator layout with 7, 8, 9 at the top (SAE ARP 571).

The detection of errors in entry is of vital importance. This can be enhanced through feedback via the visual, aural and tactual senses. The key size must be optimised as far as possible with adequate space and possibly fences between keys to prevent inadvertent operation, particularly when keys are small. Most alphanumeric keyboards for use outside the flight deck have keys and rows of keys spaced by just under 2 cm. The confined space of aircraft panels has resulted in a departure from this standard and consequently an increase in the risk of error. It may be necessary to install a special rest to steady the hand when keying in turbulent conditions – the flight deck is often a less stable platform than a computer operator's desk in the office.

Future flight deck design development may give the keyboard a more prominent role and a more favourable location. This becomes possible with the introduction of the side-stick controller and is illustrated in the so-called 'Pilot's Desk Flight Station', a concept developed by the Lockheed-Georgia Company and NASA (Sexton, 1983). This looks more like a computer operator's console rather than a conventional pilot's flight station. The Airbus A320 utilises both a side-stick controller and what is essentially a fixed position throttle that requires only pressure to modify engine output. While these features (especially the loss of cues from having a throttle that moves traditionally) have proved quite controversial, it must be admitted that most pilots who fly these aircraft, like them very much.

As keyboards are becoming the main source of data entry into flight deck automated systems, and likely to remain so, and as their design has a major impact on the accuracy with which they are used, proper Human Factors input at the design stage is essential.

Automation

There has been some confusion in the application of automation to flight deck activities. It has sometimes been mistakenly seen as an end in itself – so long as something can be automated, let's do it. This has led to consequences which were a surprise to some but which, in the light of long experience of factory automation of production, could have been anticipated. One symptom has been called the 'automatic complacency' (Ternham, 1978). It has been frequently cited in articles in aviation journals and sometimes in accident investigation reports. One such case was at JFK Airport in New York in 1984 when a DC10 overran the runway causing substantial damage. Excessive reliance on automation was noted in the accident report (NTSB-AAR-84-15). The most plausible hypothesis (proposed amongst others by ICAO) explaining the navigation error experienced by the Korean 747 in 1983, which led to it being shot down by the USSR, cites the crew interface with the automated navigation system. After this tragedy NASA ran a search of its ASRS data base and found 21 other cases of Inertial Navigation System (INS) errors previously reported (Wiener, 1985). Other automated systems, such as altitude alerting, also generate a reliance which is sometimes unjustified (ASRS Callback No. 86). No doubt, far more cases outside the USA remain unreported and not included in the ASRS data base. The Mt Erebus DC10 disaster in 1979 was examined and reported by a New Zealand Royal Commission following publication of contentious accident investigation findings. This accident apparently arose from a human error in entering incorrect data into a ground-based computer which supplied the flight plan for the airborne INS. This accident provided the subject of a book which severely criticised the original accident report (AAR-79-139) which had unjustifiably placed all blame on the dead pilot (Vette, 1983). The very reliability with which automated systems normally perform provides the foundation upon which over-confidence and complacency are built.

Numerous studies have addressed the problem of monotony and boredom which result from the under-stimulation associated with automation. Some see this as a source of stress, though evidence does not suggest that this has the same effects on health as the stress induced by over-stimulation, such as cardiovascular disorders and psychoneuroses. Concern about this in connection with the work of ATC staff led the FAA to make a review of the subject (FAA-AM-80-1). It is generally recognised that boredom and monotony in the working environment are negative factors which can affect morale and performance.

Justification

There are three broad objectives which can be cited to justify the introduction of automation into the flight deck, though there is some interaction amongst them. The first of these is related to aircraft and system performance. For many years automatic yaw dampers have been incorporated into flight control systems because the aerodynamic characteristics of the aircraft make manual flying impracticable in terms of fuel cost, passenger comfort and pilot fatigue. Many high performance and experimental aircraft are inherently unstable and effective manual control is simply not attainable. Automation of landings and low-level go-arounds in bad visibility have been developed because those particular tasks cannot be done manually with a sufficiently high level of accuracy and thus safety. Automatic altitude capture and the associated automatic altitude alerting systems have been introduced because with the high rates of climb and descent over long vertical distances, human monitoring is not adequate to assure reliable operation. Man's performance of vigilance and monitoring tasks is generally poor and this was discussed in an earlier chapter. Warning systems must therefore be automated if they are to be effective. A stall warning may activate a stick-shaker and sometimes a stick pusher or slats. So the first justification for automation is that the task is such that it cannot be performed manually with sufficient safety and reliability, if at all. It is outside the normal envelope of human skill required to ensure safe flight.

A second justification for automating a flight deck task corresponds with that of introducing a computer into the flight scheduling office – it is not essential but it permits the job to be done more efficiently. Automatic coupling of an autopilot to an inertial navigation system simply allows the desired track to be flown with greater accuracy, resulting in fuel economy and reduced aircraft separation on airways. Flight management systems (FMS) permit more efficient conduct of the operation and save aircraft time and fuel. It has been estimated that the fuel saving with such a system could amount to 3–5%, depending on the flight length (Wilson *et al.*, 1979), though service experience suggests that this may be optimistic. Aircraft data recording systems permit more cost-effective maintenance. So the second justification is purely one of economics.

A third reason why automation may be justified is to relieve the crew of certain tasks so as to reduce workload. This raises a fundamental question as to what level and type of workload is desirable during the various phases of flight. It raises questions of task priorities, crew complement and distribution of duties. And the perennial problems of how to measure workload and of the relationship of arousal to performance. There are certain phases of flight such as in the terminal area where workload is generally high and at times, particularly with a two-man crew, can reach the limit of operational acceptability. At other times in long-range cruise conditions, workload may fall to such low levels that crew members, working out of phase with their biological rhythms, fall asleep unintentionally (Fig. 11.7). High workload, and the stress associated with it, can influence fatigue and performance. Excessively low workload and low arousal will also influence performance, particularly of vigilance tasks. Automation as a means of crew workload reduction can therefore have a reflection on operational safety as well as job satisfaction. When it is applied to reduce workload to the extent of reducing crew complement, then it also has clear economic connotations.

A great many people have not realised the evolutionary process of automation in aviation. As Billings (1991) has aptly noted, aviation automation, as probably started with Sir Hiram Maxim's 1891 British patent for a gyroscopic stabiliser with a servo control loop. Automation, of one sort or another, has been with us ever since. Even before the advent of our newest aircraft, it was possible to take a regular Boeing 747 from Chicago to Honolulu, and after taking off from Chicago, not touch a single primary aircraft control until turning off the landing runway in Honolulu. The latest aircraft can accomplish the same feat by programming the entire flight (except take-off) before the aircraft leave the gate at Chicago.

Determining categorically that the increasing level of automation found in our latest aircraft is responsible for the generally increased level of safety in the operations

Fig. 11.7 Workload levels can vary extensively in different phases of flight (cartoon: J H Band).

of most airlines is difficult because there are so many variables in the air safety equation. However, there is little question that the increasing level of automation in current transports has made possible both increased efficiency and economy in these aircraft. Quite probably it has also increased safety, particularly when problems associated with training and the development of appropriate procedures have been solved. These aircraft, like all others, operate best if they are operated correctly.

As early as 1983 Wiener had called for a clear understanding of the role of the human operator in the aviation system (Wiener, 1983) and in 1985 lamented the lack of a clear philosophy of automation and the fact that most systems were still being developed 'one box at a time'. Fortunately, things seem a little better now. A full discussion of these and other automation issues is beyond the scope of this chapter, but a full discussion can be found in the occasionally controversial *Human-Centered Aircraft Automation: A Concept and Guidelines* (Billings, 1991a). A condensed version can be found in *Toward a Human-Centered Aircraft Automation Philosophy* (Billings, 1991b).

Human Factors input

It is important to keep in mind that the need for Human Factors input into the design and operation of aircraft systems is not reduced by the introduction of automation. New areas requiring Human Factors attention are created, some more complex than those of the past. The role of the pilot as a monitor, problems of boredom, sharing authority between man and machine, job satisfaction, changing pilot selection criteria and so on, all call for fresh and enlightened study.

In automating tasks – taking them away from man and giving them to the machine – another aspect must be considered. Machines function within a physical environment to which they must adapt. Man functions within a psychological as well as a physical environment. This introduces the question of motivation which is of fundamental significance relative to performance; the machine, it is said, doesn't care if it gets home safely. Man does not function like a machine and the task which is allocated to him must contain the ingredients to provide the challenge and satisfaction necessary to ensure the high performance and sense of responsibility required. While the machine does not require motivation to perform its allocated task, man, in his new managerial role over automated systems, still does.

While automation has made a positive contribution to air safety, concern has been expressed by government agencies, airlines, pilot organisations and scientific bodies at the rapidity with which it has been introduced and the occasional undesirable consequences which have followed it. What appears to be lacking is a well considered rationale or philosophy of automation (Wiener, 1985). It has not been adequately understood that while automation eliminates one potential for human error, it introduces another. The new potential is associated with the human programming, control and monitoring of automated systems. New psychological and behavioural factors are introduced. While the frequency of human errors may be reduced, their consequences may sometimes be increased. It has often been assumed that automation inevitably reduces flight deck workload. But here again one form of work, possibly manual control, has been replaced by another, the management of automated systems, and it cannot simply be assumed that the overall workload has been reduced.

The continued advance of digital technology and the increased use of integrated microprocessors will allow expansion of flight deck automation. It is essential that this expansion takes place within a proper Human Factors framework.

Advanced concepts

The use of voice as a means of communicating with systems was discussed in Chapter 7 though development and application with respect to the flight deck has been slow and is likely to remain slow.

Touch-sensitive CRT screens seem to offer potential for flight deck application, particularly at the systems station, though use in turbulence might give rise to difficulties. Fly-by-wire techniques, for long utilised in military aviation, are likely to be introduced increasingly in future civil aircraft. Already the Concorde is flying with a full authority fly-by-wire control system with an emergency mechanical back-up. Reliability has been demonstrated to be such that in about 100 000 flight hours, no complete reversion to a mechanical mode has been reported in airline service. The A310 is equipped with fully electrical controls with no manual back-up on all the moving surfaces of the wing except the all-speed ailerons. The use of fly-by-wire, that is, electrical rather than mechanical or hydraulic flight control, provides a reduction in overall weight, better control accuracy and improved aircraft handling qualities through increased stability and manoeuvrability. Current fly-by-wire development is associated with the replacement of the traditional control column and wheel by a side-stick controller, a change which can provide considerable benefits in the flight deck.

A NASA study on pilot tasks on the flight deck of airline aircraft referred to indications of high workload leading to decreased performance. This was manifested by errors in the operation of systems and mistakes in navigation. Some of the difficulties, it concluded, resulted from deficiencies in the design of flight decks and instrumentation and it reminded the industry of the importance of applying proper Human Factors input into their design (Ruffell Smith, 1979). With the extensive new technology now becoming available for application to the flight deck, the need for such input is greater than ever.

12

Space and Layout

Design of working and living areas

Matching of components

Matching of working and living areas to human characteristics is one of the primary tasks of those specialised in Human Factors. Some of the basic characteristics of human beings are those associated with size and shape and the movements of various parts of the body. Human Factors is concerned to take proper account of these in order to optimise performance, comfort and safety.

Such data can be applied in several locations on board an aircraft. On the flight deck they are used in the basic geometry, in the provision of adequate inside and outside visibility, in the location and design of controls, in seat design and so on. In the cabin similar basic data are used in the design of galleys, seats, doors, overhead luggage containers and toilets. Maintenance areas take into account human dimensions to assure adequate access to equipment and working space. In cargo compartments such data are applied to provide proper access for loaders and sufficient room to work. And also to ensure that the human forces required for operation of doors, hatches and loading equipment are realistic. The force applied by a loader to close a cargo door was a factor cited in the DC10 Paris accident in 1974 in which the door blew out and the aircraft was destroyed (UK AAR 8/76).

The design of much equipment on board makes use of information about human measurements; life-jackets, life-rafts, emergency exits, oxygen masks, meal trolleys, wash-basins, seat-belts, and so on. In the location of equipment, knowledge of the user's height, reach and sometimes girth is needed to determine accessibility. The design of controls makes assumptions about the size of the hands and the force which can be applied. In some instances handling of equipment has been shown to be impractical for one person. Such a case is the cabin door on certain wide-body transport aircraft, which slides upwards. In the event of a failure of all sources of power in an emergency, a stewardess would need to call the assistance of a strong male colleague or passenger to open the door, which typically might weigh as much as 200 kg. There are sometimes other occasions when a stewardess may need to call for the assistance of a male colleague when the force required to move or operate a piece of equipment is greater than that usually possessed by a female cabin attendant. Some smaller stewardesses also have difficulty in reaching certain switches or controls and need to call for the assistance of taller colleagues. A fundamental concept of ergonomics is that, as far as possible, equipment should be designed to match the characteristics of people, rather than the reverse.

Anthropometry and biomechanics

In order to ensure optimum effectiveness of aircraft equipment, the human dimensions and forces pertinent to its design should not be a matter of guesswork, simply reflecting the subjective feeling of a particular design engineer. In Chapter 1 it was noted that Human Factors draws upon several disciplines in order to carry out its task of matching hardware, software and the environment to the characteristics of man. We are concerned here particularly with two of these disciplines, anthropometry and biomechanics. Anthropometry is concerned with human dimensions such as the size of different limbs, weight, stature (height) and more specific measures, like seated eye height and reach when seated with and without a restraining shoulder harness. From such data it is possible to calculate requirements such as the optimum height of a work surface, location of controls, height and depth of stowage areas such as overhead luggage containers, minimum knee room between seat rows, the width of a seat, the length of an armrest or the height of a headrest. Life-raft and seat cushion design must make assumptions as to the weight of the occupant. When movement of the body is involved such as with reach requirements and working space, then the information source is called dynamic anthropometry.

A new consideration, especially with long-range aircraft, has been the provision of both flight and cabin crew rest bunks (see page 79). This is an important factor in the certification of these aircraft for they cannot be certified for long-range operations without them. A suitable bunk requires valuable space that is no longer available for passengers or cargo. However, in spite of extraordinary technical accomplishments, no manufacturer or state can alter the human disposition and its need for rest. It is unfortunate that there are still significant variations in state provisions regarding required time on-duty and rest facilities, for this is an area where no airline should find a commercial advantage.

The provision of suitable on-board rest facilities is a particular problem with the cabin crew because of the number required in some of these very large aircraft. For example, while one version of a popular, modern long-range transport includes four bunks and four resting seats in a separate space specifically allocated to the cabin crew, there is not yet consensus on their real-world efficiency. This allocated space, of course, is in addition to the provision of separate bunks and space for the flight crew.

Closely related to anthropometry and of equal concern to the Human Factors practitioner, is the discipline of biomechanics. This is concerned not so much with body dimensions as with the movement of parts of the body and the forces which they can apply. It is apparent from such data that it is not sufficient simply to assume that a certain force to operate a control is satisfactory; it will depend on where it is located relative to the body and the direction in which it has to be moved. From a biomechanical viewpoint the location of the landing gear handle on the main instrument panel, as on most large aircraft, is poor. Only limited force can be applied in this position, particularly from across the cockpit, but fortunately only limited force is needed to operate the landing gear in most aircraft.

Human dimensions

Data collection

It might first be useful to examine how this kind of data is collected and then look at the way it is used in a selection of aircraft applications. Various methods can be employed to collect information about human dimensions but to obtain meaningful data, skill and special equipment are required. One such piece of equipment is called the anthropometer and this has been developed to facilitate the measuring of stature, arm reach, sitting eye height and various other body dimensions. Photographic techniques have been developed which require the subject to sit in his normal posture against the background of a grid, the measurements of which are known.

To be of relevance to the design of equipment the data must be collected from a representative sample of the people who are to use the equipment. And the sample must be sufficiently large to cover the realistic range of potential users and to achieve reliability in the estimation of the population parameters. For specialised jobs, human size limitations are sometimes specified as a qualification for the job. One European air force has limits for pilot applicants of 0.865–1.010 m for sitting eye height and also stipulates the range of acceptable leg length and reach. Further restrictions are then later applied for pilot assignment to different aircraft types. One European airline only employs stewardesses between 1.62 m and 1.82 m tall. Some Asian air forces and airlines would not be able to use the same criteria as those used by certain European and American countries and this can create problems, as we shall see in a moment.

A further caution necessary in applying data concerning physical dimensions is that they are slowly changing. Anyone who has used a wash-basin or a mirror or who has bumped his head passing a doorway in an old house will be aware that they are all too low, though once they had suited the population of the time. People have been growing. An Australian population survey some years ago showed a general increase in mean height over half a century of 1.3 mm a year for males and 0.9 mm for females. And this is not simply due to the kangaroo steaks; the trend is universal. Equipment produced for an aircraft could still be in operational use 30 years or more after the initial design criteria were established. And the anthropometric data could have been 20 years old at the time they were used. In using human measurements in design, then, it is necessary to take into account the date when they were collected and the number of years the hardware is expected to be used. The most comprehensive source of such data is a three-volume edition published by NASA and this covers various ethnic groups (NASA, 1978).

Designing for human differences

Not only may the physical dimensions of people within one ethnic group vary from one generation to another, but wide differences can be expected between ethnic groups at any one time (Chapanis, 1975). These differences occur not only in the overall size but also in bodily proportions. Persons of African descent, are relatively long-legged compared with white Caucasians. Asians, on the other hand, are smaller overall but have relatively long trunks and short legs compared with Europeans.

In addition to differences between ethnic groups there are also differences between men and women within one ethnic group (Fig. 12.1). As increasingly men

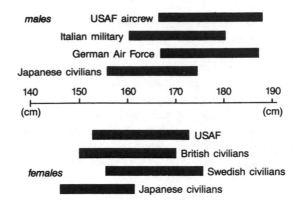

Fig. 12.1 Examples of variations in stature (height) between males and females and different ethnic groups (from NASA, 1978).

and women are doing the same jobs and using the same equipment, this raises certain practical questions which will be illustrated later in this chapter in connection with vision from the cockpit. Already it has been found advisable to ensure that aspiring female pilots can apply adequate rudder force to handle engine-out manoeuvres. Airlines differ in the way this test is done. For example, Air Canada has used a flight simulator while United Airlines has used the gymnasium. There is a relationship between eye position and foot/rudder position and it is necessary to ensure that adequate foot load can be applied while sitting at the design eye position level of seat adjustment. Rudder forces to maintain control at the minimum control speed with an engine failure are required not to exceed 68 kg (150 lb), according to US airworthiness standards (CFR 23.149d & 25.149d). Any applicant for a pilot's multi-engine rating in the USA must demonstrate an ability to control the particular aircraft in this critical engine-out condition. A study by the FAA reported in 1973 that the control force limits then stipulated in CFR 23.143 for general aviation aircraft were too high for the majority of US female pilots (FAA-AM-73-23).

In relating the variations in body dimensions to aircraft and equipment, a problem is immediately apparent. Much of the world's aviation hardware is manufactured in the USA but it is sold and used all over the world. Should the pilot's seat and rudder pedals be optimised for the American, the long-legged African or the short-legged Asian? Should seat cushions in the cabin be optimised for the 76 kg Canadian or the 56 kg Thai? A UK CAA survey published in 1983 revealed that the average weight of passengers varied considerably according to route. Men on routes to Eastern Europe averaged 84.6 kg, some 15 kg heavier than those on flights to Japan. Women to Germany averaged 68.9 kg while those to Japan averaged only 57.2 kg (CAA Paper 83003).

In the USA, regulations currently require that transport category aircraft be designed for operating by crew members of physical stature ranging from 1.57 m (5 ft 2 in) to 1.90 m (6 ft 3 in) (CFR 25.777c). At Boeing, for example, all aircraft from the 747 onwards were designed to meet these criteria. And starting with the 757/767 aircraft, special attention was given to the strength differences between male and female pilots. Numerous studies have been made on the strength differences between men and women. It has been suggested that the judicious designer should avoid making assumptions about such strength differences and should rely on empirical investigation (Pheasant, 1983). It is the view of Boeing that pilot strength requirements for aircraft control, as currently discussed in CFR 24.143(a), are well within the capability of a 1.57 m (5 ft 2 in) female.

Manufacturers may utilise various sources for anthropometric data, including the US military. With respect to ethnic variations, past experience of acceptability has a greater influence on design than the use of specific ethnic anthropometric data sources.

Compromises are clearly necessary and in order to make these, use must inevitably be made of another of the disciplines which support Human Factors, statistics. We should be aware of some of the more important elements of statistics so as to understand how design decisions are made and data interpreted and applied.

Distributions and percentiles

Once anthropometric data have been collected, they must be summarised and interpreted for practical application. In using data, it is not enough, of course, simply to quote the average or mean value of each variable measured. We also need to know something about the range and distribution we are likely to meet in the particular population with which we are concerned. The average wealth of a population is the same whether it is evenly distributed or largely held by just 5% of all the people. The use of an average in such a case tells little about the level of wealth or well-being of most of the inhabitants of that country.

Fortunately, from a statistical standpoint most human body dimensions approximate to the shape of a familiar mathematical distribution. This is known as the normal or Gaussian distribution, which has a typical bell shape (Fig. 12.2). Providing that the values obtained from the data collection fit this distribution, then only two indices are required for its description. These are the mean, a measure of the centre, and the standard deviation, a measure of variability. Using these indices it is possible

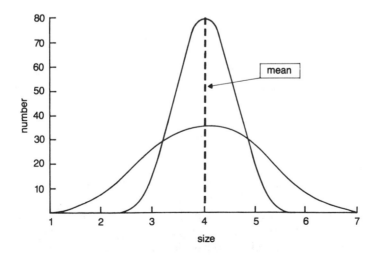

Fig. 12.2 Two normal or Gaussian curves, showing the same mean but different distributions.

273

to calculate the frequency of occurrence in the population of any point within the range. The definitions of some statistical terms related to distributions are given in Appendix 1.5.

This is perhaps an appropriate time to destroy the myth of the average man. If a doorway were designed for the average person, then 50% of the population would strike their heads when passing through it. In this case we would need to design for the top end of the distribution. If we designed shelves for the average man, then 50% would have difficulty in reaching them. Here we must use the bottom end of the distribution of human dimensions. In general, it is the larger people who determine clearances and the smaller people who determine the limits for reach.

In many cases it may not be possible to find a single design solution to suit everybody. We must then provide a range of adjustments so that most can be accommodated. Rudder pedal and seat adjustments on the flight deck are examples of where this is done, for operational safety as well as working comfort. Even then it is unlikely that the range of adjustments will accommodate everyone so a decision must be made on who is to be included and who left out. This introduces the concept of percentiles, which is a means of expressing the range of sizes to be accommodated in a particular design.

If we decide that we want to accommodate 90% of the particular population, we could decide to exclude the top and bottom 5%. We would call this designing for the 5th to 95th percentile. In other situations such as the cabin overhead luggage container height or the height of the highest container in the galley, we might exclude just the bottom 2% when deciding to accommodate 98% of the relevant population.

Accommodating a very large percentage of the population could become very expensive. We can use the vertical adjustment of a pilot's seat as an example. One study demonstrated that to accommodate 90% of the selected pilot group would require a vertical adjustment range of 11 cm. Accommodating 98% could be achieved with a total range of 15 cm. But to cover the remaining 2% would require an additional 11 cm, that is, as much as required to satisfy the original 90% (Hertzberg, 1960). It could therefore involve a disproportionate penalty in trying to accommodate extremes when designing equipment. There will be some, therefore, who will be too wide to enter an aircraft toilet door or use an emergency escape hatch, too short to reach the luggage rack and too tall to avoid striking their heads when entering the aircraft door. The decision on how many are to be included in this disadvantaged minority is a fundamental one in any design process.

Flight deck design

The flight deck as a system

The flight deck to some in the manufacturing industry is simply the front end of numerous aircraft systems, each designed under a different specialist engineer. Hydraulics, pressurisation, electrics, etc., all have components located on the flight deck. Their controls and their displays were determined largely by the individual system designers who were unlikely to have had an overall view of the task of the flight crew. Also, in the past, they were unlikely to have had any formal education in Human Factors, though in a few aircraft factories such specialists were available on call.

As the importance of flight deck Human Factors has been increasingly recognised, it has become the practice in most factories to have displays screened for Human Factors acceptability. When such a system is in use the commentary may be of an advisory nature only, although this too is changing. A strong case can be made for requiring all design engineers to undergo some basic education in Human Factors so that areas which may give Human Factors problems can be recognised at a conceptual stage in design and long before any hardware is produced.

The problems associated with trying to modify flight deck hardware after the design stage, in an environment of rigid engineering deadlines and production schedules, may become insurmountable. The design engineer should be able to identify potential problem areas, avoid most Human Factors errors in design and know when and from where to call upon further Human Factors expertise. It is most heartening to see that there is growing recognition that training requirements also should be recognised at this stage, for the design engineer is in an optimum position to identify the skills and knowledge that will be required to operate his product well.

An example of the problems of demands for late modification of a design is the overhead panel of the Boeing 747. Several evaluation pilots objected to the proximity of this panel to the pilot's head (about 8 cm at its closest point) which made vision and operation of the panel less than optimum. Head proximity to the overhead side structure (about 6 cm) was also criticised, taking into account possible body movement in turbulence. The major structural changes which would have been necessary to meet these comments were simply not practical at that stage of design and production. On the other hand, airline pilot requests to reduce the distance to reach the main flight guidance controls, made at the same design stage, could be met. Early recognition of the importance of the operational element in Human Factors expertise can save costly design changes further downstream in the development process.

Individual aspects of the flight deck should not be evaluated in isolation. The flight deck must be seen as a system with the *Liveware*, *Hardware*, *Software* and *Environment* as its components. In order to optimise the system, expertise is required as in the design of any system. As the flight deck is a workplace, with man as its central component, the expertise must be that concerned with matching the other components to the characteristics of man, combined with a specialised knowledge of the job to be performed. Earlier reference to flight deck display problems was made in Chapter 11.

Constraints

Even when the designer is in possession of valid data on human dimensions, he is regrettably required to face a number of constraints which often limit the extent to which he can produce optimised hardware.

The aerodynamic characteristics of the aircraft have a fundamental relationship with the cross-section of the fuselage and the shape of the nose. These sometimes present the flight deck designer with a difficult framework in which to work and inhibit the creation of an optimum workplace for the crew. The Concorde is a good example of the narrow and relatively cramped environment created by aerodynamic constraints, with a flight deck width of about 148 cm compared with the 191 cm of the Boeing 747. The supersonic aerodynamics of the Concorde also places severe limitations on windscreen design. A flush surface, resulting in minimal outside vision, is necessary in cruise flight and the complexity of a 'droop' nose facility is then required to provide adequate visibility for subsonic operation in the terminal area.

In addition to aerodynamic constraints, the designer must also face commercial pressures. Space in the cabin can be sold; flight deck and galley space cannot. The early DC10 flight deck was planned to be longer until the space was encroached upon by the cabin. This finally resulted in a flight deck in a wide-bodied jet which many crew members on long flights complain is cramped with inadequate space for flight bags, coat stowage and other essentials.

Geometry and visibility

Aerodynamics not only has a fundamental influence on the size and shape of the flight deck as a workplace. It also influences aircraft attitude on approach and thus forward, downward visibility. The downward visibility requirement influences the design of the windshield and the location of the design eye position. This in turn determines the seat location which then establishes the position of the rudder pedals and the control column. The installation of the main instrument panel and other control panels is dependent primarily on the location of the seat.

The flight deck designer has no control over the basic aerodynamics and is simply presented with the geometrical chain reaction, starting with the aerodynamics, in designing the workplace for which he is responsible. But he has a little flexibility. Small adjustments can be made in the distance between the pilots, for example. The closer they are to each other, the easier is cross-monitoring and the more common controls and displays can be installed. However, the trade-off in moving pilots closer together is loss of outside lateral vision, a reduction in pedestal panel space and a restriction to inboard access to the pilots' seats. In large jets the distance between the two control column centres is usually about 105 cm, though on the Concorde it is only 89 cm.

Access to pilots' seats has sometimes created problems. In early Boeing jets this resulted in a decision to move the seats slightly outwards, while leaving the control column and instruments in their original position. This applied to the Boeing 707, 720 and 727 series of aircraft. The consequence was that the pilot was not aligned with the control column and centre of the flight instruments when flying. Clearly, it was the conclusion of Boeing engineers that in this case the misalignment would not cause significant operational difficulties though the peculiarity disappeared in the 737 and later Boeing models.

Nevertheless, misalignment of pilot and controls can in certain circumstances be hazardous. In 1983 a report was made of a small aircraft crashing as a result of a design-induced error originating from such a misalignment (CHIRP Feedback No. 2). The rudder pedals were not installed directly ahead of each pilot, no doubt based on engineering and geometric constraints presented to the flight deck designer. The flight instructor pilot was accustomed to flying from the right seat and in that position was used to the rudder pedals being offset to the right. However, he was called upon occasionally to fly the aircraft from the left seat. On one such occasion, to correct a student-induced deviation, he applied the wrong rudder on landing resulting in loss of control and an accident. It later transpired that this was not an isolated case.

One important aspect of flight deck geometry concerns viewing distance for displays. The distance from the pilot's eye to the main instrument panel on large aircraft is about 71–78 cm. Depending on the particular operational requirements, an instrument may be installed in one of various locations on the flight deck, each with a different viewing distance. If installed on the overhead panel, it could be as close as 20 cm to the pilot's eye; on the lateral systems panel nearly 2 m (Fig. 12.3). The flight deck designer will need to know in which location the display is to

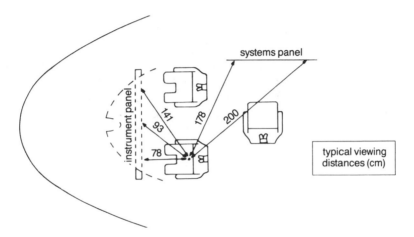

Fig. 12.3 Typical viewing distances from the pilot's design eye positon to various panels on the flight deck of a large jet.

be installed and who will be required to use it, before he can determine the size of display details including alphanumerics (see the discussion on page 114).

Panel design and layout

Historically, the panel to which most attention has been given is that incorporating the flight instruments. As mentioned in the previous chapter, it was not until the 1930s that the first serious consideration was given to the way the displays were organised. In the USA, James Doolittle played an important role during that period in making instrument flying acceptable to a largely sceptical piloting community. He also began to take an interest in the arrangement of his blind-flying instruments on the panel.

In 1937 in the UK, the RAF published details of what was then described as a standard blind-flying panel and this was installed in wartime RAF aircraft. Extensive studies of visual scanning patterns later resulted in a small change to this panel to convert it into the basic T layout which is still the core of the most advanced electronic flight instrument panels in current aircraft (Fig. 12.4). This basic T panel is configured on the need for fast and accurate scanning of four basic parameters – speed, attitude, altitude and heading – with priority being given to attitude. All these are directly related to flight.

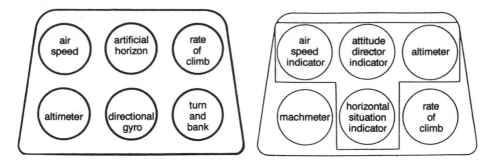

Fig. 12.4 The pre-war RAF 'Standard Blind-Flying Panel' on the left, and the later 'Basic T Panel' which forms the core of modern flight instrument panel layouts (Chorley, 1976).

A second function of an instrument panel may be to display system quantitative information, mainly for check reading. An engine instrument panel might come into this category. Several different layouts have been examined with conclusions being reached, amongst others, that the normal position of the pointer on round dials should be at the 9 o'clock or 12 o'clock position and that pointers with an extended tail are better than simply short ones (McCormick *et al.*, 1983). Applying the *Gestalt* concept to this particular function of check reading provides an additional clue as to how the layout may be optimised. This concept is based on the belief that people perceive a complex collection of separate images as a whole or single entity. The block or bank of instruments, then, is monitored as a whole rather than as individual displays. A disturbance of the symmetrical pattern of that whole as a result of a deviant indication on one instrument, should then be quickly detected.

Another kind of panel is often used for the mounting of displays and controls for fuel, electrical, pneumatic and hydraulic systems. This is called a synoptic panel and displays the system in a schematic form with the controls and displays appropriately placed in the system. The same synoptic concept is applied to CRT displays of systems where faults can be annunciated at the appropriate location in the system. Current technology also permits the system controls which are displayed on the CRT to be operated through the touch-sensitive screen.

Flight guidance control panels, now generally mounted on the glareshield, demand a different set of design criteria. The panel needs to be seen for digital read-outs and reached for operation of controls. Certain controls may need to be operated 'blind' or in peripheral vision and so questions of spacing and coding are important. The layout may be directly influenced by the operating procedures of the crew – who does what – and these could vary with different airlines. A detailed review of Human Factors of such a panel would cover a list of more than 60 items; this has been discussed in more detail elsewhere (Hawkins, 1976).

There are various other control panels on the flight deck and in the cabin which require optimisation. These include those for radio and interphone controls, circuit breakers, galley equipment and door operation. All require layout optimisation depending on their particular function.

Switches

One common element of most panels is the toggle switch. On the flight deck, two distinct methods have been used to determine switch movement – the so-called forward-on and sweep-on concepts (Fig. 12.5). The earlier forward-on concept had two clear disadvantages. Firstly, an ambiguity arose with panels mounted vertically or close to the vertical. And secondly, a module mounted overhead could not be moved to a pedestal location, or vice versa, without contravening the switch movement concept. Such relocation has occasionally been found necessary when new equipment is installed in airline service and some shuffling of panels becomes necessary to accommodate it. This latter disadvantage also meant that panels could not be standardised amongst airlines unless the locations were also standardised. In the past, this problem has usually resulted in the introduction of conflicting switch movements in a single area when the panel had been removed from its original location.

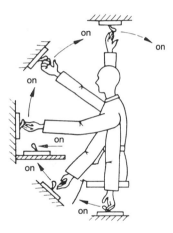

Fig. 12.5 The 'sweep-on' switch position concept which is slowly replacing the earlier 'forward-on' arrangement (Hawkins, 1976).

The sweep-on concept solves both of these difficulties and is now being increasingly applied, though multi-type fleets may find both concepts in use. Aircraft which have passed through different owners or those owned by less well-endowed companies, may reveal considerable confusion in this respect.

The question of standardisation of panel layout is not simply a theoretical or aesthetic one. Numerous cases have appeared in CHIRP reports of errors occurring as a result of inconsistent layout, sometimes involving inadvertent reversion to an operating practice appropriate to an aircraft flown previously. In the 1983 A300 accident in Kuala Lumpur the investigation report concluded that variations in panel layout between the different aircraft flown by the pilot could have affected crew performance leading up to the crash. The aircraft was on lease but operated by local crews (ICAO Summary 1984-2). It is of considerable interest that this could have been another form of negative transfer.

Sight and reach

An important principle which resulted in a fundamental revision of design policy was related to the different location required for displays (for viewing) and controls (for reaching). It was once (and in some cases still remains) normal practice for hardware vendors to design and supply to the aircraft manufacturer a display and its controls in a single module. This is still standard practice in conventional instruments such as the altimeter, with a pressure sub-scale and control knob. Yet while this serves engineering and logistical convenience, it is frequently not optimum from a Human Factors standpoint.

Starting in the 1950s, KLM in The Netherlands pioneered the concept of separating flight guidance controls from their displays so that the location of each could be optimised. This also permitted a single or common control (e.g. heading or mode selector) to be operated by both pilots, thus allowing rationalisation of flight guidance management (Hawkins, 1966). In principle, this concept involved locating the flight display on the instrument panel and the controls on a glareshield panel, until then conventionally used for engine fire controls (Fig. 12.6). This arrangement not

Fig. 12.6 Locating flight guidance controls on a glareshield panel allows both pilots to reach them and avoids the disturbing effect on instrument scanning and manual flight control from having to lean over the control column to the instrument panel (Hawkins, 1966).

only allowed both pilots to be able to reach conveniently commonly used controls. It also avoided the disturbing effect on instrument scanning and manual flight control which arose from having to lean over the control column to make settings on instruments. Originally, this concept was applied only to KLM aircraft but later became an industry standard and after the Boeing 737 all new transport aircraft were offered with this flight deck configuration. It made an important contribution to flight deck resource management and the glareshield panel came to be considered by all manufacturers as high priority flight deck real estate (SAE ARP 4053).

In the case of heading, speed and altitude, a digital display of the selected rather than the actual value is now usually installed beside the relevant control on the flight guidance panel. While this seems at first sight as a contradiction of the concept of separation of control and display, these digital read-outs are used primarily for the preselection of desired values which will be required later and are not included in normal instrument scanning. When an approximate heading command is required, such as with a large turn, then the analogue bug on the instrument is usually seen as an adequate visual display of the heading command. When a specific heading in degrees is required, then it is likely to be set using the digital display beside the control knob. These remarks are dependent, of course, on the particular aircraft display and flight guidance configuration.

Crew complement influences on flight deck layout

On large transport aircraft operated by three crew members, two basically different designs have been used. These two designs involve either the third crew member facing forward, seated at all times between the two pilots, or facing laterally at a separate station.

Even in the early post-war period, manufacturers could not agree on the optimum arrangement. At that time most third crew members were professional flight engineers often recruited from ground-based maintenance staff. The Lockheed family of transports in the Constellation series (L049 *et seq.*) favoured the lateral-facing crew member, sitting behind a curtain at night with his own set of displays and throttles. Douglas, on the other hand, in their DC4, DC6 and DC7 series, favoured the forward-facing third crew member, using the same displays and controls as the pilots. Boeing at the time with their Stratocruiser used the lateral configuration.

In subsequent series of aircraft, Lockheed with their Electra and Douglas with their DC8 reversed policies, the former switching to the forward-facing configuration and the latter to the lateral. In advancing to the wide-body family, Douglas (DC10), Lockheed (L1011) and Boeing (747) all selected the lateral configuration.

The fact that manufacturers have alternated between forward- and lateral-facing stations for the third crew member suggests that there may not be an overriding preference for one or the other configuration. In fact, a number of advantages and disadvantages for both can be proffered and these are summarised in Appendix 1.22. Once system complexity is such as to require very extensive flight deck instrumentation and controls, then a separate, lateral systems station probably becomes inevitable. But with more automation and fewer displays, compromise has been found possible by mounting some little used controls and displays on a small lateral panel which can be reached from the position of forward-facing crew members.

Some manufacturers, such as Douglas with the DC10, have tried to meet the differing demands with a compromise. A full lateral systems panel is provided but the seat tracks allow the systems operator to move and face forward between the two pilots for phases of flight when this is felt to be desirable. This also allows some flexibility for airlines to apply their own operating procedures.

The question of configuration is not simply one of technical merit and operational effectiveness. For example, the original Airbus A310 design was for a three-man crew and incorporated a lateral panel for the third crew member. By 1979, under pressure from airline managements, the third crew member was dropped and the layout redesigned. Little serious airline interest was shown in the optional three-man crew forward-facing configuration offered by the manufacturer which was widely seen as only a transitional phase before removal of the third crew member altogether. Air France as an exception to the general airline practice operated the aircraft with a three-man crew.

In order to achieve the conceptual evolution from a three-man configuration with a lateral systems panel to a two-man crew, it was necessary to create a very extensive overhead panel to accommodate many of the controls which would formerly have occupied the lateral panel. This was divided into a forward section for frequently used items, and a rear, inconvenient section for items used less frequently. The total overhead panel contained just over 200 controls and nearly as many lights. It can be expected in most two-man crew, modern, large transport aircraft, with a high degree of system automation, that little operational difficulty will be encountered in normal, routine operation. It is in abnormal and emergency conditions that difficulties may be expected.

Whether the aircraft is to be flown by a crew of two or three is of fundamental importance in the basic design process if flight deck resource management is to be optimised. This involves not only a question of relocating displays and controls. A quite different approach to system design must be adopted in each case, particularly in connection with emergency modes of operation. In a two-man crew operation, flight guidance and communication can be expected to occupy fully both crew members in busy traffic environments, leaving no significant spare capacity to handle aircraft systems. This requires, in principle, that after a first stage system failure, no crew action must be needed to continue safe operation. One means of achieving this is to provide an automatic back-up for an automatic system, instead of a manual back-up which might have been acceptable with the availability of another pair of hands. Activity which may demand prolonged head-down operation (excluding instrument flying) in busy areas, precluding visual look-out, will also need to be restricted in a two-man operation. This can result in a considerably different technical approach to crew system design and flight deck layout for a two- and three-man crew.

To achieve an optimum working environment for efficient resource management, the crew complement and in the case of the three-man crew, the location of the third crew member, must be established at the drawing board stage of development. A design which is changed during development must be reviewed with the utmost care. It is inconceivable that a design which has been optimised for a three-man crew will also be optimum for a two-man crew.

This aspect of flight deck design also has an interaction with the distribution of crew duties and thus with operating efficiency and safety. It therefore requires detailed and skilled attention. Much argument and industrial conflict was generated during the development and certification of the Airbus Industrie A310, which during its evolution appeared first in one configuration and then another.

The two- or three-man crew decision not only has an influence on the design of systems but will determine the amount of secondary duties, such as administration and log keeping, which an airline can safely allocate to the crew.

Crew seats

There have been few items on the flight deck which have given rise to more complaints than crew seats. Yet it is probable that the back troubles causing the complaints, notably lumbar region pain, are aggravated rather than created by the seat deficiencies.

Some general principles of seat design are given in Chapter 13 in connection with passenger accommodation on board. But the problem for crews is rather different. They are required to spend long hours strapped in the seat; the limited facility for moving the legs on current transport aircraft with a control column, aggravates blood circulation difficulties. Working frequently during the body night, pilots tend to slump in their seat taking up a very unhealthy posture. And, in contrast to the experience of most passengers, the exposure is likely to continue routinely for some 30 years or so.

The airline medical officer can confirm the frequency with which pilots report lumbar disc disorders. But in addition to the lost production time which this may entail, the effect of nagging back pain or discomfort in flight on behaviour, motivation and performance should not be overlooked. It creates a distraction and source of irritation which is not conducive to an optimum working situation.

Crew seat development during the aircraft manufacturing process is not always given a very high priority and is likely to be impeded by delivery date and engineering deadline pressures. Crew seats have usually been developed as a part of a new aircraft project. However, one small UK manufacturer specialising only in crew seat construction has had considerable success in breaking into air transport with a seat that has been well received by pilots (Bryanton, 1985). This particular manufacturer was very responsive to professional Human Factors input in its formative years and has since been rewarded with a steadily increasing share of the civil and military crew seat market.

Evaluation of a seat intended for use in long-range operations must take place in the appropriate environment, that is, long-range operations, and not on a short-range aircraft. Seat design aspects which require optimisation include the configuration of the seat pan and back structure, armrests, headrest, lumbar and thigh supports, upholstery characteristics and controls. In addition, attention must be given to associated hardware such as seat-belt, shoulder harness and footrest (Fig. 12.7). More complete details of the Human Factors aspects of crew seats can be found elsewhere (Hawkins, 1973).

An important result of a poor sitting posture is an incorrect curvature of the lumbar region of the spine. This places an uneven and undesirable pressure on the discs in that area. The slumped position of the shoulders also has a physiologically detrimental effect.

To avoid adverse effects on fitness and performance, proper attention to crew seat design should be assured. Appropriate recommendations should be made available to crew members on correct sitting posture and the physical activity necessary to maintain the back in a healthy condition (e.g. Finnair, 1971; Mohler, 1982).

Other applications on board

The cabin

This discussion has centred on Human Factors aspects of flight deck space and layout. But this is not, of course, the only area on board requiring this kind of attention, as indicated at the beginning of the chapter. The galley in the cabin is an important area demanding Human Factors optimisation but historically less expertise has been applied here than on the flight deck. While it is true that deficiencies in the cabin work areas are less likely to lead to such dramatic consequences as

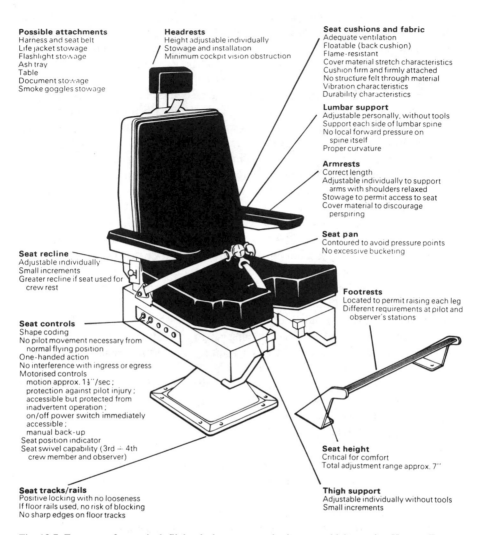

Possible attachments
Harness and seat belt
Life jacket stowage
Flashlight stowage
Ash tray
Table
Document stowage
Smoke goggles stowage

Headrests
Height adjustable individually
Stowage and installation
Minimum cockpit vision obstruction

Seat cushions and fabric
Adequate ventilation
Floatable (back cushion)
Flame-resistant
Cover material stretch characteristics
Cushion firm and firmly attached
No structure felt through material
Vibration characteristics
Durability characteristics

Lumbar support
Adjustable personally, without tools
Support each side of lumbar spine
No local forward pressure on
    spine itself
Proper curvature

Armrests
Correct length
Adjustable individually to support
    arms with shoulders relaxed
Stowage to permit access to seat
Cover material to discourage
    perspiring

Seat pan
Contoured to avoid pressure points
No excessive bucketing

Seat recline
Adjustable individually
Small increments
Greater recline if seat used for
    crew rest

Footrests
Located to permit raising each leg
Different requirements at pilot and
    observer's stations

Seat controls
Shape coding
No pilot movement necessary from
    normal flying position
One-handed action
No interference with ingress or egress
Motorised controls
    motion approx. 1½"/sec ;
    protection against pilot injury ;
    accessible but protected from
    inadvertent operation ;
    on/off power switch immediately
    accessible ;
    manual back-up
Seat position indicator
Seat swivel capability (3rd ÷ 4th
    crew member and observer)

Seat height
Critical for comfort
Total adjustment range approx. 7"

Seat tracks/rails
Positive locking with no looseness
If floor rails used, no risk of blocking
No sharp edges on floor tracks

Thigh support
Adjustable individually without tools
Small increments

Fig 12.7 Features of a typical flight deck crew member's seat which require Human Factors optimisation (Hawkins, 1973).

those up front, efficiency and the exposure to personal accidents are involved. Cabin attendants with extensive experience will be conscious of the vulnerability of cabin staff seat positions on some aircraft to flying 'missiles' from the galley in the event of an emergency landing and of the risk of scalding by water from hot-cups, burns from oven doors and abrasions. They will also be aware of the risk of back injury from badly designed container stowage facilities. Insecure doors on stowage cupboards, inadequate work surface areas and inappropriate switch and control movements will also have been experienced.

The Boeing 747 registered a significant step forward in galley area design and many of the deficiencies which had plagued cabin staff for so long were then corrected. It should be a requirement that Human Factors expertise be introduced at an early phase of cabin crew work area design and continued throughout development. Cabin crew staff associations, with perhaps the exception of those in the USA, may have neglected this aspect of their activities. Useful information is published periodically by the Flight Safety Foundation (FSF) in Cabin Crew Safety Bulletins (see Appendix 2.3).

Space in the cabin which is allocated to crew work or rest areas or which is provided for stowage of crew luggage is, like the flight deck, not available to sell at the ticket desk and so is naturally restricted as far as possible. Complaints by crew members on long-range flights on the inadequacy of stowage space for their personal hand luggage are very common, reflecting a situation which often also becomes apparent to passengers by the cabin clutter. As unstowed luggage can take on the nature of a missile and obstructs evacuation in emergency landings, this problem also has a safety dimension.

Correct assumptions about cabin utilisation must be made at the design stage if the end product is to be adequate in meeting demands. This does not always occur. The UK survey quoted earlier in this chapter (CAA paper 83003) revealed that although it was officially assumed that passengers each carried only 3 kg of hand luggage, in reality the average weight was 5.5 kg. For passengers travelling with hand luggage only, this weight averaged 8.2 kg. With more than 300 passengers on many flights, this excess and often unstowed load involves safety as well as comfort.

Safety considerations

Stowage is but one of numerous safety aspects of space and layout optimisation on board. Escape routes may become blocked by seats or other furnishings which have become detached in an emergency landing (see Chapter 13). The size of overwing emergency exits may give problems for the obese and physically handicapped passenger – the question of percentiles is relevant here. The location of emergency equipment throughout the cabin can have an influence on safety though where cabin staff fly different aircraft types, consistency in location between types may be more important than the location itself. The problem of inconsistency has always been a source of danger and has never been totally resolved. In the fatal DC9 accident at Cincinnati in 1983, mentioned in Chapter 13, which resulted from a fire originating in the rear toilet area, the smoke was so thick after landing that it would have been impossible to see the location of any emergency equipment. This has similarly been reported in the Paris Boeing 707 accident in 1973 and several other accidents. Certainty about the location of emergency equipment is of the utmost importance.

This problem is greater with smaller and non-scheduled carriers who may find it necessary to acquire their fleet of aircraft second-hand from a variety of different sources, each with its own layout. Some such companies may try to standardise the equipment, including flight deck displays, within the fleet while others may feel it is economically not justifiable to do so.

The optimisation of space and layout is important in any workplace. But in the confined working areas of an aircraft, special attention is required if safety and efficiency are to be preserved.

13

The Aircraft Cabin
and its Human Payload

The need for Human Factors in the cabin

Safety and efficiency

It is unusual to read an analysis of an aircraft accident or emergency evacuation without noting Human Factors deficiencies which cost or threatened lives. There are cases of passengers who could not find the exits or could not understand the use of emergency equipment, and seats and other cabin furnishings coming loose and causing casualties in otherwise survivable accidents. Instances are often quoted of toxic gases from burning cabin materials killing those on board who were otherwise uninjured. Passengers suffering from physical handicaps or those encountering language barriers have suffered by being unable to follow life-saving instructions. Inappropriate crew management of emergency controls or of procedures due to illogical or inconsistent design are cited. Lack of leadership or a failure to understand human behaviour in emergencies have also been mentioned. All can be found in accident reports and all reveal Human Factors deficiencies. Yet human performance and behaviour are largely predictable, as are the consequences of neglecting them in the design of cabin hardware and procedures. It is rare to find new sources of error or injury; the same cards are simply re-shuffled and dealt again in a slightly different pattern for the next accident.

But the need for a sound application of Human Factors is not limited to safety aspects of flight. Efficiency is also radically influenced by cabin layout, meal serving facilities, documentation design and other cabin elements in which Human Factors plays an important role. In a service industry such as this, where business depends on customer satisfaction, the interface between the cabin attendants and the passengers – the human payload – is of fundamental importance. Psychology plays a role here, helping to understand and manage human behaviour and the many emotions which are encountered on board.

Human Factors requirements for passengers and crew

A problem faced in trying to optimise application of Human Factors in the cabin is that two basic requirements are presented which are sometimes in conflict. The cabin is a workplace for the active crew, but it is a rest place for the inactive passengers (Fig. 13.1). This conflict does not reduce the need for optimisation; it just makes the task more difficult and frequently leads to compromise. Although such a conflict is not confined to the air transport environment, it is of particular relevance here as a result of the confined space and the lack of control by the individual over his own environmental conditions. When comfort and convenience are concerned, priority is normally given to passenger requirements. But when safety is involved, crew requirements should be considered paramount as the survival of all on board may depend on the skilled guidance of a cabin attendant; this applies particularly in the case of fire (FAA-ASF-300-1J).

Fig. 13.1 There is sometimes a conflict in interests between the passengers, for whom the cabin is a place of rest and relaxation, and the crew, for whom it is a workplace. Temperature, light and noise levels are significant here. Sometimes passengers who are unable to sleep on board congregate in the galley for a drink and to chat with the cabin attendants (cartoon: J H Band).

Understanding human characteristics

In Chapter 1 the *SHEL* model was used as a convenient basis for explaining the applied technology of Human Factors. As this is now a familiar concept to the reader, it can usefully be taken as a basis for discussing the role of man in what might be called the cabin system. In the centre of the model is man or the *Liveware*. In the aircraft cabin this includes crew and passengers. In order for the cabin system to function effectively it is essential for all the other components in the system to be matched to the characteristics, the capabilities and limitations, of man.

Anthropometry and biomechanics

A first concern here is with the criteria derived from anthropometry and biomechanics, that is, body measurements and movements. Such data are involved with most aspects of workspace and layout and with design of equipment. Whether or not seat cushions are comfortable depends initially on the weight of the occupant assumed by the designer. The height of overhead luggage racks and doors, the space between seats (the so-called seat pitch), the design of door controls and escape facilities, should all use data derived from these disciplines.

Currency of the data is particularly important when designing equipment which could remain in use for the next 20 years or so. In the Western world the population now appears to be increasing in height by about 8 mm every ten years, and so data obtained from old studies could lead to less than optimum design. This subject has already been mentioned in the previous chapter devoted to questions related to working space and layout.

Behaviour, performance and individual differences

Information on human behaviour and performance is often less tangible than data derived from anthropometry and biomechanics, but it is no less significant in establishing safety and efficiency in the cabin. This information is derived largely from psychology and its various branches. An understanding of motivation, for example, is of fundamental importance when discussing crew activities on board, which are necessarily performed far away from direct company supervision.

There are many different motives which may lead a cabin attendant to take up flying as an occupation. But unless a primary motive is related to satisfaction derived from working with people, then it may be expected that working effectiveness will deteriorate progressively. Boredom then replaces the enthusiasm associated with the novelty of the new living and working environment. Few regular passengers will not have encountered some cabin attendants whose primary motivation would appear to be seeing the world at the airline's expense and socialising with other members of the crew. On the other hand, they will also have met, particularly on long-range flights on reputable airlines, cabin staff who appear totally dedicated to the service of those in their temporary care. The difference lies in their level of motivation.

From physiology and psychology we learn that people are different. Upon these basic differences are superimposed the differences of culture. Physiological differences appear in the individual tolerance to circadian rhythm disturbances which are inevitable in long-range flying. Psychological differences are reflected in tolerance to the routine family separation inherent in the life of the intercontinental crew member and the international businessman, and in the degree of anxiety which flying generates. Cultural differences become apparent to the stewardess when the South American passenger snaps his fingers at her for service and when the Israeli rejects the standard meal and demands Kosher food. A vast range of individual differences must be accommodated in the cabin if the safety and efficiency of the operation are to be maintained. To achieve this requires a sound understanding of the *Liveware* component of the system.

Cabin hardware

The first interface between components in the cabin system requiring optimisation is that between the *Liveware* and the *Hardware*. This applies, of course, to passenger as well as crew areas. These two components must be properly matched in order for the overall system to function effectively. The *Hardware* component must be matched to the characteristics of the *Liveware* rather than the reverse. The amount of cabin hardware requiring Human Factors optimisation is so extensive that only a few examples can be cited here.

Equipment design

The galley is a major cabin region demanding Human Factors attention and this will be directed at both safety and efficiency aspects. Injury risk can be associated with general fires, ovens and hot-cups (burns), electrical appliances, and lesions (from sharp protrusions and inadequately secured equipment). In a more unusual accident a cabin attendant was killed in 1981 in the galley service lift on a DC10 flight from Baltimore/Washington to London. The investigation allocated blame to the switch system design, the service trolley retention and release system and inadequate pre-flight inspection (NTSB-AAR-82-1). This was clearly an accident involving deficiencies in the *L–H* interface and the investigation revealed nine other cases in the previous ten years with this kind of lift, all causing injury such as lacerations and broken bones.

Efficiency in the galley is dependent on the general space and layout, as in any workplace. It is also dependent upon the design of individual components such as meal and tray stowage arrangements and serving trolley design. Cabin crew injuries have been attributed to poorly designed trolleys (FAA, 1980) and useful studies have been made on what is required to make them safer and more efficient (FSF, 1982; Winkel, 1983). The number and location of the galleys play a major role in efficiency and staff requirements on board.

Control panel design has traditionally attracted more skilled attention on the flight deck than in the cabin and the cabin staff have often had to adapt to deficiencies which their pilot colleagues would never have accepted. There is no justifiable reason why the cabin should be neglected in this way. Fortunately, especially with some of the larger manufacturers, this is changing. Cabin panels are mainly associated with communication, galley equipment, lighting and in some airlines, temperature control.

This equipment is all of direct concern to the crew, though passengers no doubt benefit indirectly from its optimisation. Other *Hardware* aspects are of equal relevance to crew and passengers. One such aspect is delethalisation. This is the process applied to remove or guard sharp objects or protrusions and ensure adequate retention of loose equipment which could cause injury, particularly in an accident. Units of galley equipment can become lethal flying objects with rapid deceleration, unless properly secured. The body is able to withstand very great deceleration forces providing it does not strike or get struck by a hard or penetrating object. The FAA requires that all galley equipment, including serving trolleys, as well as crew luggage, must be secured so that it does not become a hazard by shifting in emergency conditions (CFR 121.576).

Emergency equipment

Another class of cabin *Hardware* equally relevant to crew and passengers is emergency equipment. The application of current anthropometric data is obviously a basic requirement but much more is needed. Human behaviour in situations of stress must be well understood and account taken of this in design. Two examples may illustrate the need for this attention.

In large transport aircraft there have been many cases of inadvertent deployment of escape slides during taxying. Because of the skilled work required to repack the slide and the difficulty often encountered of having this done at the local airport, flights have been inevitably delayed. Sometimes passengers have been off-loaded to meet restrictions applying when an escape slide is not available. A cost of $5000 to $15 000 might typically be the penalty for this error. While the inadvertent operation in these cases arose from human error, this error was design-induced; the door slide controls did not properly take into account normal human expectation and performance. The *L–H* interface had not been optimised.

Such examples are not simply confined to obsolete aircraft reflecting an earlier state-of-the-art. On a certain wide-bodied aircraft delivered new from the factory in 1983 it was possible to find an escape slide control designed in such a way that sooner or later an error in operation could reasonably be predicted. The control required moving up for automatic operation and down for manual operation on one door, while on another door the direction of required control movement was reversed. There is a limit to the extent to which training can, or should, go in adapting people to perform in a manner contrary to natural expectations.

In trying to optimise cabin equipment for emergency use, several problems arise. Firstly, there may be no need for use of such material until after an impact which may result in total or near total darkness in the cabin. This limits the use of colour coding and placarded operating instructions. Secondly, it may need to be used by passengers quite unfamiliar with its operation and so handling must be self-evident, as far as this is possible. A third problem is that the cabin environment at the time may be chaotic, with highly charged emotions, incapacitation and irrational activity following the accident. Evaluation of the effectivity of such material in the peaceful and rational atmosphere of a designer's office requires imagination, but more than this, it requires a sound understanding of human behaviour in such emergency situations. While good routine passenger safety briefing is of great value, most passengers seem to pay little attention to these mandatory briefings, and this is discussed later in the chapter. It is noteworthy that in 1983 and 1984 the FAA carried out tests which showed that about one in three passengers failed to don their life-vests successfully, even after viewing a demonstration. Research some ten years earlier had found a similar failure rate (FSF, 1986). The fault is not simply in the demonstration.

Medical kits

In principle the contents and use of medical kits on board come within the bounds of aviation medicine, and so outside the scope of this volume. However, certain aspects of serious medical emergencies transcend medicine and so some reference should be made here to the subject.

All cabin attendants on long-range flying are likely to encounter at some time the challenge of a passenger suffering a sudden and serious medical condition. The passenger will, in the isolated environment of an aircraft cabin, look to the crew for immediate assistance. If the passenger is unconscious, the immediate relative or colleague will call upon the crew for help.

The condition could be sufficiently serious to result in death on board – a traumatic experience for many cabin staff who may never previously have been directly confronted with death. It has been reported that 577 in-flight deaths occurred in the period 1977–1984 (Cummins *et al.*, 1989).

The immediate response of the cabin attendant to a medical emergency will be to report the situation to the captain and try to determine whether a medical doctor or nurse is on board. If so, and the medical practitioner is prepared to administer to the passenger – medical liability penalties may inhibit some – then the cabin attendant will make available the doctor's kit and await the diagnosis.

A reasonably sophisticated medical kit has been required in the USA (FAR 121.309) since 1986 though many other airlines outside the USA had accepted the need for this much earlier. The kits may not be opened or used by other than a qualified medical practitioner.

Proper use of the medical kits, which are placed on board for emergency use only, raises several problems, including the proper identification of individuals who are qualified to use the kit. They should be used only by an identified medical doctor or an osteopath. Other definition problems can also be created by the statutory restrictions. Finally there are problems such as the basic security of the kits, specific responsibility for the freshness of some of the drugs, the liability (if any) of the doctors who are asked to volunteer their services, language difficulties, etc.

The more serious challenge to the cabin attendant arises when no passenger with medical qualifications can be found. There is no universally applied requirement for cabin attendants to have any first aid knowledge whatsoever. At one extreme are airlines like Air New Zealand, where cabin crew members are all required to hold a recognised first aid qualification and receive annual refresher training (Thomson, 1987). Such training is essential to maintain a skill which is not routinely practised. At the other extreme are the many airlines in the world – including some of the best known – where there is no such requirement or training and the crew member's paramedical skill does not extend beyond the application of a sticking plaster.

Whether having a first aid qualification or not, the cabin attendant must make some kind of diagnosis and provide as much assistance as possible. The question of legal liability again arises, as the trend to sue, whenever the prospect of cash appears, seems to be increasing.

The most troublesome conditions occurring in flight are gastrointestinal, unconsciousness, shortness of breath, and chest pains (Cummins *et al.*, 1989). Even when the captain decides to land at the nearest airport, this could still leave the care of a seriously ill passenger in the hands of a totally unqualified cabin attendant for some hours. It has been concluded in a US study that improved cabin attendant training, closer to the basic level of emergency medical technicians may improve the timeliness and quality of in-flight emergency care (Cummins *et al.*, 1989).

Flammability of materials

It was estimated in 1980 that 30–40% of fatalities in survivable accidents are related to fire and its effects (FAA-ASF-300-1H). This applies to clothing as well as cabin furnishings. A major hazard is the toxic smoke which rapidly generates following the burning, in particular, of the polyurethane foam material used for various purposes in the cabin. While the flight deck crew have protective smoke masks, these are not universally available in the cabin. It has been estimated that over 80% of all transport aircraft fire fatalities are the result of the inhalation of toxic gases rather than thermal injury (Crane, 1985).

One of the many examples of the hazard of toxic smoke in the cabin was the 1980 accident in Riyadh in which all 301 occupants died. The L1011 returned to Riyadh after cabin smoke had been detected which had originated in a cargo compartment. The aircraft landed safely, turned off the runway and stopped. Some fire had been seen by this time in the rear of the cabin. After the aircraft had stopped and the captain announced an evacuation, no doors were opened from the inside and no evacuation was apparently initiated. No doors were opened from the outside either until about 26 minutes later, by which time the interior of the cabin was seen to be engulfed in smoke and flames – there may have been a flash fire shortly after the aircraft had stopped. All deaths occurred as a result of toxic smoke and fire and it seems that occupants were probably incapacitated by smoke before the fire took its toll (Saudi Arabian accident report analysed in Flight Safety Focus 4/85).

The Manchester Boeing 737 accident in 1985, in which 57 passengers died, again demonstrated the extremely rapid spread of fire which can occur in the cabin, even on the ground with fire services immediately available. It also demonstrated the lethal effect of toxic fumes on killing otherwise uninjured cabin occupants. Human error was shown to have played a role in the less than optimum performance of the fire service and deficiencies were apparent in all the *SHEL* interfaces.

A DC9 made an emergency landing in Cincinnati in 1983 with a cabin fire which had originated in the toilet area (NTSB-AAR-83-9). In 1984, following this accident, the airline concerned decided to re-furbish all of its 17 500 seats with an improved form of material with safer fire characteristics.

There has been much criticism of regulatory agencies for failing to establish adequate requirements for the utilisation in civil aircraft of non-toxic and non-flammable materials. These criticisms culminated in the inclusion of the subject in a major FAA study of cabin safety initiated in 1980 and concluded in 1985. One result of this study was that by 1987, large aircraft built or operated in the USA (and in the UK under directives of the British CAA) were required to install fire-blocking seat cushion covers. These covers retard access of a fire to the polyurethane cushion which generates toxic hydrogen cyanide and carbon monoxide when burning, and thus provide at least an extra 40 seconds for evacuation (FAA, 1983). The extent to which other countries adopt such safety measures depends on their own certifying authorities.

While mandatory regulations are established by governments, the airline operators themselves must assume some degree of responsibility in this area. A notice of proposed rule making was issued by the FAA in 1975 to amend CFR 121 to require that clothing worn by cabin attendants meet certain flammability requirements

(FAA, 1975). However, this was officially withdrawn in 1981 following comments received, with the FAA concluding that there was no proof that aviation safety regulations were needed in this area (FR Doc 91-18348). Nevertheless, one airline found it necessary to replace all stewardess uniforms after it was discovered that they were easily inflammable. In this case, the staff union had been active in promoting the safety of its members. For most passengers, however, such safety precautions are unlikely to form part of their pre-flight checklist. Human Factors practitioners should know that other remedies are required to meet this problem. It can create extremely frustrating situations for cabin personnel.

In principle, flight deck crew members should not have to leave their stations to fight a cabin fire; cabin attendants must be properly trained to perform this essential cabin task. They must have learned to adopt an aggressive approach to fire and to take the most effective measures to protect their passengers. They will know that breathing through wet towels or articles of clothing is much more effective in providing protection against toxic smoke than donning passenger oxygen masks, which should not be used during a fire. The flight deck crew will be aware that sustaining life requires adequate clean air and will not mistakenly try to reduce combustion by reducing ventilation. The captain will plan to land at the nearest suitable airport and ensure that relevant smoke and fire procedures and preparation for emergency evacuation have been followed. Effective communication between the cabin and flight deck is of vital importance.

Seating and furnishing

The cabin *Hardware* which concerns passengers most is seating. The seat is literally the airline's interface with its customer. He is often trapped for many hours in close confinement with very little chance to move about. He has to write, read, eat and sleep in the same place and optimisation of this place for all four functions represents a significant Human Factors challenge to the designer. Normally, a seat is optimised for a specific function. A typist's seat will be different from a car driver's seat. A seat for resting should have a back recline greater than one for reading and also be a little lower. A multipurpose seat such as that for an aircraft cabin must therefore necessarily have adjustment capability if it is to serve its various purposes adequately. In fact, the height of a passenger seat is not adjustable but the recline usually is, though this sometimes almost brings the occupant's head into the soup plate of the passenger behind.

As a basic principle for most seats, the weight of the body should be distributed throughout the buttock region surrounding the ischial tuberosities (commonly known as the sitting bones). This is accomplished mainly by proper contouring of the seat pan. The height of a seat should be such as to avoid any excessive pressure under the thighs. It is common practice to use a seat height of about 43 cm (17 in) for multipurpose chairs which have a sloping seat. Where the height is not adjustable, a footrest should be available to cater for the shorter person. The optimum depth of a seat is also about 43 cm, though different ethnic groups will have different leg lengths and this can influence what is considered optimum. A width of not less than 40 cm (16 in) has been proposed but this, too, will vary with ethnic groups. Another important basic requirement is that the spinal column should be kept in a state of balance by supporting the lumbarsacral section of the spine. Lumbar supports are increasingly being built into seats and in some pilots' seats the lumbar support itself is also adjustable (SAE ARP 4052/1). One final basic requirement is the provision of armrests, unless the occupant is involved in activities which call for free mobility of the shoulders, arms and trunk. When possible, the armrest height should be adjustable, but if not, the optimum height is about 18–23 cm (7–9 in) (McCormick *et al.*, 1983). Keeping in mind these basic criteria, which apply to most seats, the designer is also required to take into account the durability and weight of the material, flammability of the upholstery, structural integrity, reliability and maintainability, space available, certification requirements and cost. This is in addition to trying to meet the four functional requirements for comfort mentioned earlier. This is a formidable task and no doubt is reflected in the cost; one pair of seats for a modern passenger aircraft can cost as much as a family car.

One of the most common complaints of passengers is the lack of leg room. The tall person quickly discovers the fallacy of designing equipment for the average man and this was discussed in more detail in Chapter 12. Seat pitch, that is, the spacing of the seats one behind the other, is normally between 71 cm (28 in) and 86 cm (34 in). There is a practical minimum acceptable seat pitch, depending on the length of the flight. But the airline's financial director will be quick to point out that by decreasing the seat pitch from, say, 86 cm (34 in) to 81 cm (32 in), the human payload can increase by about 6%. This should be reflected in the fare level when competition and economics are allowed to play their normal role.

One aspect of seat pitch is related to delethalisation, which was mentioned earlier. It was said that the body can tolerate very high deceleration forces providing it does not strike a hard or penetrating object. It is worth noting that in a study of cabin safety in transport aircraft, the NTSB concluded that 'padding of aircraft interiors is insufficient to protect passengers from head injuries in survivable accidents' (NTSB, 1981). It will be clear why it is a requirement that the seat back must be in an upright position for take-off and landing, which are the flight phases when risk of rapid deceleration is greatest (Fig. 13.2).

Seat security
The certifying authorities of the USA and the UK have for decades required that the seat and attachments should be designed to withstand a forward load of 9g. Yet it has long been known from research that the human body can tolerate far greater loads than this. It has been said, with some justification, that 'we have 40 g people riding in 20g airplanes and sitting in 9 g seat and restraint systems' (Mohler, 1969). The NTSB has concluded that during the ten years from 1970, in 58% of the survivable accidents there were failures in the cabin furnishings which killed and injured passengers and blocked emergency escape (NTSB, 1981). Although many factors influence the tolerance of the human body in any particular case, the forces given in Table 13.1 represent a reasonable expectation of what the body can withstand. In fact, much higher forces have been experienced without permanent physical damage occurring. These figures are compared with the current FAA design requirements for the restraint of cabin seats and equipment. These maxima assume a normal seat-belt and a forward-facing seat. With the addition of upper body restraint, that is, a shoulder harness, the forward load tolerance can be doubled. The

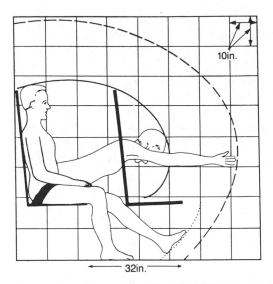

Fig. 13.2 A lap-belt-only extremity strike envelope with a 81 cm (32 in) seat pitch. With full shoulder restraint, the head would travel about 50 cm (20 in) less far forward and would not strike the seat back (US Army, 1980).

Table 13.1 Inertia forces relative to the surrounding structure found tolerable for the human body (Snyder, 1976). FAA design requirements are given in parenthesis (CFR 25.561). Tolerance is based on an acceleration duration of 0.1–0.2 sec and a rate of onset of 50 g/sec (NTSB-AAS-81-2).

| Forward | 20 - 25 g | (9.0 g) | Sideways | 10 - 15 g | (1.5 g) |
| Downward | 15 - 20 g | (4.5 g) | Upward | 20 g | (2.0g) |

shoulder harness not only spreads the load more evenly over the body but also reduces risk of injury from striking the seat in front.

Investigations have shown that in otherwise survivable accidents, the seat restraint system has sometimes failed, allowing the initially uninjured occupant, still attached to his seat, to travel at high speed within the cabin, independent of the aircraft. Many accidents have shown this pattern of injury. In the Santa Ana 737 accident in 1981, four serious and 29 minor injuries occurred as a result of failure of the seat systems and other furnishings (NTSB-AAR-81-12). The crash force levels here were below the FAA limits required for design, so it might be concluded that the seats should not have failed. In the Philadelphia DC9 accident in 1976, the g forces were well in excess of those required for design and certification and so it was predictable that the seats would fail (NTSB-AAR-78-2). Yet there were no fatalities, demonstrating that the forces experienced were within human tolerance. In this accident, 36 passengers were seriously injured as a result of the failure of the seats and other fittings (NTSB, 1981).

Recognition of the problem is not new. As far back as 1962 in the USA, following the survivable ditching in the North Atlantic of a Lockheed Constellation in which 76 people died, the CAB recommended to the FAA '... we are convinced from this accident and others in recent years that an increase in the minimum level of crashworthiness is long overdue. Accordingly, it is recommended that studies relative to crash load factors and dynamic seat testing criteria ... be expedited towards the end of achieving improved safety in this area at the earliest date' (Goldman, 1984; NTSB, 1981). The need for such activity has been routinely reinforced with each investigation of a new survivable accident. Even then, in 1962, the FAA regulations were about ten years old and remained unchanged, despite pleas such as those of the NTSB and CAB, for more than a third of a century. In 1985, the FAA published an Advisory Circular (AC 21-22) outlining a range of impact trauma which may be used in evaluating occupant survivability characteristics. A major problem here is that for certification, only static testing is required and that can be done in only one direction at a time. However, in real accidents, crash forces react with each other in complex ways, resulting in twisting movements as well as simple loads (Burnett, 1985).

Jump-seats
Cabin attendants' seats, sometimes called jump-seats, are required to be near emergency doors and must have a shoulder harness as well as a seat-belt. As far as possible they should be located to provide a direct view of the cabin area for which the attendant is responsible. They are also required to be located to minimise the risk of the occupant being struck by dislodged galley or other equipment (CFR 25.785). Sideways-facing cabin attendants' seats have not been permitted since 1980.

Rearward-facing
A number of studies have been conducted on the relative merits of forward- and rearward-facing seats. Tolerable forward loads as high as 83 g have been demonstrated using rearward-facing seats. This improvement in tolerance results from allowing a better distribution of the force over the body, providing support for the head and avoiding the jack-knifing which occurs when facing forward. Research data give overwhelming support to the increased chance of survival when using rearward-facing seats. Recommendations for such a seat configuration have appeared periodically for many years. This issue was again raised in the UK in 1986 by the Consumers Association (CA, 1986). In spite of all these findings, rearward-facing seats in civil aircraft are rare. In contrast, they have been adopted by various military organisations, where unenlightened passenger preferences carry less weight and where accident survival probability can be used as a primary criterion in design. This is interesting, because it is sometimes assumed that civil air carriers are motivated or certificated to care more for their human payload than their military counterparts. Good surveys of these and other aspects of passenger seat design are available (e.g. Snyder, 1976, 1982).

Other furnishing
While seats are the most obvious cabin furnishings requiring Human Factors attention, there are many others. In most cases the designer meets similar constraints

such as cost (which may be in the guise of material, labour, space or weight), airworthiness (such as flammability and delethalisation) and durability. Included in cabin furnishings on many long-range aircraft, is the in-flight entertainment system, with its projector, screen, individual control panels and headsets.

Lighting has traditionally been the source of design difficulty. Not only is there sometimes a conflict between cabin crew and passenger requirements, but also between one passenger and another; one wants to read while his neighbour wants to sleep or watch the movie. It is necessary to produce adequate light for reading in any seat position, while allowing the adjacent seat to remain in darkness. Ash trays must be designed so they cannot tear the passenger's clothing when he slides in and out of the seat and can be rapidly and simply removed for cleaning by unskilled workers during short ground stops. They must not shower their contents of ash and smouldering cigarette butts over the passenger behind, and his meal tray, when the armrests in which they are installed are raised.

Overhead luggage containers present several difficulties. They have to be strong but light; this is the requirement for much aircraft material. They also have to be designed so as to be operable by a non-technical passenger who has never seen the latch mechanism before. And they must be so installed that passengers do not strike their heads on rising from the seat, yet are within reach without having to call for the help of a cabin attendant. While there is nominally a weight restriction on luggage placed in the overhead containers, there is no practical way of enforcing compliance. As overhead compartments have been found to fail in 78% of survivable accidents and can then cause serious injury, this is a design area which justifies particular attention (Burnett, 1985).

Another area of cabin furnishing calling for Human Factors input is the toilet facility. Some airlines make limited provision for the handicapped passenger, others make none. In 1984, SAE issued recommendations for toilet design which referred in very general terms to the need for fire protection, facilities for the handicapped and placarding (ARP 1315). However, a more comprehensive Human Factors specification is required, including communication needs, to assure an adequate toilet area design. Toilets present a difficult challenge to the designer who is very much constrained by commercial pressure on space. The toilets are by their nature cabin areas where the passenger is not within direct observation of cabin staff and this has raised certain safety and service issues. The aircraft must be protected from passengers who may surreptitiously smoke in the toilet and deposit cigarette ends in the waste container or who may be unstable and have other more sinister motives. An attempt to at least partially control this hazard had been made in the USA with an FAA requirement to have smoke detectors in all lavatories (FAR 121.317). A completely different problem is found when some surreptitious smokers have disabled the smoke detectors. This has led to the creation of another alarm indicating that a given smoke detector has been tampered with. Such action triggers legal action when the aircraft lands and can result in severe penalties. It has been found that bottles of eau-de-cologne and other such items in the toilet are routinely stolen unless well secured, even by First Class passengers and even when the tops of bottles have been previously removed.

Passengers must also as far as possible be protected from themselves. Cases of passengers locked in and losing personal possessions are often reported; there was the example of the brave stewardess who recovered a passenger's false teeth from the flushed toilet, suggesting that service is not yet a thing of the past. Rings are often reported as having been left behind in the toilet compartment, sometimes, it is believed, quite falsely. A knowledge of the human characteristics which give rise to these situations will help the designer to optimise the toilet area, within the constraints placed upon him.

Cabin software

The second interface between components of the cabin system is between man, the *Liveware*, and the *Software*. This involves the non-physical aspects of the system such as the design of instructions, checklists and procedures. It also refers to the symbology which is used on passenger briefing cards and on various placards used in the cabin. It involves the concepts behind working methods.

Division of duties

Cabin staff have a number of discrete tasks which have to be sandwiched between boarding and leaving the aircraft. Their duties can be divided broadly into two categories, operational and service. Operational tasks may be determined by the regulatory authority and so are legally required, or by the airline. These include activities such as administrative paperwork, sealing of dutiable stores and checking and preparation of flight safety equipment. Service duties include those established basically by the marketing department such as meal provision, interaction with passengers, handling of passenger communications and sometimes ticketing. Some activities such as use of the PA system may involve both operational and service functions.

There are times when a conflict arises between these two forms of activity and judgement must be used in deciding the most appropriate action. For example, marketing management is concerned that cabin staff should remain mobile and active as far as possible during the whole flight, providing the service which they have advertised. On the other hand, operational requirements might prefer them to stow away the trolleys and loose equipment during taxying or whenever turbulence is anticipated. Analysis of NTSB data from the period 1979–1983 showed that in the USA 65% of the serious injuries on board caused by turbulence were suffered by the cabin attendants, although there may be 30 times more passengers than cabin staff (FSF, 1985). When the seat-belt sign has been put on by the captain as a precautionary safety measure, the flight deck crew themselves are tightly belted down, in fact, they are routinely belted down at all times that they are in their seats. However, cabin staff may feel obliged to continue their service activities as long as possible so as to complete their assigned programme. This may entail leaving serving trolleys and loose equipment unstowed when purely operational requirements might suggest they should be put away.

Trolley design varies, but an empty weight could be 35 kg and this could increase to some 90 kg when fully loaded. The FAA has reported that flight attendants attribute occupational injuries to various trolley deficiencies and turbulence on short flights 'when there is little time to accomplish the in-flight service functions' (FAA, 1980). Techniques are required for rapid securing of trolleys when turbulence is anticipated and some reassessment of cabin duty priorities may be needed if the frequency of such injuries is to be reduced. The NTSB and FAA are both aware of this problem area.

Human Factors considerations play a significant role in the organisation of cabin activities and involve areas such as workload, time-of-day factors and passenger preferences. While the basic division of duties is laid down by the airline management, the purser or senior cabin attendant does have a certain amount of discretion and by his own attitude and conduct has a major influence on the effectiveness of the total cabin crew. Some airlines schedule a 'purser's briefing' before departure, when the leader of the cabin attendants' team discusses the procedures to be followed on the flight and allocates individual duties. He will refer to any special passengers who may be on board – the mentally handicapped and deportee as well as the VIPs. He may also use the opportunity to make spot checks on the flight safety knowledge of his colleagues, as each aircraft has its own type of emergency door, location of emergency equipment and so on. Unlike the flight deck crew in major airlines, the cabin crew may have to fly several different types of aircraft and this can sometimes give rise to confusion. There is a fair movement to require at least the first cabin attendant (or purser) to have specific qualification for each aircraft they fly, especially if it is a long-range operation with a two-person flight crew. The opportunity may also be taken at this time to scrutinise the neatness and general appearance of the cabin attendants, as they represent the airline's shop window to the public and have a profound impact on the company's image. What goes on in the cockpit may be seen by some as being of greater significance, but it takes place largely out of sight of the customer. Personality factors are sometimes involved in the allocation of duties and the purser will at this time try to establish his own authority as team leader.

Although such briefings cost time, which is money to an airline, it is difficult to imagine how an efficient cabin working arrangement could be achieved without them. On larger aircraft there may be as many as a dozen staff to integrate into a cabin working team and they may have to serve three hundred or more passengers. The number of passengers per cabin attendant has been increasing steadily over the years. Though the comparison is certainly not without qualification, it is interesting to note that while on a large airliner the staff/passenger ratio may be 1:20, when the Queen Elizabeth II sails on a luxury cruise the ratio is about 1:2. In some sections of the cabin of a large aircraft, a single stewardess may have some 40 passengers to look after.

Communication

Software factors are involved in various forms of communication within the cabin. One such form is through the written word, by means of documentation of one kind or another and through signs and placards. Documentation has been covered more extensively in another chapter, but some particular applications are worth mentioning here.

Crew members of a large airline will be required to maintain quite considerable documentation. They will personally have one or more basic manuals which will require routine revision and this will cover working instructions, company policies and flight safety material. In addition, they can expect to receive a steady flow of bulletins from their department covering everything from the dangers of importing alcohol into Jeddah to the care of Korean babies being shipped for adoption. The basic Human Factors design of all this documentation such as the format, writing, printing and use of diagrams, will influence the effectiveness, whether commercial or safety, of the message conveyed. In documentation for cabin staff this is particularly relevant. Their individual approach to the job as a career may vary so much that different means of communication may be required for different groups of staff. What may be effective for a professionally motivated, career, purser, may be quite ineffective with a young stewardess who may be seeing the job in a different light and on a different time-scale.

In communicating with passengers, a major obstacle may exist as a result of language difficulties and in the extreme case, illiteracy. The FAA requires that passengers must be briefed on the safety equipment before flight (CFR 121.571/ 573) but does not suggest how this is to be done with such special passengers. The development of procedures to meet this requirement is left to each airline's imagination. Well designed illustrations certainly help and in-flight movie facilities offer an alternative medium for the communication of safety information to the passengers. More and more airlines are using their movie facilities to take care of passenger briefings, although there is little evidence that they then secure undivided attention from their passengers. The design of passenger briefing cards has been the subject of a number of studies and recommendations by bodies such as IATA (SAFAC 17), SAE (ARP 1384) and the FAA (AC 121-24). However, there is no international standard and airlines are free to design their own cards, approved by their own state regulatory authority.

Even were an optimum and standardised briefing card to be adopted by all airlines, a fundamental difficulty still remains that most passengers pay little or no attention to either the cabin attendant's mandatory demonstration of emergency equipment or the briefing card, usually kept in the seat pocket. There is evidence that those who pay attention to the briefing have a greater chance of surviving an accident than those who do not. Yet the problem still remains as intractable as ever and has given rise to some serious thought (Johnson, 1979). The Chairman of the NTSB tells the story of the stewardess who had become so frustrated at this inattention that she modified her PA announcement. 'When the mask drops down in front of you', she instructed, 'place it over your navel and breathe normally'. Not one passenger apparently noticed the discrepancy (Burnett, 1985). The airline position is equivocal. While of course being interested in a safe operation, they are naturally sensitive not to create any new anxiety about flight which might be commercially unfavourable.

Further aspects of communications will be discussed later when the fourth interface, that between people, is covered.

The cabin environment

The third interface with which we are concerned in the cabin is that between the *Liveware* and the *Environment*. This is not only concerned with the more obvious environmental factors such as noise and temperature, but also with less commonly recognised ones such as time zone changes. This interface was one acknowledged in the early days of flying when the pilots of open cockpit machines found they needed helmets to protect themselves from the cold and to enable communication to take place above the noise.

Noise

Factors associated with noise inside an aircraft are mostly relevant to the flight deck as well as the cabin and so both areas will be considered in the following discussion.

Sources

A major source of noise in civil fixed-wing aircraft originates from boundary layer turbulence and this increases with higher forward speed and increasing air density. In this case we are fortunate in that air density decreases with altitude which is an ameliorating factor at the higher cruise altitudes. A further contribution comes from the air conditioning and pressurisation systems and at times from the hydraulic system. External sources of noise also include the engines, which in jets have a characteristic whine from the compressor and turbine.

Within the aircraft, noise is generally of a broad-band type and is less along the aircraft centre-line than nearer the sides of the fuselage, where it can rise by as much as 6 dB. The flight deck is particularly vulnerable to the interaction of the airflow with the fuselage surface and this was demonstrated in the BAe 125. On this aircraft a modification of the windshield, top of the canopy and fuselage lines (between the -700 and -800 models) made a significant contribution to a reduction of overall noise level. This fell from 100 dB to 93 dB and at the same time the Speech Interference Level (see Chapter 7) was reduced from 91.5 dB to 83.5 dB. Removing the windshield wiper on the Trident reduced the flight deck noise level by 2 dB and the Speech Interference Level by 3 dB (Rood, 1985). The perceived noise level doubles and the actual intensity trebles with approximately each 10 dB increase in measured noise level. The higher noise level at the front of wide-bodied jets will also be noticed by passengers, particularly as speed increases.

Much progress has been made over the years in reducing noise and vibration in civil aircraft, though helicopters in this respect remain unsatisfactory. Here the source of noise is different from that of fixed-wing jet aircraft, emanating from the rotors where it is aerodynamic in nature, and the gearbox and drive mechanism where it has a mechanical origin. In propeller aircraft, some of the characteristics of a rotor appear in the noise spectrum.

In aircraft, noise can be substantially reduced by soundproofing though this has a weight penalty which is commercially unpalatable. Tackling at source a main cause of noise – the airflow over the fuselage – raises considerable problems. The BAe 125 and the Trident examples indicate that something can be done though in general, progress should be anticipated through slow design evolution rather than modification.

Effects

In broad terms, the effects of noise can be described as physiological or psychological. The most familiar physiological consequence is impairment of hearing, which has already been mentioned in Chapter 7. Long exposure to noise can permanently damage the human hearing mechanism. However, individual tolerance to this environmental stress, as to ageing, differs widely so that it is difficult to be precise in specifying safe limits of noise level or exposure time. Furthermore, the noise may be steady or intermittent and of various frequency bands. Nevertheless, some countries have established industrial limits for noise level and exposure time, the product of which is called immission. Over large groups of people, where individual differences can be discounted, it is possible to predict hearing damage as a function of immission.

In long-range aircraft, constant monitoring of HF (high frequency) radio once created a significant increase in noise exposure, but with automatic call systems (e.g. SELCAL) which obviate continuous monitoring by the crew, this additional hazard has now largely been eliminated. Even so, it remains probable that many flight crew members who are exposed to aircraft noise over a full career can expect to suffer some hearing loss in addition to the natural deterioration associated with the ageing process (presbycusis).

Verbal communication, as will have been made clear in Chapter 7, is of direct relevance to flight safety. Noise interferes with this communication by adversely affecting the signal-noise ratio and so decreases speech intelligibility. By impairing hearing on a long-term basis, it further interferes with verbal communication. In addition, noise can mask other signals such as alerts and warnings – these signals should be at least 15 dB above the masking threshold (Rood, 1985).

The second aspect of noise effect is psychological and this is far more difficult to define, though much has been written on the subject. Noise for most people is annoying. The pop-music addict would perhaps challenge this generalisation; no doubt he would classify the sounds which engulf him as signal rather than noise, though the hair cells of the inner ear would not recognise the distinction. From the flight deck crew's point of view, the annoyance may be increased by problems which noise generates in communication and may take the form of frustration, anxiety over message comprehension and the need for repetitions. Workload and fatigue are increased. Noise is known to interfere with the performance of certain mental tasks (e.g. Green, 1978).

For cabin crew members, noise has less significance from a safety standpoint than for the flight deck crew. However, it may cause hearing impairment over a long period, though most female cabin attendants do not plan such an extended flying career. It can interfere with communication with passengers and is no doubt often a source of irritation and fatigue. Cabin attendants are often seated adjacent to doors where the noise can be substantially higher than elsewhere.

Protection

In order to assure some degree of protection against hearing impairment, all flying staff should be aware of the insidious nature of noise damage and make a point of minimising the level of noise and exposure to it. Crew members when not active should select, if they have a choice, a seat in a quieter part of the cabin; earplugs also provide protection. Flight deck crews should avoid excessive exposure to HF radio transmissions as far as possible, avoid too high volume on the cockpit speakers and consider the use of combined headset and circumaural ear protector. The ear makes no distinction between working and recreational noise and so minimising exposure to noise when off duty, whether it is a disco or a motor race, also contributes to safeguarding hearing.

Passenger reaction

Passengers can be particularly disturbed by changes in noise. This may occur during approach to land with erratic power changes, either from manual control or through the automatic throttle system. This may result from windshear, technical inadequacies or poor operating techniques. Noise from gear and flap operation sometimes causes momentary alarm and on aircraft where this is characteristic it may be useful to make a PA announcement to allay the fears of the more anxious passengers. Those seated close to the galley at night may be disturbed by noises originating either from crew activities or from other passengers who cannot sleep and gather in the galley to chat and drink. Passengers who are sensitive to noise should take some care in seat selection on long flights, choosing an aisle seat away from doors, toilets and the galley. On aircraft with rear-mounted engines, they should avoid this part of the aircraft where on some types the noise can be quite excessive. On propeller-driven aircraft, there is a high noise level in the plane of the propellers. But as the speed of the aircraft increases, so the boundary layer noise rises and the propeller-generated noise plays a relatively smaller role in the total cabin noise. Nevertheless, the noise-sensitive passenger should avoid sitting anywhere near the plane of the propellers on such aircraft.

Temperature, humidity and pressure

Temperature

Cabin temperature control is perhaps the most significant environmental area which involves a conflict between the needs of the cabin attendants and passengers. Working crew members, particularly those active in the main deck galley preparing or clearing up a meal service, will find the cabin temperature far too high if it is set at an optimum level for the inactive passengers. It is not unusual to find passengers nestling under blankets while the cabin attendant working in the galley is sweating in shirt sleeves.

The problem is sometimes aggravated by inadequate heat dissipation from galley heat sources such as ovens and hot-cups. But even when the galley is located in a lower deck remote from the passenger cabin, it is inevitable that a basic conflict in environmental temperature requirements will exist between the active and passive cabin occupants. This difference is also reflected in the quantity of fluid which should be drunk during flight. A passenger needs about 110 ml an hour, while a very active cabin crew member could possibly need four times as much (Randel, 1971). The need is related to the rate of fluid loss and this varies with physical activity, temperature, air flow and relative humidity (RH). The effect of activity on fluid loss can be judged from the rate of sweating, though loss also occurs in other ways. Sweating has been measured for a clothed man on the ground in an air temperature of 27°C as 0.05 litres/hour sitting reading, 0.24 litres/hour driving a car and 0.38 litres/hour walking (Adolph *et al.*, 1947).

Humidity

High altitude jet flight brought with it a new environmental problem, that of very low RH. This results from the very low water content of the air at the altitudes which are the most economical for jet operation. The RH also depends on the number of passengers on board, as these are a major source of moisture in the cabin atmosphere. It is possible for the cabin RH to fall as low as 3% after several hours in flight (Fig. 13.3). For heated rooms, an RH of 40–45% has been said to be com-

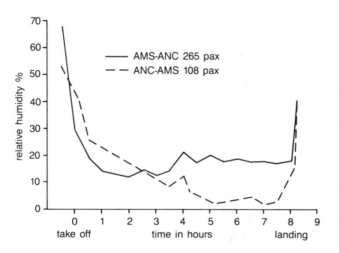

Fig. 13.3 Cabin relative humidity on a DC10 flight Amsterdam–Anchorage–Amsterdam (Hawkins, unpublished data).

fortable and in fine weather, when heating is not required, 40–60% is generally considered agreeable. Observations have shown that viruses and bacteria survive longer with an RH lower than this; less than 30% is quoted as being hygienically undesirable (Grandjean, 1973).

The discomfort which arises from a sustained exposure to low RH does not necessarily presage a serious physical indisposition. It involves drying out of the upper respiratory tract, notably the nose and throat, and dehydration of the skin generally. The nose and air passages are designed to act as a self-cleaning filter. If the mucous membrane becomes too dry through prolonged breathing of dry air, irritation and an uncomfortable feeling occur. The efficiency of the self-cleaning system then becomes reduced, bacteria may find a favourable environment in which to develop and inflammation may follow.

Provided that adequate fluid is taken during the flight, overall body dehydration can be prevented. Coffee and alcohol, which act as diuretics and thus reduce the overall water content of the body, should preferably be avoided. Incidentally, the sensation of thirst is not a good indicator of the amount of fluid needed as a replacement to avoid dehydration. One of the first things beginning marathon runners learn is that if you wait until you are thirsty before taking a drink, it is too late. It is necessary to learn how much to drink and then take it whether thirsty or not.

One side-effect of dry air is that static electricity increases. Anti-static techniques are used in passenger aircraft to reduce the minor electric shocks which are often experienced when touching metal objects such as door knobs in the cabin.

The RH in flight can be raised by the installation of humidifiers, but the amount of water and the equipment needed incur a weight and payload penalty which are unpalatable to commercial airlines. To raise the RH in the passenger cabin of a 747 from 8% to 15% would use about 20 kg/hr of water (assuming a passenger load of 50% and an altitude of 35 000 ft). On the same aircraft a separate flight deck system for the crew would maintain 25% RH and use 3 kg/hr in the same conditions. It would not be technically desirable to raise the RH substantially higher than this, as condensation and frost problems can arise. These can have an adverse influence on electrical equipment and, over a period of time, the aircraft insulation material can deteriorate due to the absorption of moisture.

Pressure

The cabin of most modern passenger aircraft is normally pressurised to the equivalent of 5000 ft to 8000 ft (1500 m to 2400 m), with the aircraft flying up to a little over 40 000 ft (12 200 m). The Concorde is an exception, cruising at between 50 000 ft (15 200 m) and 60 000 ft (18 300 m), though the cabin altitude is maintained within the same range as for subsonic jets.

While removing certain problems at altitude, pressurisation introduced others, not least, the risk of rapid decompression. The time of useful consciousness (TUC) in such circumstances depends on the aircraft altitude, the rate at which pressure falls and the level of physical activity of the individual at the time. Studies by the FAA were made on the TUC which could be expected from cabin attendants in a simulated aircraft decompression approximating to the actual decompression profile experienced by a DC10 over New Mexico in 1973 (NTSB-AAR-75-2). The decompression was from a pressurised cabin altitude of 6500 ft (2000 m) to an unpressurised altitude of 34 000 ft (10 400 m) followed by an aircraft descent at 5000 ft/min (1500 m/min).

It was found that the average TUC fell from 54 sec at rest to 33 sec when doing light to moderate work (Busby *et al.*, 1976). There was no significant difference found in the TUC of male and female subjects. Occupants of aircraft flying above 34 000 ft (10 400 m) can expect a shorter TUC – at 40 000 ft (12 200 m) only about half of the times quoted here (FSF, 1970). This FAA study emphasises the critical importance of the immediate availability of supplemental oxygen to cabin attendants if they are to continue to play any role in the cabin during a decompression emergency.

A second time interval relevant to decompressions is the time of safe unconsciousness (TSU). This is the time a person can safely remain unconscious without risk of brain damage from lack of oxygen and is about two minutes (Gaume, 1970).

An awareness of these figures could stimulate passengers to pay more attention to the cabin attendant's briefing on the use of oxygen masks. While these masks are required to be presented automatically from their stowage in the event of sudden loss of cabin pressure, some prior knowledge of their handling is very desirable to ensure adequate protection. The technical reliability of these automatic presentation systems has often left much to be desired. A further difficulty facing unbriefed passengers may be generated by the fact that the rapid decompression can be accompanied by an unusual noise, fogging of the cabin due to rapid cooling and a sudden change in breathing. All of these can increase anxiety, hinder effective use of the oxygen system and prejudice survival.

Ozone

As aircraft cruising altitudes have increased, so has the risk of ozone being present in the cabin atmosphere. Ozone is created by the action of the sun on oxygen. It absorbs ultra-violet rays which come from the sun and protects life on earth. But it is a toxic gas which damages the lungs through emphysema, an irreversible condition involving loss of elasticity of the structure of the lung. Initially, dryness of the throat and nose are felt with irritation causing coughing and chest discomfort. Sometimes eye irritation is reported. Ultimately the results may be breathing difficulty, overwork of the heart and sometimes death.

The concentration of ozone below the tropopause, the boundary between the troposphere and the stratosphere, is not likely to be high enough to create any problem. However, above this level concentration increases rapidly. In general, just above the tropopause the concentration is highest in the winter and spring months. It increases with latitude and (up to about 35 km) with altitude. For example, an aircraft flying at an altitude above about 34 000 ft (10 400 m) on a northern polar route in February could be exposed to cabin ozone concentration above the maximum established for safety. This limit has been set by the FAA at 0.25 ppmv (parts per million by volume) (CFR 121.578). Occasional peaks of 0.57 ppmv have been reported on the polar route (Preston, 1979).

It appears that the damage from ozone is far more related to the concentration level than the period of exposure. From available evidence it would seem that for healthy people an indefinite exposure to levels below the maximum specified by the FAA is harmless. Inadequate evidence is available to predict with any confidence the effect on people with pulmonary disorders. Experiments in this area with humans are impractical and it is unsafe to apply to humans conclusions derived from animal studies (FAA, 1980; Melton, 1980).

Ozone is partly destroyed as it passes through the aircraft pressurisation system, but each aircraft has its own characteristics in this respect. The high compressor temperatures of the Olympus engines of the Concorde result in the dissociation and transformation of ozone into molecular oxygen for most phases of the flight, even at the high altitudes at which this supersonic aircraft is flying. In addition, the use of nickel brazing in the heat exchanger produces further catalytic action and reduces the ozone. No other filters are needed on this aircraft type. For other aircraft where the ozone is not destroyed in the pressurisation process, carbon filters and catalytic materials can be used, though these are expensive, involve a permanent weight penalty and are therefore unattractive to commercial airlines.

Humidity might possibly have a connection with the experience of ozone sickness. Breathing through a damp cloth has been reported to provide relief from the typical coughing and chest discomfort associated with the inhalation of ozone. Evidence from British Airways' Medical Department of an increase in symptoms reported when the cabin humidifiers were switched off supports this observation. It also indicates a fringe benefit which may be attributable to the installation of cabin humidifiers.

Cabin crew may be more at risk from ozone than seated passengers. They spend more time standing and so are breathing drier air which has not yet mixed with the moister air at lower levels. Furthermore, because they are moving and working they have a higher respiratory rate.

It is often possible to detect the characteristic smell of ozone by the time it reaches harmful concentrations, but peak levels much above this have sometimes been recorded without any such detection by smell (Preston, 1979). Even when it is detected by smell, this sense seems to diminish after a few minutes, giving the illusion that the ozone level has fallen.

Smoking

Smoking is one of the main causes of emotional friction in the cabin (Fig. 13.4). Many passengers no longer feel obliged to breathe air polluted by the smoking of others. But the others feel just as strongly that they should have the right to indulge in their chosen habit and the tobacco industry gives them full support.

Fig. 13.4 Smoking is one of the main sources of emotional friction in the cabin. This may originate with the smoker who cannot secure a seat in the smoking section or the non-smoker who is irritated by having to breathe the smoke of others (cartoon: J H Band).

As the anti-social and health-damaging aspects of smoking have become more widely recognised, attitudes have become less tolerant and conflicts on board related to smoking have become sharper with physical violence sometimes following.

The Air Transport Users Committee in the UK reports increasing complaints about the constant variation in the smoker/non-smoker seating ratio, differing air circulation patterns and the problem of smoke drift. Dozens of in-flight brawls have been reported, in at least one case in 1979 on the Washington to New York shuttle causing the captain to make an emergency landing. In another example, the steward on a flight from Brussels to New York who politely advised a passenger that it was not permitted to smoke in a non-smoking part of the cabin, had a cigar stuffed in his face. The passenger spent the following night in a New York jail.

Complaint origins

There are several reasons why complaints about smoking may be more frequent in aircraft than elsewhere. Eye, nose and throat irritation is worse with low humidity and, as discussed earlier, this can be very low in flight. The normal concentration of ions in a room falls sharply when a cigarette is smoked and does not recover for some time. A disturbance in the normal level of ions may be a cause of discomfort. Although in most modern aircraft the ventilation system is such that a change of cabin air can be achieved about 20 times an hour when the system is operating at full capacity, it is not always so operated. Furthermore, it is not always technically fully efficient. Some systems recirculate a substantial part of the cabin air, including any gases present. In addition, there are some peak smoking periods, such as after take-off and after a meal, to which the ventilation system does not adapt. Inadequate airflow, even in modern aircraft, is a common complaint (Fulton, 1985). A characteristic of commercial transport aircraft is that, with the exception of the First Class cabin, occupants are packed tightly together. Consequently, those sitting close to smokers, where there is a far higher concentration of smoke than elsewhere, will inevitably be exposed temporarily to localised peak ETS (Environmental Tobacco Smoke) levels. Frequently, it is either impossible or impractical to move to another seat with a lower exposure risk.

We tend to view smoking problems in the cabin as essentially those of passengers, but this is not a faithful reflection of the current situation. In a survey conducted by the Association of Flight Attendants in the USA, 95% of respondents reported experiencing eye discomfort because of cabin ETS and 60% said they used some kind of opthalmic solution (Eng, 1979, 1979a). In a 1980 study of cabin staff in Sweden, 69% responded that they were bothered to a great extent in their working environment by smoky air; at a discomfort level of complaint, this was reported by 96% of those responding. In long-range flying, these proportions for female cabin staff increased to 84% and 99% respectively (Ostberg *et al.*, 1980).

It should be noted that air conditioning systems in aircraft were originally designed to cope with an average number of passengers smoking throughout the cabin. The concentration of smokers within confined areas of the cabin has already resulted in an accumulation of tar deposits in the air ducts of these smoking sections requiring additional maintenance. It also creates an atmosphere in these sections which is found obnoxious by many working cabin attendants.

For a number of years the cabin space allocated to smokers has shrunk, in line with public smoking behaviour. In the UK, for example, the cigarette-smoking part of the population declined, respectively, from 52% (men) and 42% (women) in 1972 to 38% (men) and 33% (women) in 1982 (GHS, 1983). Other transportation systems are also reflecting this non-smoking trend. In the UK there is a smoking prohibition in the underground systems in London, Glasgow and Newcastle, as in the San Francisco BART, the Paris Metro and the newer underground in Amsterdam. Scandinavian Airlines System (SAS) has experimented with a total smoking ban on limited local flights, but this was abandoned following objections from smokers. Air Canada in 1986, after banning smoking totally on certain domestic flights, was faced with a boycott by a powerful part of the Canadian tobacco industry. In the USA, there is now a total smoking ban on all domestic flights (FAR 121.317).

The fact that smokers feel a need to light up a cigarette may be relevant here. Flight anxiety is a common emotion amongst air travellers and some reach instinctively for a cigarette as others reach for a drink when having difficulties in coping. The length of intercontinental flights and the risks associated with clandestine smoking may also work against a total prohibition of smoking on all passenger aircraft for some time yet. Nevertheless, we have progressed a long way since 1963 when Sir Basil Smallpiece, then Chairman of BOAC, declared in a letter to the Post-Graduate Medical School in London that it was impossible to arrange non-smoking accommodation on civil aircraft. Quite impossible, he said, to block off seats or cabin sections for non-smokers and equally impossible to determine their wishes in advance of boarding.

Airline policies
The London-based 'Action on Smoking and Health' (ASH) receives regular complaints from non-smoking air travellers. They have frequently reported distress at the lack of non-smoking provision or by the airline concerned failing to deal with someone smoking in the non-smoking area. These complaints led ASH to poll all major airlines in 1983 to determine their policy in providing for non-smokers (ASH, 1983). Of the 82 responding airlines at that time, only one (Metropolitan Airways) made no provision at all for non-smokers. Of the remainder, the seats allocated to non-smokers varied from 10% (Afghan Airlines) and 20% (Middle East Airlines) up to 75% (Olympic Airways and British Caledonian) and 100% (Loganair). A number of airlines operated a flexible seating allocation, often varying with the type of aircraft and class of travel. Half of the airlines reported a total ban on the use of pipes and cigars.

On average, the European and US airlines have offered notably more seats to non-smokers than airlines from the rest of the world, even before the total ban on domestic flights in the USA. Airlines of the Near and Middle East on average appeared to be significantly less sympathetic in space allocation to non-smokers, though some of these totally prohibited alcohol on board. This survey is up-dated periodically. Linjeflyg, the domestic airline of Sweden, has had completely smoke-free flights since 1983.

Non-smokers who may be interested in checking on an airline's policy before they travel can contact ASH in London or Washington. This may be particularly helpful for those who suffer an allergic reaction to tobacco smoke or a respiratory disorder. The figures quoted above can be expected to change with time.

Some airlines like KLM, made the smoking/non-smoking division laterally across the cabin, often allocating a complete fuselage section to one or other category. Others such as Iberia, made the division longitudinally with the left or right seat rows catering for the different categories. Configurations varied widely. Air Algerie granted non-smokers just four to six seat rows in the rear, while Guernsey Airlines adopted the more common practice of placing its six rows of non-smoking seats in the front.

While objections from non-smokers having to breathe smoky air is a primary reason motivating airlines to restrict smoking on board, it is not the only one. Burn damage to carpets and upholstery is common. The cabin pressurisation outflow valve becomes clogged with tobacco tar, requiring maintenance which costs money.

Fire hazard

There is also a real fire hazard from smoking on board (Brenneman, 1985). Many toilet compartment fires have been reported. In 1973, 124 people died following a toilet area fire in a 707 near Paris, possibly as a result of passenger carelessness. The smoke generated was so severe, even on the flight deck, that the pilot had to make a forced landing away from the airport as he was unable to see his instruments any longer (ICAO Aircraft Accident Digest No. 21). At least two similar fires were reported to the NTSB the following year. In cases such as the 1983 DC9 accident in Cincinnati in which 23 passengers died, the exact cause of the fire could not be determined (NTSB-AAR-84-9). In other cases like the Tampa-bound DC9 flight the same year, it has been shown to be due to burning material in the toilet waste container – paper towels and tissues – usually attributable to cigarette disposal (Fulton, 1985).

For a decade, the NTSB had been calling unsuccessfully for FAA action to make smoke detectors and fire protection in toilets mandatory. Had comprehensive action been taken in response to these recommendations, the NTSB has declared, the severity of the Cincinnati accident would have been lessened. Nearly one year after this accident, the FAA gave notice of a rule change proposal (NPRM-84-5) which would have the effect of requiring the installation of smoke detectors and waste container fire extinguishers in aircraft toilets. The rule became effective in 1986.

In 1984 PAA, which had on its own initiative installed toilet smoke detectors, reported about 40 warnings in the short period after installation. Two of these were caused by an actual fire in the waste container, ignited by a cigarette. Most of the others resulted from the detection system being set off by passengers smoking in the toilet. While this is specifically prohibited, control is virtually impossible. The belief of the FAA in the efficacy of such prohibition may explain its earlier reluctance to take more effective protective action.

Those addicted to smoking are no less determined than those addicted to other drugs. They may go to inordinate lengths to satisfy their craving and this could lead to attempts to conceal cigarette ends and thus create a greater hazard. It must be recognised that, firstly, some passengers will smoke in the toilet regardless of any prohibition or warning placards. And secondly, that burning cigarette ends may be deposited in any location in the toilet and not only in ash trays provided for the purpose. Fire protection measures must take this into account.

The problems associated with smoking in aircraft had reached sufficient proportions by 1983 for representatives of WHO, IATA and ICAO, together with those of certain airlines and other experts, to meet for two days at the WHO headquarters in Geneva. Working papers of the meeting were published under the reference of WHO/SMO/84.3.

While these problems have been discussed in this chapter as a *Liveware–Environment* issue, it is apparent that they can generate problems of a *Liveware–Liveware* nature, more about which in a moment.

Circadian and time zone effects

The problem associated with the disruption of the pattern of day and night on transmeridian flights was covered in more detail in Chapter 3. Reference here should just be made to the impact of this environmental influence on the aircraft cabin.

It is assumed that cabin attendants coming on duty in the middle of the night will be fresh and will perform as though they were starting work at the normal time in the morning. This could be far from the truth. They could have had difficulty in sleeping before departure, due to an out-of-phase condition of the biological rhythms. They may be working at the lowest phase of their daily cycle of performance (see Fig. 3.2). Some may have used a sleeping tablet inadvisedly and be experiencing a hangover effect. Of course, the same could be said for passengers, although their role on board is likely to be a passive one in which high performance is not required.

The scheduling of various activities on board such as meals, in-flight movies and commercial tax-free sales, should take into account not only the local time, but also the body phase of the majority of passengers. Forcing a passenger to take lunch when his body phase is in the middle of the night is not likely to endear the airline to its paying customer. Even the opening and closing of the window shades should involve an educated consideration of these circadian factors. The timing of tax-free sales activity is also likely to affect their commercial success. Much depends on the knowledge and attitude of the purser or chief cabin attendant in planning an optimum activity schedule on board.

The interface between people

In the first chapter of this book it was said that Human Factors is about people. Nowhere is this truer than in the cabin of passenger aircraft. This is reflected in the last of the four interfaces, that between people, the *Liveware–Liveware* interface. In this section it may also be convenient to discuss certain passenger characteristics which often interact with other passengers and with staff.

Communication

A major part of the activities in the cabin involves communication of one kind or another, between crew members, between crew and passengers and between passengers themselves. We should first look at some routine Human Factors aspects of this communication.

There has been a trend over the years to erect barriers to communication between the flight deck and the cabin crew. As a protective measure against hijacking, the cockpit door in many airlines is kept locked. Since the introduction of jet flight it has been compulsory for both pilots to remain strapped in their seats throughout the flight. The palmy days when the captain, as on a ship, would return to the cabin for a leisurely dinner with his passengers has long since gone, except in special circumstances when an additional crew is on board. The purser on large aircraft in many airlines has developed into a passenger handling or cabin manager and has taken over some cabin responsibilities from the captain. As the complexity of cabin service has steadily risen, with several different cabin sections, separate galleys and meals and a dozen or so staff, such a change in the role of the purser was inevitable. Furthermore, there are times during flight, notably during take-off, climb, descent and landing, when the flight deck is considered 'sterile' and when even cabin staff are not permitted access. This policy developed from the findings of accident investigations which revealed the part played by distraction in reducing the performance of vital cockpit tasks. In the USA the sterile cockpit is now enforced by law (CFR 121.542).

But in spite of these barriers it is essential to maintain the communications link in both directions. The pilot needs to know when meals are being served or the movie shown. He, and the flight engineer if carried, need to be aware of cabin technical problems and when oxygen is needed by a passenger. And the captain would want to know of any sickness occurring or of any time the medical kit had been utilised without his prior knowledge, particularly when medical assistance might be required on arrival. Commercial information may also need transmission by radio to the airline. In the other direction, the purser requires to be kept advised of schedule changes, the weather to be expected, the technical status of the aircraft, particularly if delays are involved, and he may want the captain's advice on rule interpretation. In certain circumstances, such as with a serious illness or death occurring on board, this two-way communication link becomes essential if efficient operation is to be maintained. And, of course, in emergency situations the link is vital.

An embarrassing example of a breakdown in this two-way communication occurred on a 727 line flight in the USA. It was not until reaching cruising altitude and a passenger knocked on the cockpit door asking when they could expect to get a cup of coffee, that the pilot discovered that he had left his cabin crew behind at the airport (FSF, 1984). Incidentally, this is not the only example of such an occurrence.

Communication between crew members on board has one aspect which makes it different from that in other working situations. In civil flying, particularly in the larger airlines, cabin attendants are required to work under a different leader – a different boss – and with different colleagues on every trip. While this has some advantages, such as the temporary nature of any problems arising from personality clashes, it does mean that the team must be re-established for each flight. Sometimes cabin or flight crew are changed at what is for them simply an en route stop. This practice, which is far from unusual, exacerbates cabin-crew/cockpit-crew communication problems. Personality clashes do sometimes occur and occasionally they may become apparent to passengers. The purser's skill in moulding the cabin attendants into an effective team on any particular flight is an important and often underestimated element in cabin safety and efficiency.

One of the most apparent and most criticised instruments of crew-to-passenger communication is the public address (PA) system. Technically it seems difficult to design or maintain a PA system adequately; in practice it seems often to be inaudible to some while at the same time giving an aural shock to others, squealing viciously whenever the cabin attendant tries to speak. Even when the system works adequately technically, many cabin attendants – and pilots, too – seem to be untrained in the art of communication through this medium. On a flight across the USA a passenger might be required to listen to a continuous historical homily on everything from the design criteria of the Brooklyn Bridge to the last stand of Davy Crocket at San Antonio. On the other hand, the pilot may see the cabin simply as the structure which holds the tail to the flight deck and provide his human payload with no information at all. The quality of speech also varies widely. It may be an automatic, rapid and impersonal repetition of stereotyped phrases or a slow and hesitant announcement by a crew member who did not start to think of what he was going to say until the microphone button was pushed.

Use of the PA system requires a certain skill which must be learned and practised. There are a few rules which can form the basis of good PA announcements and these are listed in Appendix 1.23.

Most of the communication between cabin attendants and passengers is at a much more personal level. In Chapter 7, non-verbal communication was discussed and this is very relevant in the cabin. In a way, the cabin is like a theatre with cabin attendants the actors and passengers the audience. Whenever a crew member passes along the aisle, dozens of pairs of eyes follow him or her, often drawing a conclusion about the safety, service and efficiency of the operation. This is particularly so when turbulence is experienced or when unusual noises or manoeuvres occur. The passenger is hoping for reassurance or is seeking to reinforce his anxieties and fears. When the captain, in particular, walks through the cabin, above all in a problem situation, he is communicating non-verbally with all the passengers whether he plans to do so or not. This situation is partly generated by the feeling of helplessness and inability to control their own destiny which many passengers feel acutely on board.

Special passengers; the physically handicapped

Language difficulties, illiteracy and lack of attention to the safety briefing given by the crew are not the only barriers to communication on board. There are certain special categories of passengers who would like to communicate but for one reason or another have difficulty in doing so. And there are others who can communicate but for whom relating to the crew poses special difficulties. We should look at just a few of these as examples of the kind of challenges which may be presented on board and the sort of steps which can be taken to meet them.

One in six of the population in the UK is said to suffer from a material hearing deficiency (Davies, 1983). Artificial hearing aids may help some forms of deafness, but not others. It may be expected, then, that some passengers on every flight will not hear boarding announcements, will not understand the safety briefing and will not hear the spoken instructions to fasten seat-belts. More important, they will not receive any emergency evacuation instructions if it is dark and no source of visual information is available. Some such passengers may carry a special card which draws attention to their handicap and asks for consideration. A similar sign can be placed on a check-in counter or in an airline or travel agent's window to indicate that such a person will receive sympathetic attention. There are many ways in which air travel for those with hearing difficulties can be made easier and safer and some of these are listed in Appendix 1.24. Many state or charitable organisations are available to provide guidance (e.g. The Royal National Institute for the Deaf in London).

Visual impairment is another common handicap which influences communication on board. In the USA there are said to be six million who have some visual difficulties, even with corrective lenses, and many of these are now regular air travellers (FSF, 1980). Some thought has been given to the problems created for the blind in emergency situations and research has been done on related subjects, such as the use of canes on board (Chandler *et al.*, 1980). Simple guidelines can also be valuable here in ensuring safe and sympathetic handling of those handicapped in this way (see Appendix 1.25).

In the USA, CAB regulations (CFR 382) prohibit unjust discrimination against the handicapped, so as to ensure that they have proper access to air transport. This includes the timely presentation of information of both a routine and emergency nature. Furthermore, the regulations specifically permit guide dogs and wheel chairs to be taken on board with the passenger. Various other kinds of physical and mental handicap should be covered by instructions from aviation medical departments and are beyond the scope of this volume.

A special bulletin, Ninnescah International Service, is published every other month in New York by the Ninnescah Corporation. It is targeted towards those in the aviation industry with responsibility and interest in the problems of special passengers, particularly the handicapped, and is mainly concerned with design, equipment, training and the travel market.

Special passengers; the intoxicated

Cabin attendants are occasionally faced with the problem of managing another form of handicap, the self-inflicted one of alcohol intoxication. The FAA prohibits an airline from allowing on board any person who appears to be intoxicated (CFR 121.575), but this condition is not always evident when boarding. A commercial airline might be reluctant to ban marginal cases and no breath analyses are used to determine the threshold of acceptability. To take a few drinks before a flight is, after all, a common practice for dulling the fear of flying amongst those so afflicted.

A further difficulty related to alcohol arises on special flights when passengers have been alcohol-deprived prior to the flight. Flights from Saudi Arabia to Houston, for example, or groups of seamen flying home after a stint in a 'dry' ship, present problems to the cabin attendants. Not unnaturally, every airline likes to be seen as providing as friendly a service as possible and competition is a factor in the liberal availability of alcohol on board. For certain airlines the problem does not exist. Those from countries such as Saudi Arabia and Pakistan satisfy the thirst of their customers on board only within the non-alcoholic bounds of national religious customs.

A watershed sometimes arises on board when a cabin attendant must decide whether or not to agree to demands from a well-lubricated passenger for more to drink. The hope will be that the imbiber will become more sedated, though uncontrolled loss of inhibition is also a risk. Refusal to supply more drinks can result in an aggressive response in one already enjoying his escape from reality. Responses are unpredictable and cabin staff have a difficult task in resolving the dilemma, fully conscious of the potentially hazardous environment in which they are working.

Special passengers; the fearful

Another special passenger who should now be considered is the one who only becomes special when he has to fly – the one who is afraid of flying. The degree of anxiety can vary from mild apprehension to complete phobia. Various names have been coined to give this condition a similar status to longer established phobias such as agoraphobia and hydrophobia. Words such as aviaphobia, aviophobia and aerophobia are used, but these all have impurities. Users of the word aerophobia are perhaps not aware that it already has a dictionary definition as a morbid fear of draughts. Here we shall just use plain English and refer to apprehension, anxiety, flight fear and, in the extreme case, flight phobia.

Increasingly in commerce, politics, engineering, entertainment and sport, rapid travel is taken for granted. For those suffering from fear of flying, travelling can become a nightmare. Some will even avoid looking at magazines or television in case an aircraft picture appears. Sportsmen like Mohammed Ali and Geoff Capes, entertainers like Doris Day and David Bowie and many other well-known people who must travel have admitted to a fear of flying. In 1984 an English football manager disembarked his whole team as his fear became untenable when the aircraft returned to the departure gate after an aborted take-off.

Individuals differ somewhat in the phase of the journey which gives rise to the greatest apprehension. Some find the journey to the airport a harrowing experience, with tension rising as they approach the airport. On reaching the departure hall some will desperately search for a reason not to depart – a child at home is sick, travelling by car would be more convenient, next week is early enough (Fig. 13.5). How many 'no-shows' (passengers with reservations who fail to check in) arise

Fig. 13.5 ... on reaching the departure hall some who suffer from a fear of flying will desperately search for a reason not to depart (cartoon: J H Band).

from this cause is not known, but it could be significant.

Take-off for most of those who fear flying is a very difficult phase, but the condition which often creates the greatest concern is air turbulence, even when announced beforehand on the PA by the cabin crew or the pilot. Unusual noises (at least, unusual for the uninitiated) and manoeuvres can also be expected to create an immediate arousal condition, with a rise in heart rate, systolic blood pressure and adrenaline output. The journey can terminate in exhaustion. It can end as it started, with an escape from reality through alcohol.

Size of the problem

While accepting that this condition exists, it is reasonable to ask just how big a problem it is; what percentage of the public suffers anxiety when flying. Certainly the attitude of the travel industry has been one, with few exceptions, of not recognising the existence of a problem and in this way hoping it will go away; a temporary headache that will simply disappear if you do not think about it too much. Even passengers themselves are often reluctant to admit to such fears in a technology-oriented society where fear of the instruments of technology may be seen as failure or weakness.

Several studies have been made on the incidence of flight fear in the population. A semantic difficulty exists here in trying to define precisely the terms used in these studies, such as apprehension, anxiety, fear and phobia. Such definitions are necessary in making precise comparisons between different studies and for the more academic analyses. But in spite of such qualifications, the extent of the problem has become somewhat clearer, although one must avoid placing too precise an interpretation on the figures quoted.

A review sponsored by the Boeing Airplane Company concluded that one in six of the US population is afraid of flying (Fig. 13.6). These people, it was concluded,

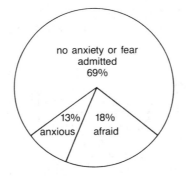

Fig. 13.6 Responses to a survey in the US population regarding anxiety and fear related to civil flying (Dean *et al.*, 1982)

make only one third of the airline trips of those who are not afraid. In the study it appeared that 24% of actual airline passengers were anxious or afraid of flying (Dean *et al.*, 1982). In a Swedish study, 23% responding to a questionnaire reported some degree of fear, but an additional 8% who denied any fear said they were interested in attending a course to reduce the fear of flying (Nordlund, 1983).

This suggests that the incidence of flight fear may still be underestimated. A study in The Netherlands found that 21% of those who had flown and 32% of those who had not flown had a negative attitude towards flying with feelings of apprehension (Peterson, 1979).

We are primarily concerned here with Human Factors in aircraft operation. But those more interested in the marketing aspects of commercial aviation will surely note the very substantial sales potential represented by these reluctant travellers and those too fearful to fly at all.

Flight phobia treatment
In Europe until the 1980s little interest had been shown by airlines in encouraging treatment for flight phobia. One airline medical officer even claimed he had never met anyone afraid to fly and another official of the same airline said 'perhaps once a year we get someone who is frightened'. Another major airline says it has 'no official policy' on the subject. Another has discouraged any support for flight fear reduction programmes.

In the USA there has been a far more progressive and realistic approach. In 1976, PAA began a series of 'Fearful Flyers' courses and by 1980 had processed 2000 people. Two thirds of these were women, but this might reflect a greater concern of men to maintain a fearless image amongst their peers than a lower incidence of male fear. These courses involve typically seven, once-weekly, two-hour classes and have been conducted by ex-Captain 'Slim' Cummings, a former PAA pilot with interest in psychology. A number of other courses are also available in the USA and the number of these courses has increased considerably. In Europe, psychologist Dr Maurice Yaffé at the York Clinic in Guys Hospital, London, has run courses for some time and these, like some in the USA, include a 'graduation' flight, usually to Amsterdam or Paris. In The Netherlands and Germany courses are also available. Each course adopts a slightly different approach and the cost can vary substantially. A useful unpublished survey of the subject was made in Switzerland by Dr Marc (and Dominique) Muret.

A common characteristic of the person afraid of flying is the simultaneous presence of other phobias, though this is not always the case. Muret cited 45% of the normal population having a fear of heights (acrophobia) while amongst those afraid of flying the figure was 59%. For claustrophobia the figures were 38% and 41% respectfully. There is often present a marked ignorance of elementary theory of flight and meteorology. One marketing manager said he was afraid that one day they would encounter an 'air pocket' which would be so deep that the aircraft would simply fall into it and down to the ground. A few minutes of education and he felt considerably relieved on this point, though other irrational fears still remained to be tackled.

The individual origins of this condition are rarely evident initially and the behavioural characteristics of the person are sometimes complex. A multi-dimensional approach to treatment has therefore seemed appropriate and most likely to succeed. This is likely to involve systematic desensitisation (a technique to reduce the sensitivity of an individual to the problem), knowledge about flight, and a graduated confrontation with the fearful situation. Systematic desensitisation consists of contemplation of the fearful situation when in a completely relaxed condition. Techniques such as bio-feedback, Autogenic Training or Progressive Relaxation may be used to induce the relaxation. Aversion relief has also been used in the multi-dimensional therapy. The principles involved in treatment are listed in Appendix 1.26.

The courses can be rather expensive, particularly when an actual flight is incorporated into the programme, but high success rates have been claimed. Not only have careers been salvaged but ticket sales will have been reinforced, so cost-effectivity has probably been high.

The scope for airline activity in this field can be divided into two distinct categories; firstly, for those who feel able to fly sometimes and secondly, for those who do not. Those who are able to fly are likely to come in contact with ticket sales staff, the airport check-in agent and cabin staff on board. Some develop a fear of flying only after their first flight. Each of these groups of airline staff therefore has a role to play not only in preventing the development of anxiety but also in the relief of it. Recommendations for recognition of the symptoms of flight fear and their handling by staff are also given in Appendix 1.26. On short flights there may be little that can be done on a personal basis on board, though enlightened use of the PA system can help. On longer flights, more can be done through an educated approach by cabin staff.

The second category of activity concerns those who through fear feel totally unable to fly. Airline operational staff can initiate little action to help these people, though bringing flying staff into contact with them as part of a fear reduction programme is usually considered helpful. Initiation of programmes in this category, or at least sponsorship or encouragement of them, must be left to marketing management in the airlines or the travel or aircraft industry who stand to benefit from an expansion of the air travel market.

Sufficient studies have been completed and published to indicate that there is a sizeable, untapped reservoir of traffic, untapped because of flight fear. This fear is real, persistent and of substantial proportions. However, something can be done about reducing it.

Passenger behaviour

Cabin staff sometimes remark that after several hours of friendly, sympathetic and personal service to an irritable and apparently discontented passenger, he may disembark without the least indication of satisfaction with the effort expended on his behalf. Of course, there are those who, having paid what they consider an excessive fare for the journey, particularly in Europe, consider that the airline owes them something anyway. But cabin staff, who usually join the flight after a statutory rest period, may not be aware of the hours of tiresome events which could have preceded the passenger's boarding. Perhaps a full day's work at the office, home problems to resolve before departure, ticketing and check-in procedures, possibly luggage and over-booking confrontations and schedule delays. And this may well be superimposed upon some degree of anxiety about flying itself.

This introduces, though it will not be the sole cause of, a situation which is giving increasing concern; the abuse and assault of cabin attendants by passengers on board. This may be a reflection of the trend in society generally, but in aviation, flight safety gives it another dimension. Many cases are routinely reported such as attacking a stewardess with a knife, throwing a meal tray at the stewardess, refusing to extinguish smoking material and failing to fasten the seat-belt (both leading to abuse), theft of other passengers' property leading to kicking and hitting the flight attendant and the aggressive consequences of drunkenness. These few examples were selected, in fact, from the crew reports of one US airline in one ten-day period (FSF, 1979). In a major European airline, one purser suffered two physical assaults in 12 months, on both occasions getting punched in the face by a passenger.

In some such cases the flight crew radio ahead for police to meet the aircraft and remove the offending passenger. Assault at 35 000 ft (10 700 m) is inevitably to be taken seriously, but there are some practical measures which can be taken and these are outlined in Appendix 1.27. Airlines differ in the extent to which they recognise the problem and make provision for it. The fact that in the USA the FAA specifically forbids interference with the crew (CFR 91.8) provides a legal basis for action against such offenders but there often appears to be a reluctance to proceed with prosecution. In 1983, one European airline experienced an assault by a drunken sea captain, travelling as a First Class passenger. He took the stewardess by the throat and lifted her from the floor. No subsequent legal action was taken against him. It is likely that the majority of assaults, like this one, are not reported to any law enforcement authority for action, though organisations like the Association of Flight Attendants in the USA are encouraging a more vigorous response. On the whole, airlines prefer to avoid what they see as the adverse publicity which such a vigorous response would generate. Recommendations are available for cabin crew handling of another difficult problem, that related to emotionally disturbed passengers (Mason, 1984).

Group influences

These remarks have referred particularly to individual behaviour which may involve an interface problem with other occupants on board. But there is another aspect of interaction between people which calls for understanding. When people come together, their behaviour is often different from when they are acting individually. We see this in hooligan behaviour after a football match, with violence and the destruction of property. And in foolish behaviour at a party which causes embarrassment when looking back next morning.

As far as aircraft occupants are concerned, the groups are of circumstance rather than of choice. Exceptions to this generalisation would be when crew members can influence their own scheduling and so arrange to fly together and when passenger groups such as an orchestra or a sports team fly together, often on a charter flight. A second factor which distinguishes aircraft groups from most others is that there is usually an element of fear somewhere in the group.

For a group to be recognised as such in the psychological sense, there should be norms within the group and there will be pressures to conform to those norms. The flight crew can be seen as such a group. A new stewardess may feel obliged to follow the norm and go with the rest of the crew to the bar after arrival, although she may feel tired and prefer to go to her room. She feels under pressure to conform.

In the case of passengers, they generally cannot be classed as a group in the same sense as the crew as they usually do not have common norms. One may pull out a prayer mat, ask which way is Mecca, and periodically prostrate himself in the aisle. Another may refuse standard aircraft food and demand a Kosher meal. One may be seeing the flight as a pleasurable part of a vacation, another as a necessary evil on the way to a business meeting. Yet another, oblivious to the service on board, may be on the way to plant a bomb at a foreign embassy. Yet the fact that they are together on an aircraft can still influence individual behaviour.

Behaviour in emergencies

It is particularly important to understand the kind of passenger behaviour which can be expected in an aircraft emergency. The situation facing passengers in such a case has certain characteristics which make it different from most other emergency situations. The occupants are enclosed in a confined space and are generally rather densely packed. In flight there are no escape routes. On the ground they are severely limited and may involve the somewhat athletic use of escape slides. The generation of fire and toxic smoke can be extremely rapid and visibility can be quickly reduced to near zero. All of this may take place in a chaotic and structurally damaged cabin with widespread disorientation and incapacitation following an impact. The passenger throughout feels powerless with little or no control over the events determining his destiny. It has been suggested that the consumption of alcohol by passengers may also contribute to their inability to respond quickly and appropriately (Burnett, 1985).

Dependency

In most non-aviation emergencies people generally use their own initiative to reach safety. But in the unfamiliar and often fearful environment of an aircraft emergency, experience has shown that passengers look to the crew for leadership, guidance and instructions to ensure escape and rescue. Without these instructions from the crew, panic, freezing into inaction and other inappropriate behaviour can be expected to develop, reducing the chance of survival (FAA-ASF-80-4). It was already noted in Chapter 6 that members of a group become more dependent on a leader in situations creating stress. And at such times they also tend to transfer some of their own responsibility to him.

There has been considerable study of human behaviour in other emergency situations such as hotel fires (e.g. Canter, 1980), but not much for aircraft. This may be partly due to the very large number of variables involved in an aircraft accident and the very short time available for escape after the crash. Another factor may be that aircraft accident investigators have often in the past appeared to lose interest in the case after legal liability or blame was established. However, some limited analyses have been made to contribute to a better understanding (e.g. NTSB, 1977; Snow *et al.*, 1970).

Reliance placed by passengers on crew guidance is a general observation. Without such guidance there are cases reported of passengers trying to return to their seats to rescue personal belongings, sometimes at the cost of their lives; taking a distant escape exit rather than one close because that was where they came in; opening the wrong emergency exit directly into a fire; and blocking a cabin door in desperation and thus preventing it from being opened. Crew procedural deficiencies have also been recorded, in particular, related to communication. The PA system and megaphones may be the only way to provide guidance to passengers when time is short. In the accident to a DC8 at Phoenix, Arizona in 1980, the failure of the crew to use the megaphones to control the emergency evacuation could have resulted in a drastically reduced chance of occupant survival. The NTSB added that the failure over many years to require an effective communication system for the rapid evacuation of passengers in emergency had placed in jeopardy countless lives and resulted in many serious injuries (NTSB-SIR-81-4).

Yet much remains unknown. Seat location seems not unreasonably to be a factor in survivability; the nearer the emergency exit the greater the chance of survival. The 1970 study by Snow *et al.* quoted above found that age and sex may also be factors which influence the chance of successful escape. Age is not surprising, with the very young and very old being particularly vulnerable and requiring additional attention from cabin staff. Sex is less clear. Although females statistically seemed to face a greater survival risk than males, it is not clear whether this is because they are less strong, wear more flammable clothing or whether they react differently in such situations due to psychological, social or cultural reasons. There remains much scope for research and the passenger and crew member might be excused for pondering on an interpretation of the Hippocratic aphorism, 'Life is short and art long: the occasion instant, decision difficult, experiment perilous'.

Recognition

At the beginning of this volume we said that the 1970s were the decade when the operational significance of Human Factors as a technology began to be more widely recognised in aviation and this was reflected in the 1975 IATA Istanbul Conference. The same can be said about the 1980s with respect to the aircraft cabin and here two conferences in 1984 reflected the new recognition. The first of these was the First Annual International Aircraft Cabin Safety Symposium at the Institute of Safety and Systems Management at the University of Southern California. This three-day meeting was attended by more than 300 participants representing 55 organisations from 25 countries. The problem confronting delegates was made clear in the address by the Vice-Chairman of the NTSB. She cited as examples the fact that comparing aviation in 1984 with 1954, seats still broke loose with even light impact, life-vest design remained virtually unchanged and passenger safety briefing after 30 years was still seriously deficient. Passenger protection had not progressed with the same speed as other technical aspects of flying. The NTSB had been calling for improvements for decades and they were long overdue. It was worrying that it required major tragedies, such as the Cincinnati DC9 fire with its associated publicity, to bring about improvement; dangers should be anticipated and we should not wait first for disasters to occur (Goldman, 1984).

The second event was particularly significant in that the FAA, with power to legislate, was the sponsoring body with the FSF handling the organisation. This conference was attended by 370 delegates from 18 countries, again reflecting increasing world-wide concern at many aspects of cabin safety. The FSF/FAA meeting lasted four days and gave the FAA the opportunity to survey a wide-ranging programme of cabin safety improvements involving areas such as cabin crew shoulder harness, evacuation slides, oxygen masks, fire extinguishers, flammability of materials and emergency lighting (McSweeny, 1985). In spite of this programme, numerous areas of concern remained such as passenger safety information facilities, crashworthiness, flight attendant duty times, excessive cabin baggage, galley equipment, and training standards. All these have long been recognised as important factors in survivability. This meeting also revealed some of the areas of conflicting interests of manufacturers, airlines, government agencies and the travelling public. The same year as these conferences took place, the FAA published a Cabin Safety Subject Index to promote awareness of FAA documents related to cabin safety (FAA-AM-84-1).

Cabin seizure with violence: hijacking

Aspects of Human Factors are extensively involved in security screening of passengers and luggage before boarding. The vigilance effect, discussed in Chapter 2, will also apply to those concerned with this task. But more amenable to discussion than pre-flight protective measures is the violent seizure of the cabin, better known as hijacking.

This problem poses a more frequent hazard than is generally realised. In the USA alone, for example, there were 22 hijack attempts during 1980. The year 1985 was also bad – in June alone there were three hijackings. Sixty lives were lost in trying to rescue hostages from an Egyptian aircraft. The following year 18 people died trying to escape from a PAA plane and a fight between hijackers on an Iraqi aircraft resulted in it crashing with the loss of 65 lives. In 1987 hijackings averaged just over one a month worldwide. By 1990 reported hijackings continued to be a major problem and were a much too familiar occurrence. In many countries such attempts will not voluntarily be reported. The consequences relate not only to the safety of those on board but may also involve a legacy of long-term psychological stress symptoms for the surviving victims.

The chances of a passenger being hijacked or taken hostage are very small but crew members, being exposed continuously throughout a career, are far more vulnerable. Every cabin attendant in a large airline is likely to know of a colleague who has faced such a terrifying experience. They would therefore be wise to take advantage of the studies which have been made of the behaviour of hijacker and hostage to enhance the probability of survival. The following is good basic information but is not a substitute for procedures developed for specific airlines.

Those who originate attempts to take over the control of the aircraft cabin – and thus the aircraft – can be divided very broadly into three classes, viz, the terrorist, the criminal and the mentally unbalanced. We can briefly examine each of these and the cabin circumstances created by them.

The terrorist
The most ruthless, skilled and dangerous of these is the terrorist, who has no hesitation in killing quite indiscriminately. He may value his own life less than his cause, and value the lives of others still less. Many such terrorists are now graduates from sophisticated schools of terrorism, are usually well aware of the measures currently in use to foil their activities and are knowledgeable in aircraft operational matters – sometimes even piloting. Crew members must become equally skilled in their own task of providing as much protection as possible for themselves and their passengers.

What the captain says to the terrorist and the way he acts towards him can play a significant role in the safety of all on board. A careless, hostile, provocative or sarcastic word from a crew member can seal the speaker's fate. On such occasions the only sensible objective of those on board is survival. Generally, behaviour should be that of compliance with the demands of the terrorists and avoidance of any action which might be interpreted as provocative. The atmosphere will be highly charged and tense and a word or action out of place could spell disaster.

The captain's role in such a seizure of the aircraft is crucial. He is usually the channel of communication between the terrorist and negotiators in the outside world. Without him the terrorist's means of success – the aircraft – cannot function. He is seen by the terrorist as having the authority to maintain orderly and manageable behaviour amongst the hostage crew and passengers. The crew should see resolution of the crisis as coming from outside, while their own role generally is a passive one of avoiding any provocation and concentrating on survival.

The criminal

The second category of hijacker is the criminal and while crew and passenger response to this type of menace should also be one of caution and compliance, he can usually be seen as less dangerous than the terrorist. He is usually more rationally concerned with his own safety and future and is not blindly dedicated to a political cause for which he is prepared to die. Martyrdom is not a common characteristic of the criminal.

The criminal may be escaping from the law, attempting to extort money, defecting or fleeing from political persecution (the term criminal may be debatable here). He may simply be trying to obtain publicity for himself or some cause. In fact, the first hijack occurred in 1930 and was of a Fokker FVII by a group of Peruvian revolutionaries who wanted to scatter propaganda leaflets amongst the population. During the half century that followed this incident, several hundred hijacks were reported – certainly the full total is not known – and 63% of those were carried out by defectors or escapees (Dorey, 1983). Most of the 113 reported hijackings between 1947–1969 were politically motivated, but would be classified as criminal rather than terrorist in nature.

Generally, both terrorists and criminals will want to maintain the effectivity of the flight deck crew to fly the aircraft, communicate with outside negotiators and to operate aircraft systems such as air conditioning and lighting. Flight deck crew members therefore may be in less peril than passengers. Cabin staff may be useful as hostages but are also of particular use to the hijacker to manage sometimes hysterical passengers, organise food distribution and generally to aid in maintaining order in the cabin.

The mentally disturbed

The third broad source of serious violence associated with attempted seizure of the aircraft is from the mentally disturbed passenger. This type of person is unpredictable and could, for example, receive 'a message' at any time to take a certain action. The dangerously mentally unbalanced passenger could be a paranoic schizophrenic, a psychopath or could be suffering from a depressive or neurotic illness.

He will usually be acting alone, which makes resolution of the crisis less complex than in the case of the terrorist who is likely to have accomplices.

Airline ground staff have a particular responsibility to report immediately to the senior passenger handling officer any passenger displaying unusual behaviour before boarding. A staff member with special training in approaching and evaluating such passengers should, if possible, always be available on call. If any doubt exists, the passenger should be prevented from boarding or allowed to board only under escort and with the captain's approval. Briefing of the cabin staff would then be necessary.

Situations which could create a notable increase in emotional pressure, such as forecast en route bad weather, could justify the captain refusing to accept such a passenger. Many airlines routinely carry on board restraining equipment, such as a rope and handcuffs.

Passenger response

From studies of hostage behaviour, it has been suggested that the victim tends to go through four phases (Strentz, 1985). First of these is the alarm phase, where the victim perceives a life or death situation. Some will seek refuge in a defence mechanism, denying reality. Those who are psychologically unprepared may panic.

The second phase is considered of critical importance for survival. This phase is the transition from initial shock to the acceptance of the reality of the situation – a return to some normality in behaviour. The passengers may begin to feel a sense of isolation and claustrophobia in a seized aircraft cabin and experience a loss of sense of time.

The third phase in a long hijacking case is one of accommodating to the situation. This frequently generates a feeling of boredom, perhaps interrupted with moments of terror when facing sudden threat of violence. During this phase some hostages may display signs of the 'Stockholm Syndrome', named after a notable incident in Sweden. This syndrome involves the subconscious development of positive feelings of compassion or sympathy by the hostage towards the hijackers with negative feelings against the authorities who are opposing them. The hijackers sometimes reciprocate this positive feeling. While there are some favourable aspects of this syndrome there are also negative aspects if it should be allowed to impede the activities of security staff.

The fourth and final phase of the experience is that covering the period of resolution. This may generate mixed feelings of hate for the distress caused by the hijackers and thankfulness to them for allowing survival.

The behaviour of passengers – who are likely to be quite unprepared for the traumatic experience – can influence their chance of survival as well as the long-term psychological damage they will suffer. Without specific guidance, however, their behaviour is likely to be a natural reflection of their individual personalities (Strentz, 1986).

Crew response

In many ways, the basic pattern of crew response to violent seizure of the aircraft is likely to be similar to that of passengers. However, with proper training, they will be far better able to cope with the situation and, to a considerable degree, will know what to expect and the most effective way to act.

With all such forms of violence on board, the primary task of the crew is to lower tension, be unprovocative and to establish credible, reasonable and non-controversial communication with the hijacker. It is essential that they avoid development of what has been called the 'London Syndrome' (Barthelmess, 1988). This was named after a situation in the siege of the Iranian Embassy in London when a hostage passionately argued the 'righteousness' of his case, finally getting killed. There is no sense in inviting martyrdom as a hostage in an aircraft hijacking and crew members should avoid such an attitude and encourage passengers to avoid it. It is the responsibility of airlines and regulatory agencies to ensure that there is a high level of security awareness amongst crews, through specialised, confidential training programmes. Some general guidelines are noted in Appendix 1.28.

As with the passenger, the extent to which a crew member may suffer long-term psychological damage as a result of a hijacking trauma depends very much on individual personality. However, the need for professional counselling in this as in other traumatic events is being increasingly recognised (Johnstone, 1985).

This chapter has discussed the need for a proper application of Human Factors to the design of equipment, software, the aircraft interior environment and the interaction between people in the cabin. The passenger will be aware that he can enhance his personal safety by seat selection and by paying attention to the safety information given by the crew. Enlightened observation may help him to select an airline when he travels which not only provides the most exotic drinks and has the most lithesome stewardesses, but which takes greatest account of the application of Human Factors in the cabin.

If some of the information in this chapter gives rise to concern, then this can only be in the best interest of air safety and the air transport industry. But in feeling concern, it is necessary to keep the matter in perspective. Most crew members and other air travellers will never need to make serious use of the aircraft emergency equipment or procedures throughout a whole lifetime of flying. And even a cursory examination of the safety precautions taken in other forms of transportation will reveal that other houses also need putting in order, though their failures tend to attract less media attention.

Nevertheless, lives are lost and efficiency reduced unnecessarily as a result of the inadequate application of researched, published and available information from the applied technology of Human Factors. Responsibility is not restricted to a single sector of the industry but must be shared by the airlines, manufacturers, regulatory authorities and staff unions, in whom the powerless and packaged passenger must place his trust.

14

Education and Application

Education

Selective omissions

These chapters have served as an introduction to Human Factors in transport aircraft operation. But within the covers of one modest book they cannot claim to be comprehensive and the discriminating reader will certainly find gaps in the total coverage. For example, personnel selection has not been discussed. This is a specialised field and while it is certainly a human aspect of the operation of aircraft it is not in the mainstream of our definition of Human Factors and has therefore been omitted.

We have also tried to exclude physiological aspects of flying which might come within the scope of aviation medicine. There are, of course, some grey areas here, as physiology is involved in Human Factors as well as medicine. But we have avoided where possible sickness induced by the flying environment and given more attention to performance. While Human Factors is also concerned with the abnormal person and the exceptional environment, we have intentionally avoided these in a general introduction and in view of the nature of the aviation industry. The reader is asked to be tolerant of our selection of the most relevant material with the most practical application.

Industry and academia

The late Professor Hywel Murrell, who in 1949 created the term 'ergonomics', wrote that the lift between the ivory tower of academia and the shop floor appears to have got stuck halfway. There are several reasons for this communication breakdown and the resulting inadequate application of Human Factors. From an organisational point of view there has been a lack of knowledge of the technology of Human Factors in the design office and behind the manager's desk. As a result there has been a lack of funding and support for Human Factors activity and a lack of proper Human Factors effort during system development. Things are getting better now, but there is still a long way to go. Meaningful understanding and adequate funding are still major problems.

From the user end of the industry – the airlines – inadequate enlightened demands have been made to the manufacturer to ensure proper and timely Human

Factors input. Here too the record is spotty but getting better. Formulation of such demands, of course, requires a level of airline Human Factors expertise which is lacking in many airlines and still available in only a very few. Of consumer products generally, it has been said that 'whenever the consumer consistently makes errors in the operation of a product, he has been cheated in the design process' (National Design Council, 1975). It is usually assumed that the designer will have made the necessary Human Factors input, but experience has demonstrated that educated pressure from the consumer may be necessary to ensure this.

Technical problems also exist. Academia does not usually translate research findings into operationally usable language or techniques. Indeed, it does not generally accept that it is its role to do so and has little interest in practical issues in industry unless invited by industry to become involved. There have been too few operators to keep Murrell's lift moving between the sources of research knowledge and those who are faced with the practical problems which they have to solve with inadequate background or resources. Furthermore, techniques are not always adequate to resolve human behavioural problems, even when the problems are recognised and the right questions presented. In a highly cost-conscious and basically safe and successful industry such as air transport, quantification of the cost-benefits which can be expected from investment in better Human Factors, demanded by financial controllers, is often very difficult to calculate.

And so there are technical as well as organisational problems in getting Murrell's lift moving. We can see the lift stopping at five different floors – academia, the design office, the manufacturer, the certifying authority and the operator – and a contribution should come from each one to ensure its proper functioning. It is unrealistic to expect such a contribution unless an appropriate level of Human Factors expertise is assured throughout. Education to provide such expertise is now available, and today the problem is beginning to be recognised.

Degree qualifications

Although it is possible at certain institutions to obtain a degree in ergonomics, several other degrees may also provide the basis for the profession and practice of Human Factors. In the same way that physiology provides a necessary foundation for medicine, so psychology provides the necessary foundation for Human Factors (see Appendix 1.1). Having made that basic distinction, it may be useful to look in more detail at the branches of psychology involved, sometimes more a question of semantics than of major practical significance.

In the UK, ergonomics was institutionalised as a technology in its own right with the foundation of the Ergonomics Research Society in 1949. Occupational and industrial psychology are branches of science not far removed from ergonomics. The term ergonomics is now in common usage in much of the world except in the USA, where acceptance is progressing only slowly. There, the term generally used is Human Factors, with the most common academic background being in industrial or experimental psychology and in engineering .

At degree level, then, Human Factors expertise can be founded on a variety of psychology-based academic courses. A breakdown of the academic background of members of the Human Factors Society emphasises the psychological orientation of Human Factors education. However, it should be recalled that the degree titles largely represent the US academic scene rather than the European one (see Appendix 1.2). Directories of Human Factors/Ergonomics educational programmes in the USA and internationally are available from the Human Factors Society in Santa Monica, California.

In order to illustrate the level and sources of Human Factors expertise which would be considered appropriate in any major organisation within the air transport industry, it might be helpful to make use of a model. This could be called the 'Company Human Factors Wigwam' (Fig. 14.1).

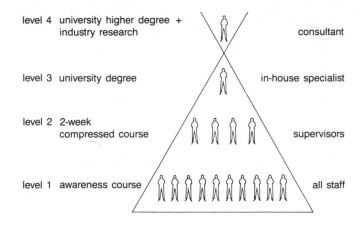

Fig. 14.1 The Company Human Factors Wigwam, indicating the various levels of expertise and education required within an organisation to ensure proper application of Human Factors (Hawkins, 1984).

Level 4: the Human Factors consultant

The highest level in the model is the Human Factors consultant. He is nominally outside the company structure though he may have a long-term association with it.

There are various advantages in a company using an independent consultant. He is not on the normal permanent payroll, with its vacation, pension and social security costs, and is only called upon when needed. He is able to make analyses and proffer advice free from internal organisational pressures and biases. He will have routine contact with industry and research developments which may be of benefit to the company he is advising, as the problems presented may already have been encountered elsewhere.

He will hold a higher degree (Master's or Doctor's) in one of the relevant psychology-based fields mentioned earlier and will have knowledge of the Human Factors problems of the air transport industry, including flight operations. Different consultants may have specialised knowledge in different aviation fields such as operations, engineering, or marketing, and a large organisation may make selective use of more than one such specialist.

Level 3: the in-house specialist

There seems to be little justification for any large organisation not employing, in-house, one or more degree-qualified Human Factors specialists. In fact, without some level of in-house expertise, Human Factors problems are unlikely to be recognised adequately to generate a call for reference to an external consultant. Even with the availability of the in-house specialist, there will still be a need to maintain links with an outside consultant.

Within an organisation, the Human Factors specialist will be available on demand to solve Human Factors problems wherever they occur. He should also be able to carry out research studies to provide answers to questions when the answers are not readily available from elsewhere; some assistance from the consultant may be needed here.

A modern aircraft is a workplace with a unique environment and with its own distinctive problems. The Human Factors specialist must acquire a full understanding of this work situation before the airline flight operations department will develop sufficient confidence to make extensive use of his skills.

Those who have been applying Human Factors on a DIY basis for decades may see the Human Factors specialist as a threat to their security and so may resist his input. Even when the specialist has ample operational experience, there still appears to be an entrenched attitude to avoid asking his advice or assistance. This attitude is reflected in the statement of one senior DIY official of a major operator at a European airline Human Factors seminar in 1984, which we quoted at the end of Chapter 1. 'In our view', he said, 'Human Factors is just an excuse for incompetence'. Perhaps this type of official must be seen, in the context of Max Planck's analysis quoted in the last few lines of this book, as a lost case. However, it does emphasise that without some degree of Human Factors education at supervisory level in a company, the task of the in-house specialist is likely to be a frustrating and largely fruitless one.

In the summer of 1988, a NASA/FAA/Industry sponsored workshop entitled 'Flight deck automation: Promises and realities' was held in Carmel Valley, California (Norman *et al.*, 1988). The workshop dealt with many of these issues. Representatives of European and American airlines, European and American pilots, the manufacturers, independent Human Factors consultants, academia, and NASA and the FAA all participated. There was little doubt that the importance of Human Factors was recognised in all groups although there was not always agreement on its implications (Norman *et al.*, 1988). It would have been difficult to have achieved such broadly based attendance had such a workshop been held in the previous decade.

Suitable degree courses for this Level 3 of expertise are available in the USA, Europe and elsewhere. In the USA it is possible to specialise in aviation Human Factors at universities such as Embry Riddle and Ohio State, though such specialisation is not available in Europe.

Level 2: supervisory staff

Some Human Factors education is required by supervisory staff to enable them to carry out their own tasks effectively. They should be able to recognise Human Factors problems as such and understand the kind of support and assistance which is available. They should be able to solve the more straightforward Human Factors problems and know when and from where to get help.

The level of expertise suitable for supervisory staff is likely to involve about a two-week intensive course provided by a university, technical institute or organisation specialising in aviation Human Factors education. General, non-aviation, ergonomics courses of this kind have long been available (e.g. that offered by the West of England Management Group). However, such courses, covering a wide range of working environments, including mass production manufacturing industries, are not really cost-effective for air transport staff. For this purpose, an industry-dedicated course is needed.

In 1972 such a course was established at Loughborough University, England, and later transferred to the University of Aston in Birmingham until 1983. The course, 'Human Factors in Transport Aircraft Operation' (HFTAO), was the responsibility of the Applied Psychology Department of the university and the course consultants represented the airlines and operational engineering side of the aircraft manufacturing industry. In addition to conventional lectures, the course provided for valuable interaction among supervisory staff. In the first ten years, staff from 30 different countries had attended this annual course. For smaller countries, where no similar expertise is available, the graduate from the course will have returned home, relatively speaking, a specialist and certainly with a new attitude to human problems. The course has also been conducted away from the university by arrangement with local organisations outside the UK; taking the course to the student rather than the reverse has proved a very cost-effective way of providing this kind of education.

A Human Factors course lasting four and a half days ('Schnittstelle Pilot/Flugzeug') was founded in 1979 and is offered annually at the Carl-Crans-Gesellschaft in Heidelberg. The course, which is given in German, is designed specifically for those working in the design, development and evaluation of aircraft systems rather than for those involved in operating aircraft.

The University of Southern California (USC) provides a five-day Human Factors course as part of the Aviation Safety programme offered by the Institute of Safety and Systems Management. It is less comprehensive than the HFTAO course and allocates more time to physiology and biomedicine. Both the Heidelberg and USC courses are rather short for the supervisory staff in Level 2.

The task of the supervisor in introducing new attitudes towards Human Factors, however, is likely to be a daunting and frustrating one unless all staff with whom he has to deal have a general level of awareness of Human Factors.

Level 1: all staff

Operations management may publish a crew bulletin on idiosyncrasies of a particular engine or navigation system and this is likely to be an effective communication. All flight deck crews have had formal basic education in the principles of engines and navigation upon which to found their comprehension and response. A similar bulletin on a Human Factors topic is likely to generate less comprehension and response if there is no formal Human Factors education upon which to draw.

Many deficiencies in the operational environment are corrected as a result of feedback from staff in the form of reports, letters, questionnaire responses or verbal communication. The quality and thus effectiveness of that feedback will be greatly enhanced if it is made with some basic knowledge of the technology involved, in this case, Human Factors. This is true whether the feedback is coming from a pilot on a question of an instrument display or from a check-in agent on the colour coding of luggage labels.

It is reasonable to assume that in time all basic technical education will include in the syllabus aspects of human performance and behaviour reflecting this technology. In fact, in 1981 in The Netherlands, the State Flying School (RLS) acquired from KLM the course described below and has integrated this into its two-year flying training programme. While ICAO's Flight Safety and Human Factors Programme has done pioneering work in furthering interest among its 166 Contracting States, it will probably be many decades before we can expect to see the air transport industry fully staffed with operational personnel qualified in this way. In the meantime it remains the responsibility of industry to fill this basic educational gap.

A number of criteria can be set down for the design of such a course for use on a large scale in aviation:

- it must provide coverage at an awareness level of the full range of Human Factors applicable to air transport operations
- it must provide education for large numbers of staff at low logistical cost per head
- the hardware must be such as to provide mobility to permit course presentation in different locations with minimal logistical cost or inconvenience
- the credibility must be high with very high Human Factors quality in content and hardware
- the presentation, format and content must be suitable for use by all staff with responsibility for the safe and efficient conduct of aircraft operations.

These criteria provided the basis for the course commissioned by KLM Royal Dutch Airlines in 1977. The 'KLM Human Factors Awareness Course' (KHUFAC) consists of 15 separate, self-contained audio-visual units, each on a different aspect of Human Factors and each lasting about 40 minutes. Single- and twin-screen tape-slide versions are available as well as video-tape. KHUFAC is the compact, audio-visual companion of this volume with unit titles corresponding with the book chapter titles.

In KLM, the course is supported by a library of relevant Human Factors books which are available for purchase by course participants. Some 800 were bought by staff during the first five years of the KHUFAC programme, reflecting a notable interest by them in the subject.

While this course utilises air transport examples, the principles discussed are relevant to any industry or activity. Interest in the programme has been shown by shipping, oil industry and medical representatives, all concerned with human performance in their own spheres of activity.

Further details of the education at these four levels are available elsewhere (Hawkins, 1984).

Selected Human Factors training

In addition to these general Human Factors educational facilities, a number of airlines have adopted training programmes to tackle specific Human Factors problems which they have detected within their own operational environments. An airline which mixes ex-military pilots with those from a civil flying school could find that this created certain problems which required resolution through training. Some countries which may have characteristically a generally lower level of achievement motivation (see Chapter 6) may need to make a special training input to adapt to the demands of an occupation requiring a high level of achievement motivation. The task distribution amongst flight deck crew and the professional background of the crew members may have an influence on job satisfaction, motivation and performance and the resulting situation might give rise to a need for special training for a particular airline.

While all such programmes can be seen as tackling Human Factors problems, they are designed to meet the specific needs of one airline and may or may not be suitable for adoption elsewhere. Furthermore, the 'language' and level of the course may or may not be optimum for other countries and cultures.

One training philosophy which has evolved over the years and has now been widely adopted is the training, and sometimes testing, of flight crews as a team rather than as individuals. This development is the outcome of a realisation that individual skill acquisition is not sufficient when the skill must be practised as a group activity and where personality and situational factors interact to create overall performance. It also follows from the conclusion that the safety-through-redundancy concept, which has for long been the *raison d'être* of multiple-crew complement, does not appear to have been sufficiently effective.

An example of this type of training has been called Line-Oriented Flight Training (LOFT) and has had particular application in the USA, where it has received FAA approval (Lauber *et al.*, 1983). A further refinement is in the use of video-taping of the training sessions so that the crew, as a team, can later review their own performance in a seminar-type environment, with an emphasis on self-analysis rather than instructor criticism. This type of programme is particularly oriented towards the *L–L* interface described in Chapter 1.

Flight deck resource management

A fashionable term with respect to aviation Human Factors is flight deck or cockpit resource management (CRM). A further refinement is to call it crew resource management to ensure that the cabin crew are included, and also to ensure that credit for what are largely semantic modifications (in the eyes of many) accrue to those responsible for the new terminology. These rather intimidating expressions simply mean the management and utilisation of all the people, equipment and information available to the aircraft. It is in principle no different from the management and utilisation of all the people, equipment and information available in any other workplace involving skilled activities in a technological environment. It could equally be applied to a hospital operating theatre where the surgeon is in command. It is appropriate to a television studio, where the producer is the leader and manager and is coordinating use of actors, electricians, scripts, script writers and cameramen and trying to do as an effective a job as possible. In fact, flight deck resource management can be seen as a special case of *SHEL* system management.

The thesis which generated this term with respect to aircraft operation originated from two conclusions. First, that accidents occur frequently because the resources available are not adequately managed by the aircraft commander and second, that sufficient resources were available to have avoided the accident had they been utilised in an optimum way. Very few air transport accidents cannot be categorised in this fashion.

As management is a human activity associated with work, this concept is directly within the scope of the technology of Human Factors. Elements of resource management on board include leadership, communication, task distribution, setting of priorities, monitoring of information sources and individual perform-ances and extraction and utilisation of data. These elements can also be applied to the operating theatre and television studio.

It is important to understand just what can and cannot be achieved by training in resource management. Personality traits may be innate and certainly are acquired and well established early in life. They are very stable and resistant to change. The problems and uncertain success of even long-term psychotherapy in trying to modify personality distortions is well recognised and was mentioned in Chapter 8. It would therefore be quite unrealistic to expect the captain who displays, say, aggressive tendencies, to experience a significant change in this underlying personality trait as a result of such airline training. This emphasises the need for applying effective and skilled selection techniques on entry to the airline. Neither will such training change the effect on performance of, say, domestic stress, fatigue, circadian rhythm disturbance, drugs or medication. Nor will it see the end of errors in information processing, eliminate visual illusions or do away with design-induced errors. What training can do is help better control these very real problem areas.

On the other hand, various management skills can be taught very effectively and these can be largely dissociated from personality. This kind of training has an important role to play. Attitudes should be seen as distinct from personality. They have been learned during life and although they can be irrationally resistant to change, they are less stable than personality traits and many can be modified through enlightened influence.

Analyses of aircraft operating problems which have been reported, notably through the FAA/NASA ASRS scheme, show that many can be classified in terms which are familiar to an ergonomist. These would include social and communication skills; planning, problem-solving and decision-making; role perception; workload distribution; human performance; controls and displays; documentation; human error; and so on.

Attention has been focused on developing training programmes which address these particular areas. They aim primarily at refining the *Liveware* to match the other components in the system – the *Software, Hardware, Environment* and, of course, other *Liveware*. While this is naturally a necessary activity in training, it is important not to lose sight of the need, as outlined in Chapter 1, of refining the task, including all system components, to match the characteristics of man. We need to ensure that not only are resources available, but that they are available in a way to facilitate processing and management. For maximum effectivity, all such training programmes must be built upon the foundation of a sound understanding of the technology of Human Factors.

Application

Aircraft hardware

Whose responsibility is it to ensure that Human Factors is properly applied in aviation? In the past there has been a fair amount of 'buck-passing' which, together with the educational gap just discussed, has resulted in years of inadequate action. Perhaps aircraft hardware is a good place to start in reviewing areas of responsibility.

It used to be commonly said of equipment manufacturers that their approach to the customer was 'you name it, and we'll make it'. When challenged on the lack of proper Human Factors input, the manufacturer might have replied that such input on their part would be largely a waste of time and money, as the customer would in any case modify the design to suit himself. And if there were 10 different customers for an aircraft, there would be 11 different cockpits and cabins, the basic model being unacceptable to anyone. The customer, on the other hand, would say that when he buys a product he expects it to be optimised from a Human Factors standpoint; he only uses it and does not design it.

While there is an element of truth in both these postures, they were too often adopted to avoid the necessity and the initial cost of doing proper Human Factors homework and applying proper Human Factors expertise to design. Things are better now – perhaps to a large extent because Human Factors are involved in a great many air transport accidents and because air transport accidents are very expensive both morally and financially.

Responsibility in this area is, in fact, shared amongst the designer, the manufacturer, the operator or airline and the state certifying authority. There are certain basic criteria derived from ergonomics which should be applied at the design stage, and no input should be required from the user for these. For example, optimum instrument scale progressions and graduations are well documented. Similarly, there are basic rules concerning controls such as direction of movement, coding and operating forces.

On the other hand, it is the user who has the experience of his own operating environment and this could be reflected in equipment layout based on the number and qualification of crew members, route structure, or commonality with other company equipment.

In spite of this division of responsibility, it has been the common experience of airline pilots to encounter design-induced errors such as those resulting from controls in too close proximity to each other, poor instrument displays, unsatisfactory seat design and inadequate visibility. User input at the early design stage, even in areas which should be the responsibility of the manufacturer, has proved to be essential to reduce the number of such deficiencies in a new product.

Certifying authorities employ specialists in system and hardware performance in order that they are able to make educated assessments of the quality of the product before certification. They should also have available Human Factors expertise to predict the influence of man in the system in the operational environment. This is gradually becoming a more standard practice for notable deficiencies have sometimes been discovered well after the aircraft has entered service.

Unfortunately, Human Factors input generally cannot usefully be left until a late stage of development or certification. Input after manufacturing engineering dead-lines have passed or, worse, after aircraft delivery, is likely to be found prohibi-tively expensive. Six stages where Human Factors input must be made have been proposed – system planning, pre-design, detail design, production, test/evaluation and operations (Meister, 1982). These, and the training required, were among the factors discussed at the previously mentioned NASA/FAA/Industry Seminar in Carmel Valley, California (Norman *et al.*, 1988).

In trying to make this input, the Human Factors specialist will meet many problems. The designer, who may have had no formal Human Factors education, may not take kindly to being guided in his design unless immediate and obvious benefits are apparent to him for the changes proposed. He faces engineering deadlines, which would rarely be modified on Human Factors grounds. When money is tight, it is usually this expertise which is abandoned first. The opera-tional requirements may not be specified in adequate detail at an early stage to permit proper Human Factors input. An example here is the Airbus Industrie A310 which started its design process as a three-man crew aircraft. During development, as a result of customer commercial pressures, it was required to be certified for a two-man crew operation, calling for some fundamental flight deck redesign to ensure an acceptable work environment.

A further and significant element in the amount of Human Factors input made by the manufacturer is that it will reflect the customer's, that is, the airline's, demand for such input. This is no different from the way in which seat-mile costs, noise levels and maintainability reflect customer requirements. Without a structure of Human Factors expertise within the operator's organisation such a demand is unlikely to materialise.

The cost of making a different aircraft model for each customer has become prohibitive and this is having an influence on Human Factors input. Groupings of airlines such as ATLAS (Alitalia, Lufthansa, Air France and Sabena) and KSSU (KLM, SAS, Swissair and UTA) have arisen to standardise aircraft. This has tended to increase the power of the moderately sized airline relative to the manufacturer but has inevitably led to compromises, not all of which are favourable from a Human Factors standpoint. The cost of making changes from a basic aircraft model has increased the advantages of getting the design right in the first place through timely Human Factors input. Nevertheless, Human Factors deficiencies are still being designed into new aircraft, partly as a result of inadequate specification requirements from the customer.

Training

Flight crew training is largely the responsibility of the operator, though the regula-tory agency – the national CAA – is normally required to approve the training methods of operators within its jurisdiction. The manufacturer is usually responsi-ble for training an initial group of instructors from the airline. And so, once again, the responsibility for adequate Human Factors input is shared.

As was apparent from Chapter 9, Human Factors is involved in the techniques of instruction as well as in training hardware and software. In view of the fact that training represents a major cost item in an operator's profit and loss account, it is important that optimum cost-effectiveness and training transfer are achieved.

Operating procedures and documentation

It will be apparent from earlier chapters that the design of operating procedures and documentation can be expected to have a significant influence on safety and efficiency in aircraft operation.

Usually, basic operating procedures are designed by the manufacturer and then approved, firstly, by the certifying authority of the state of manufacture and, secondly, by the customer's own CAA. The certification process is not simply a formality or a rubber-stamp exercise. However, it is unrealistic to expect too much of the limited staff of that authority and at least some of the information provided by the manufacturer must be taken by them for granted. Responsibility is shared, as illustrated by the FAA acceptance of a proportion of the legal liability claims which followed the Paris DC10 crash in 1974, caused by design deficiencies in the cargo door.

The primary responsibility, however, for Human Factors input in operational procedures and documentation must reside with the airline or operator, under the supervision of its national CAA. It is usual for national regulatory agencies to accept the operating standards established by the authority of the state of certification, subject to minor additions. A notable human aspect of certification relates to crew complement, that is, the number of crew members and their function (e.g. pilot, flight engineer, pilot/engineer). The manufacturer's state authority will certify the aircraft with a minimum crew complement but this can be increased by the operator's national authority, which may also be more specific with regard to the function of the crew members concerned. The same principle applies to the approval of aircraft operating documentation.

Inadequate Human Factors input in the early stage of system and equipment design may finally necessitate the use of complex normal and emergency operating procedures which in turn may increase the risk of human error. While the true responsibility for the error in such cases may go back to the original design, it is the operator who is usually left 'holding the baby'.

Staff/management relations

In spite of the extensive documentation available on the subject of personnel relations, practical application of this knowledge often leaves much to be desired. Frequently, there is a failure at supervisory level to understand the relative merits of various aspects of Human Factors influencing work efficiency and harmony. These include, amongst others, positive and negative behaviour reinforcement, job satisfaction, job enlargement and enrichment, motivation at work and the effects of emotional stress. Union/management relations often seem to be based on confrontation rather than cooperation. Experience suggests that with respect to the teaching and application of Human Factors, cooperation with trust between staff associations and management can prove fruitful and constructive.

Accident and incident investigation

Perhaps nowhere is the public made more aware of the existence of aviation Human Factors than in media coverage of aircraft accidents and incidents, though the content of reporting normally reveals little knowledge of the technology.

Human error sells books and newspapers. The Charge of the Light Brigade at

Balaklava in 1854, immortalised by Tennyson, resulted in the loss in dead and wounded of 303 of the 'noble six hundred' unquestioning soldiers taking part. This would never have received such literary and film attention if it had not resulted from a catastrophic series of human errors; 'not tho' the soldier knew some one had blundered ...' Similarly with the loss of 356 lives in HMS *Victoria* in 1893 when two ships on exercise in the Mediterranean collided. The London Moorgate underground disaster in 1975 killed only 43, yet the unexplained human failure involved has kept discussion alive ever since.

Closer to aviation, the Staines Trident accident, the DC10 cargo door failures and the Mt Erebus crash all generated dedicated books in which human failures at various stages were elaborated. And the milestone of aviation disasters, at Tenerife in 1977, provided a catalogue of Human Factors deficiencies for discussion. The destruction by the USSR of the Korean Flight 007 in 1983 with the loss of 269 lives, led to many hypotheses of the possible sequence of human errors on board which could have generated the navigation diversion which brought it over Soviet territory.

All this might suggest that Human Factors expertise is now recognised as an essential component in accident and incident investigation. Yet in the 1975 IATA Istanbul Conference conclusions it was declared that 'there is an urgent need for the inclusion of Human Factors experts in all accident investigation teams. Airlines should consider the use of the same type of expertise in their in-house safety programmes'. IFALPA's 1977 Washington Conference concluded 'that Human Factors aspects of accident/incident investigation should be more deeply pursued'.

In the same year it was written in the prestigious journal *Human Factors* in the USA about the NTSB that its 'so-called Human Factors sections of accident investigation are medically oriented rather bland accounts of accident survivability and physical conditions of the crew, with little or no recognition of the man-machine or information-processing aspects of Human Factors that may have prevented the accidents' (Wiener, 1977). This confusion between medicine and Human Factors was discussed in the first chapter of this book as a significant obstacle to progress and it still persists in many important aviation bodies both national and international. However, it must be added that since 1982, the NTSB interest in Human Factors has been notably expanded and more specialised staff have been recruited and engaged in accident investigations. At a later date and after considerable external pressure, the operational branches of the FAA also began to show increased interest in Human Factors.

In the UK in 1986, the Air Accidents Investigation Branch (AAIB) reported having no staff qualified in Human Factors. Its accident investigators received no formal Human Factors education either before or after they attended the seven-week Aircraft Accident Investigation course at Cranfield, even to the educational Level 1 quoted earlier in this chapter. It was the view of the AAIB that if the investigator concluded that he required specialised Human Factors assistance, he could obtain this on an *ad hoc* basis from the RAF Institute of Aviation Medicine.

As a notable exception, and in contrast to the UK AAIB, the Australian Bureau of Air Safety Investigation (BASI) has a degree-qualified Human Factors specialist in a senior managerial function. It also provides formal Human Factors education for all its investigators to Level 1 using the KHUFAC audio-visual programme as a basis. The organisation in BASI now fully conforms to the structure in Fig. 14.1. Australia's air safety performance has long had international recognition.

In 1984, the International Society of Air Safety Investigators (ISASI) felt able to allocate a half-day discussion at its conference at London Airport to the subject of 'Human Factors investigation is a waste of time'. Although the title was intended to stimulate debate, the fact that debate was even possible on such a premise reflects both the practical difficulties and the entrenched attitudes that have been associated with Human Factors investigation.

Many of the difficulties are organisational and legal rather than scientific. Public pressures and encouragement from insurance and legal sources, which are themselves tending to take a more serious interest in Human Factors, has also stimulated progress.

While responsibility to ensure skilled Human Factors input clearly resides with the body conducting the investigation, others, such as staff associations, insurers, the aviation media and international bodies, can all have an influence, if they have done their homework.

The regulatory authority

Many of the basic criteria employed by regulatory agencies are based on the Annexes of the ICAO Bermuda Convention. Until comparatively recently, neither these nor operational training programmes conducted internationally by ICAO reflected any recognition of Human Factors as an applied technology or that such expertise was available. This is not true today. Formerly, national regulatory authorities had no international basis for requiring Human Factors knowledge as an element in training. However, in 1986 the ICAO Assembly set the foundations for a programme for aviation Human Factors for all nations by passing Assembly Resolution A26-9. ICAO's Air Navigation Commission defined the objective of this resolution by stating that its purpose was 'to improve safety in aviation by making States more aware and more responsible for the importance of Human Factors in civil aviation through the provision of practical Human Factors materials and measures, developed on the basis of experience within States' (ICAO, 1989). Then three years later, the ICAO Council enacted a revision to the 1989 update of ICAO's Annex I. Annex I now requires that all pilots in the future be familiar with 'human performance and limitations' as they relate to their flying activities and the formal privileges of their licence.

In reality, of course, it was not necessary for progressive national flying training institutions to await recommendations from ICAO before taking action. As mentioned earlier, The Netherlands State Flying School (RLS) in 1982 began integrating the KHUFAC audio-visual awareness course into its normal two-year flying training programme without any statutory pressure. Today, this type of initiative is becoming less of an exception.

In spite of the vitally important role played by Human Factors in safe and efficient air transport, it has not been usual for a CAA (Civil Aviation Authority/ Administration) to employ a permanent and degree-qualified Human Factors specialist. In 1983, however, the Swedish CAA acquired the KHUFAC course described earlier and this may have been the first such authority to establish a formal programme of Human Factors education for its staff.

Of the FAA it has been said by a leading US aviation Human Factors authority that it 'is mired in politics, sluggish and seemingly unable to make changes and recognise hazardous conditions until accidents and their resultant political pressures impel them' (Wiener, 1977). This environment is slowly being modified for many of the reasons already mentioned. However, it still may not be an environment that is totally conducive to a progressive approach to Human Factors.

Marketing

While marketing is outside the field of aircraft operation, use is often made of operational information and sometimes operational staff in marketing activities and so perhaps reference should be made to it here. Marketing is a specialised activity in which an understanding of human behaviour is of paramount importance. In Chapter 13, a number of special cases were discussed which can have relevance to market potential. Various phobias, including flight phobia, and common handicaps such as diminished hearing and vision capacity can create resistance to air travel and can also have an influence on operational procedures. Considerable work has been done on minimising the effects of such handicaps and the resistance to flying which may accompany them. Marketing staff are responsible for acquiring a knowledge of this work and seeing that it is applied. This is one area where responsibility is not shared. The extent to which such factors are taken into account varies very widely from one airline to another.

Quo vadis?

Credibility

Human Factors aspects of any Human Factors teaching programme must be optimised. Credibility will be rapidly lost if the presentation technique employed and the quality of the material do not reflect the standards of Human Factors being advocated. Human Factors is a dynamic technology and any Human Factors educational programme also needs to be kept up-dated if it is to retain credibility and achieve maximum effectiveness.

Another component of credibility is trustworthiness and it is important that the programme does not come across as company propaganda or an exercise in exhortation. There is a well of natural interest of staff in this technology and care must be taken to ensure that it is properly nurtured and not abused.

Resistance to progress

Early in this volume, reference was made to some reasons why progress in the better application of Human Factors has been retarded. There are sometimes vested interests acting as obstacles to change. However, often the origins lie in inflexible and out-dated attitudes and managerial financial concepts, though these are far less likely to exist when there has been some exposure to formal education in the technology.

Chapter 1 opened with a quotation from philosopher and mathematician Bertrand Russell, emphasising the difficulty in getting people to listen to newly discovered concepts. Physicist Max Planck (1858–1947) went a stage further when he said,

'An important scientific innovation rarely makes its way by gradually winning over and converting its opponents ... what does happen is that its opponents gradually die out and the growing generation is familiarised with it from the beginning'. While practitioners of Human Factors will recognise a substantial degree of truth in this pessimistic view, it emphasises the need for more formal education in the technology, particularly amongst those destined for future managerial roles.

Managerial financial concepts sometimes present a different problem. Improved application of Human Factors, including better education, may initially involve some investment of money, though this is not always the case. It is a common and prudent accounting practice to demand an estimate of the cost-benefit which can be expected from any new financial investment. In some cases it is indeed possible to conduct an experimental Human Factors research programme so as to estimate the reduction in costs which may be expected. However, such studies are themselves not without cost and can sometimes be extensive and complex if meaningful results are to be assured.

In other cases such experimental programmes are not practical. Estimates of the future saving of life in accidents by better application of Human Factors in the cabin equipment and environment can only be made in the most general terms and such quantification cannot be applied to any single airline or operator. Similarly, it is virtually impossible to quantify the benefits in safety and efficiency which may accrue to a company from the elimination of Human Factors deficiencies on the flight deck. We must frequently rely on enlightened attitudes and policies rather than on accounting methods.

Expectations from the better application of Human Factors

It is easier to demonstrate historically the cost of human error and inadequate Human Factors than to quantify the future benefit which can be expected from applying Human Factors in a more enlightened fashion.

With more effective Human Factors in communication we would have had no Charge of the Light Brigade to write and talk about and no Tenerife collision to establish a black landmark in aviation history. The million dollar Gemini 9 space docking project would not have failed if Human Factors in documentation had been better applied. Countless hours have also been shown to have been lost in industry due to inadequate documentation and similar deficiencies on board may have played a role in aircraft accidents. The Challenger space shuttle disaster in 1986 may have been avoided if more consideration had been given to Human Factors aspects such as communication, fatigue and circadian effects on performance before the vital but misguided decision was made to proceed with the launch. And the eventual and as yet undetermined final cost in human life and suffering resulting from the Chernobyl nuclear catastrophe, also in 1986, would have been avoided but for the catalogue of human errors which preceded it.

Looking back, the list of benefits from better application of Human Factors is so extensive as to reach beyond the scope of an introductory book of this kind. Looking forward, it might confidently be expected that better application of Human Factors in aircraft operations will:

- improve user feedback and thus the quality of the tools or the product
- improve the effectiveness of documentation and other forms of communication and instruction

- improve the quality of leadership and resource management
- reduce human error as a result of increasing an awareness of human limitations and behaviour
- reduce the consequences of human error by better design of equipment and procedures
- minimise adverse environmental effects on personal well-being and performance
- favourably modify attitudes and enhance motivation
- raise the level of job satisfaction
- improve working efficiency by optimising the design of displays, controls and the workplace, including the flight deck and the cabin
- increase passenger satisfaction through a better staff understanding of human behaviour and needs and by refining the cabin environment
- improve the cost-effectiveness of training.

There is a mass of evidence to demonstrate that the good application of Human Factors, like the maintenance of good health, is intrinsically desirable and that its neglect is likely to prove costly. As with good health, better education is the key to success.

Without such education, progress will remain slow and tortuous and the human and financial penalties will continue to be paid. With it, will gradually come a proper allocation of resources and increased utilisation of expertise to solve Human Factors problems in aviation so as to ensure a higher level of safety, efficiency and individual well-being.

Appendices

1. Notes on Procedural Recommendations and Other Data

1.1. NASA-Ames 'Aerospace Human Factors Research Division'. A breakdown of the educational background of Federal and contractor employees involved in research as of July 1992.

Psychology	32	Computer Science	1
Experimental psychology	8	Aviation Medicine	1
Cognitive psychology	5	Behavioural Sciences	1
Psychophysiology/Clinical psychology	2	Biology	1
Cognitive/Mathematical psychology	1	Business Management	1
Cognitive engineering	1	Education	1
Aeronautics and Astronautics	4	Formal Sytems	1
Mathematics	3	Industrial Engineering	1
Electrical Engineering	4	Law	1
Electrical Eng. and Computer Science	2	Nursing	1
Aeronautical Engineering	2	Physics	1
Architecture	3	Social Sciences	1
Communications	2	Sociology	1
Computer Music	1	Spanish	1

1.2. The Human Factors Society (HFS). A breakdown of the HFS membership by highest academic degree obtained, expressed as a percentage of the total Society membership (1984 and 1992).

	1984	1992		1984	1992
Psychology	51%	44%	Education	1%	2%
Engineering	16%	19%	Computer science	1%	1%
Human Factors/Ergonomics	6%	8%	Safety	—	1%
Industrial design	4%	3%	Others	7%	6%
Med/Phys/Life Sciences	4%	4%	No degree (incl. students)	8%	9%
Business administration	2%	2%			

1.2A. Eye witnesses; testimony precautions.

- obtain a written or recorded statement as soon as reasonably possible after the event
- prevent unskilled questioning from any source
- prevent exposure of the witness to external influence, particularly by interested parties, before official interviewing
- assess any possible influence of the location, environmental conditions and skill of the witness in making the observation
- assess to what extent expectation may have played a role in witness recollection
- assess the possible influence of trauma and the nature of the event on the witness

- assess the independence of the witness and any possible influence of loyalty, obligation or the fear of punitive consequences
- check any early statement against later ones to determine possible occurrence of in-filling and post-event enhancement of memory
- do not misconstrue witness confidence as testimony accuracy

1.3. Dietary concepts suggested as being associated with acceleration of circadian rhythm resynchronisation, though not necessarily suitable for flying crew.

- maximise the synchronising effect of mealtimes by alternating daily fasts with three full high protein meals on alternate days just before the flight
- restrict caffeinated drinks to the appropriate time in the circadian cycle on the day of departure; in the morning only for westbound flights and evening only for eastbound flights
- after arrival, no caffeinated drinks in the morning
- after arrival, take a high protein meal at breakfast and lunch to facilitate the rise in brain catecholamine synthesis associated with the active phase of the circadian cycle
- after arrival, take a high carbohydrate evening meal so as to facilitate an increase in brain serotonin synthesis which typically precedes sleep
- after an eastbound flight go to bed early; after a westbound flight go to bed at the normal time for the new location

1.4. Precautions before the use of hypnotics by flight crews.

a) Advice for crew members
- consult a physician familiar with aviation before using hypnotics
- know the half-life of the drug
- know the effect on performance
- use the lowest dose that is effective
- use only when really necessary
- do not normally carry out a skilled task within 24 hours of using the drug
- avoid alcohol within 24 hours of using the drug
- first try non-drug techniques to facilitate sleep

b) Physician's considerations
- the toxicity and the overdose risk
- the effect on performance
- the addictive characteristics
- the effect combined with other medication
- the effect combined with alcohol
- personal tolerance
- the expected frequency of use
- clinical insomnia indications requiring counselling or a non-drug technique
- personality factors

1.5. Some statistical terms used in the description of data distributions.

Mean – the average; the sum of the scores divided by the number of scores added.
Median – the score or measure which would divide the total score or measure in half.
Mode – the most frequently occurring score or measure in a distribution.
Normal (Gaussian) distribution – a bell-shaped distribution curve with diminishing frequencies about a point with greatest frequency. In an ideal normal curve, the frequency diminution is symmetrical and the peak frequency is at the mid-point of the curve.
Skewed curve – a normal distribution in which the peak frequency is to the left (positively skewed) or to the right (negatively skewed) of the mid-point.
Standard deviation – the square root of the mean of the squares of the deviation from the mean.

1.6. Optimisation of sleep potential in hotel rooms.

- emotional stress or excitement should be avoided prior to sleep; this is particularly relevant to interpersonal friction but can also arise from other causes
- a high level of cognitive activity (e.g. study) should be avoided close to the sleeping period
- the thermal environment (e.g. bed covers and room temperature) should be properly controlled
- good ventilation should be ensured
- reasonable relative humidity should be established as far as possible
- the room should be capable of complete darkening with opaque curtains
- the room should be totally quiet, so remote from lifts, concourses, car parks, musical entertainment or main entrances
- the bed should be firm but comfortable
- stimulants such as coffee or tea should be avoided prior to sleeping; smoking has been shown to have a slight stimulating effect
- the sleep situation should be approached gradually rather than suddenly

1.7. Exercises which are suitable for use on board the aircraft.

a) Body exercises
- shoulder shrugs
- high stretch
- knee flex
- neck bends and twist
- stomach pull-ins
- seated side bends
- forward bend
- straighten out
- torso twist
- bent arm circles

b) Eye exercises
- rotations – move eyes to all directions
- fixations – focus consecutively at different distances
- alternate eye blinking – 20 times
- full blinking – 20 times
- converging, diverging – repeatedly focusing near nose and far away

1.8. There are a number of conditions when it is advisable not to jog.

- when it is too cold or too hot
- when feeling over-tired
- with a minor illness, such as a cold
- when it appears excessively hard work
- if the pulse rate becomes too high
- on medical advice

1.9. There are various ways in which a tendency towards alcoholism can be recognised by close associates.

- preoccupation with the next drink
- high tolerance to alcohol
- drinking fast or gulping drinks
- often drinking alone
- use of alcohol as a tranquilliser
- loss of memory of events when drunk
- protection of alcohol supply
- loss of control of drinking

A list of publications and teaching aids for alcohol education is available from the Health Education Council in London and elsewhere.

1.10. A small selection of the life events surveyed and the Life-Change Units allocated to them (Holmes *et al.*, 1967).

Death of a spouse	100	Financial change	38
Divorce	73	Death of friend	37
Sickness	53	Mortgage	31
Marriage	50	In-law trouble	29
Retirement	45	Sleeping habit change	16

1.11. Major sources of some vitamins (WS = water soluble; FS = fat soluble).

A — fish oils, liver, butter, eggs, margarine, cheese, milk, many yellow, red or dark green vegetables (e.g. carrots, tomatoes), fruits (FS)

B1 — wheatgerm, wholegrain cereals, lentils, pork, nuts, yeast extract, potatoes (WS)

B2 — brewer's yeast, liver, meat extract, cheese, eggs, peanuts, beef, wholemeal bread, milk, fish (WS)

B3 (niacin) — bran, wholegrain cereal, yeast extract, lentils, liver, kidney, meat, fish; can be synthesised from tryptophan (WS)

B6 — meat, liver, vegetables, wholegrain cereals, bran (WS)

B12 — meat, liver, eggs, milk (WS)

C — citrus fruits, currants, berries, green vegetables, potatoes (especially when new) (WS)

D — sunlight, oily fish, butter, eggs, margarine (FS)

E — wholemeal flour, nuts, wheatgerm, eggs, unrefined vegetable oils (FS)

K — vegetables, peas, cereals (FS)

1.12. Advantages of white cockpit lighting over red.

- permits effective colour coding of displays and use of red flags
- permits colour coding in documentation, maps and charts
- reduces heat dissipation problems in incandescent integrally lighted instruments and thus improves instrument reliability and life
- reduces power requirements
- reduces eye fatigue and the effect of hypermetropia and presbyopia
- improves instrument and display contrast ratios for good readability
- provides better illumination in thunderstorm and daytime conditions

1.13. Low visibility landing categories (ICAO Manual of All-Weather Operations, AN/910, 1982).

Category I
    Decision Height (DH); not below 60 m (200 ft)
    Visibility; either not less than 800 m or
    Runway Visual Range (RVR); not less than 550 m

Category II
    Decision Height; below 60 m (200 ft) but not below 30 m (100 ft)
    Runway Visual Range (RVR); not less than 350 m

Category IIIa (fail-operational)
    Decision Height (DH); either less than 30 m (100 ft) or no DH
    Runway Visual Range (RVR); not less than 200 m

Category IIIb
    Decision Height (DH); less than 15 m (50 ft) or no DH
    Runway Visual Range (RVR); less than 200 m but not less than 50 m

Category IIIc
    Decision Height (DH); none
    Runway Visual Range (RVR); none

1.14. A small selection of motives from many which have been distinguished and studied (Murray, 1938).

Abasement – passive submission to external force
Achievement – striving for achievement for its own sake
Affiliation – wanting to be liked and accepted
Aggression – a drive to overcome opposition forcefully
Deference – admiring and supporting a superior
Dominance – controlling one's human environment
Order – putting and maintaining things in order
Play – acting for amusement without further purpose
Power – influencing the behaviour of others
Understanding – asking or answering general questions

1.15. Leadership requirements as might typically be applied to a captain or a purser in a civil aviation environment.

- cordial but business-like relations on board
- firmness with tact in pursuing company goals
- teamwork and cooperation encouraged
- sympathetic attention to personal difficulties
- reinforcement of desired performance
- setting an example of desired behaviour
- demonstrating technical competence

1.16. The ICAO standard alphabet (ICAO Annex 10).

A – alfa	H – hotel	O – oscar	V – victor
B – bravo	I – india	P – papa	W – whiskey
C – charlie	J – juliett	Q – quebec	X – x-ray
D – delta	K – kilo	R – romeo	Y – yankee
E – echo	L – lima	S – sierra	Z – zulu
F – foxtrot	M – mike	T – tango	
G – golf	N – november	U – uniform	

1.16A. Avoidance of errors in communication readback.

- request re-confirmation when any doubt exists
- maintain a 2-man listening watch to provide monitoring of the communication
- use standard phraseology
- speak the critical part of the readback more slowly and clearly and with emphasis
- always give the aircraft callsign clearly and fully
- be alert to the existence of other aircraft with a similar callsign
- do not assume lack of response to mean confirmation
- recognise that pilots and controllers are not immune to error

1.17. Computer-based training (CBT) may be desirable for individualised, self-contained training, with little involvement from an instructor.

The following conditions might suggest that CBT is suitable (Medsker, cited in Bailey, 1982).
- mastery of specific, measurable objectives by each learner is required
- active responding by the learner (practice) is needed
- job conditions are constant or predictably variable, which minimises the need for on-the-spot adaptation to local conditions
- long development time is acceptable
- subject matter is fairly stable

- a large number of people will take the training
- flexible scheduling of trainees is desired
- instructors are expensive or difficult to find

If the following conditions prevail, then CBT may not be desirable.
- a requirement to adapt training to rapidly changing subject matter or job conditions
- when time or resources are limited
- when courses are not for use over a long period of time

1.18. The qualities which should be evaluated in the selection of pilot candidates for training as flight or simulator instructors or check pilots.

- competency as a pilot on the aircraft type
- interest in teaching and the principles of learning
- ability to communicate and an interest in communication methods
- ability to analyse problems and present material in a clear and credible manner, verbally and in writing
- personality traits such as patience, tact, sense of humour, intelligence and integrity
- interest in human behaviour and performance
- appreciation of the significance of the flight deck management role of the captain
- appreciation of the significance of the management role of other crew members and their ability to work as a member of a team
- appreciation of flight deck technical development
- appreciation of Human Factors as a technology
- acceptability to all potential students
- appropriate sense of responsibility and authority
- constructive attitude towards company objectives

1.19. Recommendations for effective lecturing and instructing.

DO:
- properly prepare the lecture material
- speak with authority and confidence
- be patient, considerate and polite with students
- understand individual learning differences
- retain a sense of humour and try to make the session a pleasurable experience
- use appropriate gestures, mannerisms and movement
- use proper voice modulation and volume
- use a language that the audience can understand
- periodically assess the suitability of the pace of the lecture

DON'T:
- embarrass or humiliate students
- criticise the equipment or the system
- over-control the class
- be excessively formal or informal in appearance
- talk down to an audience
- be sarcastic or cynical
- use obscene or profane language

1.20. One example of a simple method applied to test readability of text in the USA (Gunning, 1968).

- take a sample of 100 words of text
- calculate the average number of words per sentence
- count the number of words with three or more syllables
- add the two above items
- multiply by 0.4
- the result equals the reading grade level (USA)

Note: to obtain a reading age, add five to the US reading grade level, i.e. a grade level of 5 gives a reading age of 10 years.

1.21. Typographical characteristics which reduce the clarity and effectiveness of graphic material (Hartley, 1981).

- reversed lettering (e.g. white lettering on black)
- show-through of the print from the reverse side of the page
- words set at an angle to the horizontal
- lines haphazardly connecting labels to reference points
- a wide variety of type sizes and styles

1.22. A summary of the operational advantages and disadvantages associated with the forward-facing and lateral-facing third crew member's station on the flight deck.

Forward-facing
- closer crew proximity for communication
- third man more flight orientated
- third man available for radio/navigation tasks
- less flight deck space required
- all system controls available to pilots
- possible pilot distraction by system displays and controls
- monitoring of overhead panel inconvenient
- panel space restricted
- pilot seat access restricted
- work table for third man not convenient

Lateral facing
- third man more systems orientated
- extensive and continuous displays possible
- good panel monitoring angle for system displays
- pilot seat access free in cruise
- pilots not distracted by systems
- better workspace, stowage and seating possible
- more flight deck space required
- difficult pilot access to systems
- crew communication distance greater
- radio/navigation and outside visual tasks less convenient for third man

1.23. Use of the public address (PA) system.

- think out the announcement before speaking
- avoid excessive announcements
- speak directly into the microphone
- do not use aviation jargon or slang
- speak each word distinctly

- speak at an even rate
- speak slower than in normal conversation
- remember that emotions can come over the PA system
- report technical deficiencies for maintenance action

1.24. Handling of those with hearing difficulties.

- identify those with hearing problems
- indicate where sympathetic attention is available
- speak slowly and clearly but do not shout
- reduce background noise
- face the light and the person
- leave the mouth visible for lip reading
- use plain language; rephrase if necessary
- advise on visual sources of information available
- indicate location of hearing aid plug-in station if available
- use visual aids in safety briefing
- be considerate, not patronising
- pass on information to subsequent staff member

1.25. Handling of the visually handicapped.

- offer assistance but do not insist
- the handicapped person holds the helper's arm
- offer pre-boarding, but do not insist
- know emergency requirements (FAA-AM-80-12)
- explain the menu and meal tray layout (clockwise)
- give personal, tactual, safety briefing
- understand guide dog requirements:
- dog stays with owner on board
- do not touch, pet or interfere with the dog
- in a slide evacuation, dog exits on owner's lap

1.26. Fear of flying.

a) Recognition
- admission of fear by passenger
- unusual irritability, hostility or aggressiveness
- unusual loss of inhibition
- excessive sweating
- tense face
- rigid, immobilised seat posture
- withdrawal
- excessive talking, humming or whistling
- avoidance of certain seat locations
- immediate desire to disembark after boarding
- immediate desire to smoke or drink after boarding
- excessive mobility or restlessness
- visual appealing to cabin staff after turbulence, etc.
- questions to staff revealing concern about safety

b) Handling
- display professionalism plus understanding
- consider passenger preference for male/female attendance
- avoid reinforcement of anxiety
- use caution with alcohol as response is unpredictable
- do not use oxygen to treat hyperventilation
- advise on the availability of professional therapy
- offer choice of seat if possible

c) Professional therapy
- consideration also given to other phobias
- behaviour therapy as a basis
- multi-dimensional approach used
- relaxation technique taught
- systematic desensitisation used
- habituation used
- principles of flight explained
- exposure and confrontation gradually introduced

1.27. Assault on board.

- refuse to accept boarding of drunk or violent passengers
- use caution with alcohol on board
- warn passenger of off-loading and arrest
- be aware of the location and use of restraining equipment
- notify purser and captain
- radio ahead for police to meet the aircraft on arrival
- record the passenger's name
- obtain names and addresses of witnesses
- report the incident in writing after arrival

1.28. Crew behaviour and hijacking.

Prevention
- closely observe behaviour of boarding passengers
- watch for unauthorised persons on board during transit stops
- report immediately any suspect or unclaimed package/luggage on board

Handling
- avoid provocation, argument or confrontation
- generally, comply with the hijacker's demands
- engage in non-controversial conversation if possible
- obtain as much information as reasonably possible about weapons and accomplices
- if possible, keep hijacker away from the flight deck
- establish reasonably 'friendly' relations with the hijacker
- when possible, play for time
- remain calm and lower cabin tension or excitement
- follow the captain's guidance, when available
- in principle, leave resolution to outside authorities

2. Recommended Books, Journals and Bulletins

These lists are designed to provide a student and an organisation with guidance on the availability of Human Factors literature related to the subject matter of the different chapters of this volume.

Appendix 2.1 lists further reading and attempts to select the more easily readable and more reasonably priced books which are available. In some cases, books which are out-of-print have been retained on the list where no suitable alternative is available. Such books may still be found in libraries or second-hand book shops. It is unfortunate that paperback publications tend to go out-of-print more quickly than the more expensive hard-cover issues.

Appendix 2.2 lists books which, in addition to those in Appendix 2.1, could form the basis of a Human Factors library but which can be seen as reference books and are in most cases too expensive for personal purchase.

Appendix 2.3 lists the journals and bulletins which should be available in any aviation Human Factors library. There are, of course, many more covering specialised fields of Human Factors, such as *The Journal of the Acoustical Society of America, Instructional Science* and *The Journal of Verbal Learning and Behaviour*.

2.1 Recommended reading

OP = Out-of-print
PB = Paperback
HB = Hardback

Chapter 1

Edholm O G
  Biology of Work; Weidenfeld and Nicolson; (OP)
Green R G, *et al.*
  Human Factors for Pilots; Avebury Technical (PB)
Jensen R S (Ed)
  Aviation Psychology; Gower Technical (PB)
McCormick E J & Sanders M S
  Human Factors in Engineering and Design (5th Edition); McGraw Hill (PB)
Murrell H
  Men and Machines; Methuen (PB)
National Design Council
  Design for People; Canadian Government, Cat.Id-23/75 (PB)
O'Hare D & Roscoe S
  Flightdeck Performance: The Human Factor; Iowa State University Press (HB)
Roscoe S N
  Aviation Psychology; Iowa State University Press (PB)
Shackel B
  Applied Ergonomics Handbook; IPC (PB)
Singleton W T
  Man-Machine Systems; Pelican (PB) (OP)
Warr P B
  Psychology at Work; Penguin (PB)
Wiener E L & Nagel D C
  Human Factors in Aviation; Academic Press (HB)

Chapter 2

Hurst R & Hurst L R (Eds)
 Pilot Error (2nd Edition); Granada (HB)
Norman D.A.
 The Psychology of Everyday Things; Basic Books (PB)
Perrow C
 Normal Accidents; Basic Books (HB)
Rasmussen J, Duncan K, & Leplat J
 New Technology and Human Error; John Wiley and Sons (HB)
Reason J & Mycielska K
 Absent Minded; Prentice-Hall (PB) (OP)
Senders J W & Moray N P
 Human Error: Cause, Prediction, and Reduction; Lawrence Erlbaum Associates (HB)
Swain A D
 Design Techniques for Improving Human Performance in Production; Swain (PB)

Chapter 3

Bartley S H & Chute E
 Fatigue and Impairment in Man; McCraw Hill Book Co (HB)
Benson H
 The Relaxation Response; Avon (PB)
Carrington P
 Freedom in Meditation; Doubleday/Anchor (PB) (OP)
Hartmann E L
 The Functions of Sleep; Yale University Press (PB)
Hawkins F H
 Sleep and Body Rhythm Disturbance in Long-Range Aviation; Hawkins (PB)
Jacobsen E
 You Must Relax; Unwin (PB)
Luce G G
 Body Time; Paladin (PB)
Meares A
 Relief Without Drugs; Fontana (PB)
Selye H
 Stress Without Distress; Hodder & Stoughton (PB)

Chapter 4

Carruthers M & Murray A
 F/40; Fitness on 40 Minutes a Week; Futura (PB) (OP)
Cooper K H
 The New Aerobics; Bantam (PB)
HMSO
 Eating for Health; HMSO (PB)
HMSO
 Manual of Nutrition; HMSO (PB)
Iyengar B K S
 Light on Yoga; Unwin (PB)
Morehouse L E & Gross L
 Maximum Performance; Simon and Schuster (HB)
Royal College of Physicians
 Smoking or Health; Pitman Medical (PB)
Royal College of Psychiatrists
 Alcohol and Alcoholism; Tavistock (PB)

Steen D
 Canadian Pilots Fitness Manual; Fitzhenry and Whiteside (HB)

Chapter 5

Gregory R L
 Eye and Brain (3rd Edition); Weidenfeld and Nicolson (PB)
Robinson J O
 The Psychology of Visual Illusion; Hutchinson Press (PB) (OP)

Chapter 6

Drucker P F
 Management; Harper and Row (HB)
Murrell H
 Motivation at Work; Methuen (PB)
Vroom V M & Deci E L
 Management and Motivation; Penguin (PB)

Chapter 7

Carpenter A
 Human Factors in Speech Communication; Medical Research Council (UK), PS 5/78 (PB)
OU Press
 Communication; Open University Press (UK) (PB) (OP)

Chapter 8

Brown J A C
 Techniques of Persuasion; Pelican (PB)
OU Press
 Attitudes and Beliefs; Open University Press (UK) (PB) (OP)
Reich B & Adcock C
 Values, Attitudes and Behaviour Change; Methuen (PB)
Warren N & Jahoda M
 Attitudes; Penguin (PB) (OP)

Chapter 9

Baddeley A
 Your Memory: A User's Guide; Penguin (PB)
Mace C A
 The Psychology of Study; Pelican (PB)
Stammers R & Patrick J
 The Psychology of Training; Methuen (PB)
Wiener E, Kanki B & Helmreich R (Eds)
 Cockpit Resource Management; (HC)

Chapter 10

O'Connor M & Woodford F P
 Writing Scientific Papers in English; Pitman Medical (PB) (OP),
 (2nd Edition, 1987 John Wiley) (PB)
Oppenheim A N
 Questionnaire Design and Attitude Measurement; Gower Publishing, (PB)

Chapter 12

Moroney M J
  Facts from Figures; Pelican (PB)
Reichmann W J
  Use and Abuse of Statistics; Pelican (PB)

Chapter 13

Harris T A
  I'm OK – You're OK; Pan Books (PB)

2.2 Library reference books

Bailey R H
  Human Performance Engineering; Prentice-Hall
Fitts P M & Posner M I
  Human Performance; Prentice-Hall, (OP)
Grandjean E
  Fitting the Task to the Man (3rd Edition); Taylor and Francis
Hartley J
  Designing Instructional Text (2nd Edition); Kogan Page, London; Nichols, New York
McCormick E J & Sanders M S
  Human Factors in Engineering and Design (5th Edition); McGraw-Hill
Oborne D J
  Ergonomics at Work; Wiley
Poulton E C
  Tracking Skill and Manual Control; Academic Press
Van Cott H P & Kinkade R G
  Human Engineering Guide to Equipment Design; US Government (OP)
Welford A T
  Skilled Performance: Perceptual and Motor Skills; Scott, Foresman
Woodson W E
  Human Factors Design Handbook; McGraw-Hill

2.3 Human Factors journals and bulletins

Journals

Applied Ergonomics; UK; IPC Science and Technology Press; 3-monthly
Ergonomics; UK; Taylor and Francis; monthly; official journal of the International Ergonomics
  Association
Journal of Applied Psychology; USA; American Institutes for Research; 2-monthly
Human Factors; USA; Johns Hopkins University Press; 2-monthly; official journal of the Human
  Factors Society
Aviation Space and Environmental Medicine; USA; Aerospace Medical Association; monthly; occa-
  sionally includes a Human Factors paper
The International Journal of Aviation Psychology; USA; Lawrence Erlbaum Assoc.; 3-monthly

Bulletins

Government Departments

- Work Research Unit papers; UK; Department of Employment; periodically, also includes bibliographies

Flight Safety Foundation; USA

- Cabin Crew Safety Bulletin; 2-monthly; practical and operational
- Human Factors Bulletin; periodically; mainly physiology-oriented
- Pilots Safety Exchange Bulletins; periodically; diverse and irregular

Incident Reporting Systems

- Callback; USA; NASA-Ames; monthly; bulletin of the ASRS
- ASRS Quarterly Reports; USA; NASA-Ames; 3-monthly; summaries and analyses
- ASRS Contractor Reports; USA; NASA-Ames; periodically; ASRS data analyses
- Feedback; UK; Institute of Aviation Medicine; 4-monthly; bulletin of the CHIRP
- In-Flight or APRç; CN; Transport Canada; bulletin of the CFRP
- In Australia, data and reports are included in Basic Aviation Safety Information (BASI) publications
- Flashback; NZ; Bulletin of the Independent Safety Assurance Team Programma (ISAT)

3. Abbreviations Used in the Text

AAIB	Air Accidents Investigation Branch (UK)
AAR	Aircraft Accident Report
AB	Alert Bulletin
AC	Advisory Circular (FAA)
ADI	Attitude Direction Indicator
AEA	Association of European Airlines
AFA	Association of Flight Attendants
AFD	Advanced Flight Deck
AGARD	Advisory Group for Aerospace Research and Development (NATO)
AI	Articulation Index
ALPA	Air Line Pilots Association
AM	Amplitude Modulation
ANC	Air Navigation Committee (ICAO)
APU	Applied Psychology Unit
ARINC	Aeronautical Radio Incorporated
ARP	Aerospace Recommended Practice
AS	Aerospace Standard
ASH	Action on Smoking and Health
ASR	Automatic Speech Recognition
ASRS	Aviation Safety Reporting System (USA)
ATC	Air Traffic Control
ATLAS	Alitalia/Lufthansa/Air France/Sabena
BA	British Airways
BABS	Beam-Approach Beacon System
BAC	Blood Alcohol Concentration
BAe	British Aerospace
BASI	Bureau of Air Safety Investigation
BOAC	British Overseas Airways Corporation
CA	Consumers Association
CAA	Civil Aviation Authority
CAB	Civil Aeronautics Board
CAIR	Confidential Aviation Incident Reporting
CAL	Computer-Assisted Learning
CAP	Civil Aviation Publication
CASRP	Confidential Aviation Safety Reporting Program (Canada)
CBT	Computer-Based Training
CFF	Critical Fusion Frequency
CFR	Code of Federal Regulations
CFRP	Confidential Human Factors Report Programme
CHIRP	Confidential Human Factors Incident Reporting Programme (UK)
CNS	Central Nervous System
COHb	Carboxyhaemoglobin
CRM	Crew/Cockpit Resource Management
CRT	Cathode-Ray Tube
CVR	Cockpit Voice Recorder
DHEW	Department of Health, Education and Welfare (USA)
DHSS	Department of Health and Social Security (UK)
DIY	Do-It-Yourself
DME	Distance-Measuring Equipment
DOE	Department of Employment (UK)
DSK	Dvorak Simplified Keyboard
DVI	Direct Voice Input
EADI	Electronic Attitude Director Indicator
EAP	Employee Assistance Program
EC	European Community
ECAM	Electronic Centralised Aircraft Monitoring

ECG	Electrocardiogram
ECR	Error-Cause-Removal
EEG	Electroencephalogram
EICAS	Engine Indicating and Crew Alerting System
EL	Electroluminescence
EMG	Electromyogram
EOG	Electro-oculogram
ETS	Environmental Tobacco Smoke
FAA	Federal Aviation Administration
FAR	Federal Aviation Regulation
FL	Flight Level
FM	Frequency Modulation
FMS	Flight Management System
FSF	Flight Safety Foundation
G-10	Aviation Behavourial Engineering Technology
GGT	Gamma glutamyl-transferase
GHS	General Household Survey
GMT	Greenwich Mean Time
GPWS	Ground Proximity Warning System
GSR	Galvanic Skin Response
HBT	Human Behaviour Technology
HEC	Health Education Council (UK)
HF	High Frequency (radio)
HFTAO	Human Factors in Transport Aircraft Operation
HMSO	Her Majesty's Stationery Office (UK)
HSI	Horizontal Situation Indicator
HUD	Head-Up Display
IATA	International Air Transport Association
ICAO	International Civil Aviation Organisation
IEA	International Ergonomics Association
IFALPA	International Federation of Airline Pilots Associations
ILO	International Labour Organisation
ILS	Instrument Landing System
INS	Inertial Navigation System
ISASI	International Society of Air Safety Investigators
ISAT	Independent Safety Assurance Team
ISBN	International Standard Book Number
ISO	International Organisation for Standardisation
ISSN	International Standard Serial Number
JAA	Joint Aviation Authority
JAR	Joint Aviation Requirements
KHUFAC	KLM Human Factors Awareness Course
KLM	Koninklijke Luchtvaart Maatschappij (Royal Dutch Airlines)
KSSU	KLM/SAS/SWR/UTA
LCD	Liquid-Crystal Display
LCU	Life-Change Unit
LED	Light-Emitting Diode
LOFT	Line-Oriented Flight Training
LORAN	Long-Range Aid for Navigation
LSD	Lysergic Acid Diethylamide
LTM	Long-Term Memory
NASA	National Aeronautics and Space Administration
NATO	North Atlantic Treaty Organisation
NCC	National Consumer Council
ND	Navigation Display
NMAC	Near Mid-Air Collision

NPRM	Notice of Proposed Rule Making
NTSB	National Transportation Safety Board
OAA	Orient Airlines Association
OHCS	Hydroxy corticosteroids
OU	Open University (UK)
PA	Public Address
PAA	Pan American World Airways
PAG	Pilot Advisory Group
PANS/RAC	Procedures for Air Navigation Services/Rules of the Air and Air Traffic Control
PB	Phonetically Balanced
PF	Pilot flying
PFD	Primary Flight Display
PNF	Pilot-not-flying
ppmv	parts per million by volume
QAP	Quality Assurance Programme
RAF	Royal Air Force
REM	Rapid Eye Movement
RH	Relative Humidity
RLD	Rijksluchtvaartdienst (Netherlands CAA)
RLS	Rijksluchtvaartschool (Netherlands State Flying School)
RPM	Revolutions Per Minute
RTF	Radio Telephony
SAE	Society of Automotive Engineers
SAFAC	Safety Advisory Committee (IATA)
SAS	Scandinavian Airlines System
SASAG	Special Air Safety Advisory Group
SELCAL	Selective Calling System
SFENA	Société Française d'Equipements pour la Navigation Aérienne
SHEL	*Software–Hardware–Environment–Liveware*
SI	Système Internationale d'Unités
SID	Standard Instrument Departure
SOP	Standard Operating Procedure
STM	Short-Term Memory
SWR	Swissair
SWS	Slow-Wave Sleep
TSU	Time of Safe Unconsciousness
TUC	Time of Useful Consciousness
UAL	United Airlines
USAF	United States Air Force
USC	University of Southern California
UTA	Union de Transports Aériens
VAM	Visual Approach Monitor
VASIS	Visual Approach Slope Indicating System
VDT	Visual Display Terminal
VDU	Visual Display Unit
VIP	Very Important Person
WGD	Windshield Guidance Display
WHO	World Health Organisation
WRU	Work Research Unit
ZDP	Zero-Defect Programme

Note: Recommended nomenclature and abbreviations for use on the flight deck are published in SAE ARP 425.

References

Adolph E F & Associates (1947)
Physiology of Man in the Desert
New York: Interscience Publishers

Aerospace (1986)
Goodbye to bifocals
Aerospace, 13, 6:39

AGARD (1980)
Fidelity of simulation for pilot training
NATO AGARD–AR–159

Airola P (1977)
Hypoglycaemia: a better approach
Phoenix: Health Plus Publishers

Aitken R C B (1969)
Prevalence of worry in normal aircrew
British Journal of Medical Psychology, 42:283

Akerstedt T & Gillberg M (1981)
The circadian variation of experimentally
displaced sleep
Sleep, 4:159–170

Alkov R A, Borowsky M S & Gaynor M S (1982)
Stress coping and the US Navy aircrew factor
mishap
Aviation, Space and Environmental Medicine,
53:1112–1115

American Airlines (1979)
Computer based instruction
In: *IATA 5th General Flight Crew Training
Meeting*, Miami, 15–19 October 1979, WP 3.10

Angiboust R (1970)
Discussion period
In: *Aspects of Human Efficiency*, W P
Colquhoun (Ed),
London: English University Press

Arendt J, Aldhous M & Marks V (1986)
Alleviation of jet lag by melatonin: preliminary
results of controlled double blind trial
British Medical Journal, 292:1170

Armstrong D (1991)
*Enhancing Information Transfer: The Aircraft
Perspective*
Los Angeles Aircraft Certification Branch,
Federal Aviation Administration, Long Beach,
California

Asch S E (1956)
Studies of independence and submission to
group pressure
Journal of Abnormal and Social Psychology,
41:258–290

ASH (1983)
Non–smoking provision in airlines
Action on Smoking and Health, Mortimer
Street, London

Atkinson J W (1964)
An Introduction to Motivation
Princeton New Jersey: Van Nostrand

Atkinson J W, Heyns R W & Veroff J (1954)
The effect of experimental arousal of the
affiliation motive on thematic apperception
Journal of Abnormal and Social Psychology,
49:405–410

Baddeley A D (1976)
The Psychology of Memory
New York: Harper & Row

Baddeley A (1982)
Your Memory: a user's guide
Harmondsworth: Penguin Books

Bailey R H (1982)
*Human Performance Engineering: a guide for
system designers*
New York: Prentice–Hall

Baird J A, Coles P K L & Nicholson A N (1983)
Human Factors in air operations in the South
Atlantic Campaign: discussion paper
Journal of the Royal Society of Medicine,
76:933–937

Baranek L L (1949)
Acoustic Measurements
New York: John Wiley & Sons

Baron R A, Byrne D & Griffitt W (1974)
Social Psychology
Boston: Allyn & Bacon

Barrett T R & Ekstrand B R (1972)
Effect of sleep on memory; III Controlling time
of day effects
Journal of Experimental Psychology, 96,
2:321–327

Barthelmess S (1988)
How to survive in a hijacking and hostage
situation
FSF, Cabin Crew Safety Bulletin, 23,
4, July/August 1988

Beard R R & Grandstaff N (1970)
Carbon monoxide exposure and cerebral
function
Annals of the New York Academy of Science,
174:385–395

Beard R R & Wertheim G A (1967)
Behavioural impairment associated with small
doses of carbon monoxide
American Journal of Public Health,
5:2012–2022

Becker C E, Roe R L & Scott R A (1974)
*Alcohol as a Drug: a curriculum on pharma-
cology, neurology and toxicology*
New York: Medcom Press

Bender W, Gothert M & Malorny G (1972)
Effect of low carbon monoxide concentrations
on psychological functions
Staub; Reinhaltung der Luft, 32, 4:54–60

Bennet G (1983)
Psychiatric disorders in civilian pilots
Aviation, Space and Environmental Medicine,
54, 7:588–589

Benson A J & Burchard E (1973)
Spatial disorientation in flight: a handbook for
aircrew
NATO AGARD–AG–170

Billings, C E (1991a)
*Human-Centered Aircraft Automation: a
concept and guidelines* (NASA TM 103885).
NASA/Ames Research Center, Mountain View,
California

Billings, C E (1991b)
Toward a human-centered aircraft automation
philosophy
*The International Journal of Aviation
Psychology*, 1, 4: 261–270

Billings C E, Wick R L, Gerke R J & Chase R C
(1973)
Effects of ethyl alcohol on pilot performance
Aerospace Medicine, 44, 4: 379–382

Blake R R & Mouton J S (1962)
Overvaluation of own group's product in
intergroup competition
Journal of Abnormal and Social Psychology,
64:237–238

Bonnet M H (1986)
Performance and sleepiness as a function of
frequency and placement of sleep disruption
Psychophysiology, 23, 3:263–271

Bootzin R R (1981)
Insomnia
Arts and Sciences, 4, 1:2–6, Magazine of the
College of Arts and Sciences, Northwestern
University, Ill.

Botting D (1980)
The Giant Airships
Chicago: Time-Life Books

Brebner J & Sandow B (1976)
Direction-of-turn stereotypes – conflict and
concord
Applied Ergonomics, 7, 1:34–36

Brenneman J J (1985)
Smoke detection and control, interior system
design, fire proofing, occupant protection, cabin
equipment and use
In: Proceedings of *Cabin Safety Conference
and Workshop*, Arlington, 11–14 December
1984, DOT/FAA/ASF100–85/01

Broadbent D M (1958)
Perception and Communication
London: Pergamon Press

Brown E L, Burrows A A & Miles W L (1972)
Optimisation of maintenance manuals
Paper 6065, Long Beach, Douglas Aircraft
Company

Brown J A (1975)
Safety – A systematized commitment
Safety in Flight Operations, Vol. 2
International Air Transport Association 20th
Technical Conference, Montreal

Bruggink G M (1988)
Reflections on air carrier safety
FSF Flight Safety Digest, 7, 6 June 1988

Bryanton R (1985)
IPECO. Sitting machine makers
Aerospace, 12, 1:28

Buley L E (1970)
Experience with a physiologically-based
formula on determining rest periods on long
distance air travel
Aerospace Medicine, 41:680–683

Bunnell D E, Bevier W C & Horvath S M (1983)
Effect of exhaustive exercise on the sleep of
men and women
Psychophysiology, 20:50–58

Burg A (1966)
Visual acuity as measured by static and
dynamic tests
Journal of Applied Psychology, 50, 6:460–466

Burnett J (1985)
Statement by the Chairman of the NTSB
In: Proceedings of *Cabin Safety Conference
and Workshop*, Arlington, 11–14 December
1984, DOT/FAA/ASF100–85/01

Burnhill P, Hartley J & Young M (1976)
Tables in text
Applied Ergonomics, 7, 1:13–18

Busby D E, Higgins E A & Funkhauser G E
(1976)
Protection of airline flight attendants from
hypoxia following rapid decompression
Aviation, Space and Environmental Medicine,
47, 9:942–944

CA (1986)
Flying – how safe?
Holiday Which, May 1986:134–137, London,
Consumers Association

Cameron S (1973)
Job satisfaction: the concept and its measure-
ment
Work Research Unit, Occasional Paper No. 4,
London

Canter D (1980)
Fires and Human Behaviour
Chichester: John Wiley & Sons

Carruthers M & Murray A (1976)
F/40; Fitness on 40 minutes a week
London: Futura Publications

Chandler R F, Garner J D, Lowry D L,
Blethrow J G & Anderson J A (1980)
Considerations relative to the use of canes by
blind travellers in air carrier cabins
FAA-AM-80-12

Chapanis A (1965)
Words, words, words
Human Factors, 7:1–17

Chapanis A (1975)
*Ethnic Variables in Human Factors
Engineering*
Baltimore: Johns Hopkins University Press

References

Chorley R A (1976)
Seventy years of flight instruments and displays
Third H P Folland Memorial Lecture,
Royal Aeronautical Society, London,
19 February 1976

Claxton E (1980)
Do children use verbal quantifiers in the same
way as adults?
Unpublished project paper available from
J Hartley, Keele University, Staffordshire

Cohen G (1989)
Memory in the real world
Hillsdale, New Jersey: Laurence Erlbaum Assoc

Colquhoun W P (1971)
Circadian variations in mental efficiency
In: *Biological Rhythms and Human Performance,*
W P Colquhoun (Ed), London & New York:
Academic Press

Condom P (1983)
Automatic voice recognition; a new form of pilot-
system interaction
Interavia, 38, 1:82–83

Cooper C L & Sloan S (1985)
The sources of stress in the wives of airline pilots
Aviation, Space and Environmental Medicine,
56:317–321

Cooper K H (1970)
The New Aerobics
New York: Bantam Books

Cooper K H, Gey G O & Bottenberg R A (1968)
Effects of cigarette smoking on endurance
performance
Journal of the American Medical Association,
203, 3:189–192

Crane C R (1985)
Human tolerance to toxic components of smoke
In: Proceedings of *Cabin Safety Conference and
Workshop,* Arlington, 11–14 December 1984,
DOT/FAA/ASF100-85/01

Cummins R O & Schuback J (1989)
Frequency and types of medical emergencies
among commercial air travellers
Journal of the American Medical Association,
261, 9

Cuthbert B N, Graeber R C, Singh H C & Schneider
R J (1979)
Rapid transmeridial deployment. II. Effects of age
and countermeasures under field conditions
In: Proceedings of *The 14th International
Conference of The International Society for
Chronobiology,* Milan: Il Ponte

Cutts M & Maher C (1984)
Writing Plain English
Stockport: Plain English Campaign

Czeisler C A, Weitzman E D, Moore-Ede M C,
Zimmerman J C & Krauer R S (1980)
Human sleep: its duration and organisation
depends on its circadian phase
Science, 210:1264–1267

Damkot D K & Osga G A (1978)
Survey of pilots' attitudes and opinions about
drinking and flying
Aviation, Space and Environmental Medicine, 49,
2:390–394

Davies D R & Parasuraman R (1981)
The Psychology of Vigilance
London: Academic Press

Davis A (1983)
Hearing disorders in the population; first phase
findings of the MRC National Study of Hearing
In: *Hearing Science and Hearing Disorders,*
M E Lutman & M P Haggard (Eds),
London: Academic Press

Davis D R (1949)
Pilot Error
London: HMSO

Davis D R (1958)
Human errors in transport accidents
Ergonomics, 2:24–33

Dean R D & Whitaker K M (1982)
Fear of flying: impact on the US travel industry
Journal of Travel Research, Summer 1983,
7–17

DeCelles J L & Terhune G (1979)
Flight instrumentation requirements for all-
weather approach and landing
In: Proceedings of *The International Air
Transport Conference ASCE,* New Orleans,
30 April–3 May 1979

Degani A & Wiener E L (1990)
*Human Factors of Flight-Deck Checklists:
the normal checklist* (NASA CR 177549)
NASA/Ames Research Center, Mountain View,
California

Dexter H E (1975)
Pilot fatigue study: a first look
Aeromedical Committee Research Study, Air
Line Pilots Association, Washington

DHEW (1979)
Involuntary smoking
In: *Smoking and Health – a report of the
Surgeon General,* Chapter 11, Department of
Health, Education and Welfare, DHEW
publication 79–50066, Washington

DHSS (1979)
Eating for Health
London: Department of Health and Social
Security, HMSO

Dille J R & Linder M K (1980)
The effects of tobacco on aviation safety
FAA-AM-80-11

DOE (1981)
Guidelines for environmental design and fuel
conservation in educational buildings
London: Department of Education, HMSO

Donoghue J A
Tests and trials, an accident aftermath
Air Transport World, December 1989

Dorey F C (1983)
Aviation Security
New York: Van Nostrand Reinhold Co

DOT (1985)
Transport statistics Great Britain 1974–1984
London: Department of Transport, HMSO

Drew G C (1967)
The study of accidents
Bulletin of the British Psychological Society,
16:1–10

Drucker P F (1973)
Management
New York: Harper & Row
Dvorak A (1943)
There is a better typewriter keyboard
National Business Education Quarterly, 12,
2:51–58
Eastburn M W (1987)
A management tool – the accident record
In: *Fourth Annual International Aircraft Cabin
Safety Symposium*, University of Southern
California, Las Vegas, Nev, 2–6 March 1987
Edel D H (1967)
Introduction to Creative Design
Englewood Cliffs: Prentice-Hall
Edwards E (1972)
Man and machine: systems for safety
In: Proceedings of *The BALPA Technical
Symposium*, London
Edwards E (1975)
Stress and the airline pilot
In: *BALPA Medical Symposium*, London
Edwards E (1985)
Human Factors in aviation
Aerospace, 12, 7:20–22
Ehret C F (1981)
New approaches to chronohygiene for the shift
worker in the nuclear power industry
In: *Night and Shift Work: Biological and Social
Aspects. Advances in the Biosciences*, A
Reinberg, N Vieux & P Andlauer (Eds),
Oxford: Pergamon Press
Ehret C F, Potter V R & Dobra K W (1975)
Chronotypic action of theophylline and of
pentobarbital as circadian *zeitgebers* in the rat
Science, 188:1212–1215
Ekstrand B R, Barrett T R, West J N &
Maier W G (1977)
The effect of sleep on human long-term
memory
In: *Neurobiology of Sleep and Memory*,
R R D Colin & J L McGaugh (Eds),
New York: Academic Press
Elgerot A (1976)
Note on selective effects of short-term tobacco
abstinence on complex versus simple mental
tasks
Perceptual and Motor Skills, 42:413–414
Eng W G (1979)
Survey on eye comfort in aircraft: 1. Flight
attendants
Aviation, Space and Environmental Medicine,
50:401–404
Eng W G (1979a)
Survey on eye comfort in aircraft: 2. Use of
opthalmic solutions
Aviation, Space and Environmental Medicine,
50:1166–1169
Englund C & Naitoh P (1978)
A validation study of the Biorhythm Theory
Naval Health Research Center, San Diego
(unpublished)

Evans P (1975)
Motivation
London: Methuen
FAA (1975)
Flight attendant clothing
FAA NPRM Notice No. 75–13, (40 FR 11737)
FAA (1980)
Summary of safety résumés. Cabin safety
FAA-ASF-80-3
FAA (1983)
Flammability requirements for aircraft seat
cushions
FAA NPRM Notice No. 83–14,
(48 FR 197:46250)
Farmer E W (1984)
Personality factors in aviation
International Journal of Aviation Safety,
2, 2:175–179
Farmer E W & Green R G (1985)
The sleep-deprived pilot: performance and EEG
response
In: *The 16th Conference of the Western
European Association for Aviation Psychology*,
Helsinki, 24–28 June 1985
Fast J (1971)
Body Language
London: Pan Books
Fentem P (1978)
Answer to a question on physical fitness and
higher mental function
British Medical Journal, 6150:1484
Feuer B (1984)
The AFA EAP: members helping members
In: *The First International Aircraft Cabin
Safety Symposium*, University of Southern
California, 7–9 February 1984
Fineman S & Warr P B (1971)
Managers: their effectiveness and training
In: *Psychology at Work*, P B Warr (Ed),
Harmondsworth: Penguin Books
Finnair (1971)
Coffee break gymnastics during the flight for
Finnair pilots
Helsinki, Finnair Medical Department
Finsberg G (1982)
Hansard, London, 26 January 1982, 16: col 337
Fitts P M (1951)
Engineering psychology and equipment design
In: *Handbook of Experimental Psychology*,
S S Stevens (Ed),
New York: John Wiley & Sons
Flack M (1918)
Flying stress
London, Medical Research Committee
Flight International (1984)
Airliner office equipment
Flight International, 126, 3921:128–135
Folkard S, Monk T H, Bradbury R & Rosenthal J
(1977)
Time of day effects in school children's
immediate and delayed recall of meaningful
material
British Journal of Psychology, 68, 1:45–50

References

Foushee H C (1982)
The role of communications, socio-psychological and personality factors in the maintenance of crew coordination
Aviation, Space and Environmental Medicine, 53:1062–1066

Foushee H C, Lauber J K, Baegte M M & Acomb D B (1986)
Crew factors in flight operations: III, The operational significance of exposure to short-haul air transport operations
NASA Technical Memorandum 88322

FSF (1965)
More illusions
FSF, Pilots Safety Exchange Bulletin, 65-109

FSF (1970)
What is your TUC?
FSF, Cabin Crew Safety Exchange, 70-501/502

FSF (1979)
Flight safety eroded as cabin assaults grow
FSF, Cabin Crew Safety Bulletin, March/April 1979

FSF (1980)
The deaf passenger. The blind passenger
FSF, Cabin Crew Safety Bulletin, March/April 1980

FSF (1981)
Wiper effect
FSF, Accident Prevention Bulletin, August 1981

FSF (1982)
Beverage/meal cart safety
FSF, Cabin Crew Safety Bulletin, 17, 4

FSF (1984)
Preflight crew briefings: a step towards improved communication
FSF, Cabin Crew Safety Bulletin, 19, 3

FSF (1985)
Cabin safety: what the future holds
FSF, Flight Safety Digest, April 1985; statistics compiled from the NTSB data base

FSF (1986)
Behaviour: using life vests and flotation cushions
FSF, Cabin Crew Safety Bulletin, 21, 3, (4)

Fulton H B (1985)
A pilot's guide to cabin air quality and fire safety
New York State Journal of Medicine, 85, 7:384–388

Gabriel R F (1977)
Some potential errors in human information processing during approach and landing
In: *International Symposium on Human Factors,* Air Line Pilots Association, Washington, 8 February 1977, Douglas Paper 6587

Gander P H, Graeber R C, Foushee H C & Lauber J K (1986)
Crew factors in flight operations II:
Psychophysiological responses to short-haul air transport operations
NASA Technical Memorandum 88321

Gander P H, Myhre G, Graeber R C, Andersen H T & Lauber J K (1985)
Crew factors in flight operations I:
Effects of 9-hour time-zone changes on fatigue and the circadian rhythms of sleep/wake and core temperature
NASA Technical Memorandum 88197

Gaume J G (1970)
Factors influencing the time of safe unconsciousness (TSU) for commercial jet passengers following cabin decompression
Aviation, Space and Environmental Medicine, 41:382–385

Gaume J G (1981)
Advances in physiological conditioning and exercise monitoring procedures
DC Flight Approach, 37, Long Beach, Douglas Aircraft Company

GHS (1983)
Cigarette smoking 1972–1982
OPCS Monitor Ref. 83/3, Office of Population Censuses and Surveys, Kingsway, London

Glines C V (1976)
Probable cause: pilot fatigue?
The Airline Pilot, 45, 10:19–21

Goldman P (1984)
Coffee and health – what's brewing
New England Journal of Medicine, 310, 12:783–785

Goldman P A (1984)
Aircraft cabin safety: fire isn't the only hazard
In: *First Annual International Aircraft Cabin Safety Symposium,* University of Southern California, 7–9 February 1984

Goodwin A R, Thomas S K & Hartley J (1977)
Are some parts larger than others: quantifying Hammerton's qualifiers
Applied Ergonomics, 8:93–95

Gorowitz S & MacIntyre A (1976)
Toward a theory of medical fallibility
Journal of Medicine and Philosophy, 1:51–71

Gould J D & Grischkowsky N (1984)
Doing the same work with hard copy and with cathode–ray tube (CRT) computer terminals
Human Factors, 26, 3:323–337

Graeber R C, Cuthbert B N, Sing H C, Schneider R J & Sessions G R (1979)
Rapid transmeridial deployment; 1. Use of chronobiologic countermeasures to hasten time zone adjustment in soldiers
In: *Proceedings of The 14th International Conference of the International Society for Chronobiology,* Milan: Il Ponte

Graham W (1978)
Air carrier approach and landing accidents
Prepared for the FAA Office of Aviation Policy by Questek Inc., New York

Grandjean E (1973)
Ergonomics in the Home
London: Taylor & Francis

Grandjean E & Vigliani E (Eds) (1983)
Ergonomics Aspects of Visual Display Terminals
London: Taylor & Francis

Gray M (1975)
Questionnaire typography and production
Applied Ergonomics, 6, 2:81–89

Green R G (1978)
Communication and noise
In: *Aviation Medicine*, G Dhenim (Ed), Vol. 1,
London: Tri-Med Books

Green R G (1985)
Stress and accidents
Aviation, Space and Environmental Medicine,
56, 7:638–640

Greenwood M & Woods H M (1964)
A report on the incidence of industrial
accidents upon individuals with special
reference to multiple accidents (1919)
In: *Accident Research*, W Haddon,
E A Suchman & D Klein (Eds),
New York: Harper & Row

Gregory R L (1977)
Eye and Brain
London: Weidenfeld & Nicolson

Greist J H, Klein H H, Eischens R R, Faris J,
Gurman A S & Morgan W P (1979)
Running as treatment for depression
Comprehensive Psychiatry, 20:41–53

Grether W F (1948)
Analysis of types of errors in reading of the
conventional 3-pointer altimeter
Aero–Medical Laboratory, Air Material
Command, Report MCREXD-694-14a,
16 March 1948, Dayton, Ohio

Guilford J S (1973)
Prediction of accidents in a standardised home
environment
Journal of Applied Psychology, 57:306–313

Gunning R (1968)
The Technique of Clear Writing (2nd Edition)
New York: McGraw-Hill

Hackman J R, Ed. (1990)
Groups That Work (and Those That Don't)
San Francisco: Jossey–Bass

Halberg F (1977)
Implications of biologic rhythms for clinical
practice
Hospital Practice, 12, 1:139–149

Harper C R & Kidera G J (1973)
Hypoglycaemia in airline pilots
Aerospace Medicine, 44, 7:769–771

Harris N (1986)
Positive health
Safety and Risk Management, British Safety
Council, March 1986

Hartley J (1981)
Eighty ways of improving instructional text
*IEEE Transactions on Professional Communi-
cation*, PC–24, 1:17–27

Hartley J (1985)
Designing Instructional Text (2nd Edition)
London: Kogan Page; New York: Nichols
Publishing

Haward L R C (1974)
Effects of domestic stress on flying efficiency
Revue de Médecine Aéronautique et Spatiale,
13:29–31

Haward L R C (1984)
Effect of Autogenic Training upon the flying
skill of pilots under stress
In: Proceedings of *The 2nd International
Interdisciplinary Conference on Stress and
Tension Control*, Brighton, 30 August–3
September 1983,
New York: Plenum Press

Hawkins F H (1966)
Aspects of sight and reach in a civil aircraft
cockpit
Alan Cobham Prize Paper, Guild of Air Pilots
and Air Navigators, London

Hawkins F H (1973)
Crew seats in transport aircraft
Shell Aviation News, 418:14–21

Hawkins F H (1976)
Some ergonomic aspects of cockpit panel
design for airline aircraft
Shell Aviation News, 437:2–9

Hawkins F H (1977)
Cockpit visibility
Shell Aviation News, 440:20–25

Hawkins F H (1980)
Sleep and body rhythm disturbance amongst
flight crews in long-range aviation. The
problem and potential for relief
Thesis, University of Aston, Birmingham

Hawkins F H (1984)
Human Factors education in European air
transport operations
In: *Breakdown in Human Adaptation to Stress.
Towards a multidisciplinary approach*, Vol. 1,
for the Commission of the European Communities, The Hague: Martinus Nijhoff

Hawkins F H (1984a)
Autogenic Training as a stress management tool
in air transport operations
In: Proceedings of *The International Interdisciplinary Conference on Stress and Tension
Control*, Brighton, 30 August–3 September
1983, New York: Plenum Press

HEC (1974)
Recognising the alcoholic
London, Health Education Council

Hefti H (1986)
The cost of accidents
Swiss Reinsurance Co, Zürich, April 1986

Heglin H J (1973)
NAVSHIPS. Display illumination design guide
II. Human Factors
NELC/TD223, Naval Electronics Laboratory
Center, San Diego

Helmreich R L (1980)
Social psychology on the flight deck
In: Proceedings of a NASA/industry workshop,
Resource Management on the Flight Deck,
San Francisco, 26–28 June 1979

Helmreich R L (1983)
What changes and what endures: the capabili-
ties and limitations of training and selection
In: *The Flight Operations Symposium*, Irish Air
Line Pilots Association/Aer Lingus, Dublin,
19–20 October 1983

Helmreich R L and Wilhelm J A (1991)
Outcomes of crew resource management training
The International Journal of Aviation Psychology, 1, 4:287–300

Hertzberg H T E (1960)
Dynamic anthropometry of working positions
Human Factors, 2, 3:147–155

Herzberg F (1966)
Work and the Nature of Man
New York: World Publishing

Herzberg F, Mausner B, Peterson R & Snyderman G (1959)
The Motivation to Work
New York: John Wiley & Sons

Higgins E A, Chiles W D, McKenzie J M, Iampetro P F, Winget C M, Funkhauser G E, Burr M J, Vaughan J A & Jennings A E (1975)
The effect of 12 hour shift in the wake/sleep cycle on physiological and biochemical responses and on multiple task performance
FAA-AM-75-10

Hockey G R J (1970)
Changes in attention allocation in a multi-component task under loss of sleep
British Journal of Psychology, 61, 4:473–480

Holdstock L F J (1980)
British Airways pre-command training programme
In: Proceedings of the NASA/industry workshop, *Resource Management on the Flight Deck*, San Francisco, 26–28 June 1979, NASA Conference Publication 2120

Holmes T H & Rahe R H (1967)
The social readjustment rating scale
Journal of Psychosomatic Research, 11:213–218

Hood W R & Sherif M (1962)
Verbal report and judgement of unstructured stimulus
Journal of Psychology, 54, 1:121–130

Hopkin V D & Taylor R M (1979)
Human Factors in the design and evaluation of aviation maps
NATO AGARD-AG-225

Hore B D & Plant M A (Eds) (1981)
Alcohol Problems in Employment
London: Croom Helm

Horne J A (1985)
Sleep loss: underlying mechanisms and tiredness
In: *Hours of Work*, S Folkard & T H Monk (Eds), Chichester: John Wiley & Sons

Horne J A & Moore V J (1985b)
Sleep EEG effects of exercise with and without additional body cooling
EEG and Clinical Neurophysiology, 60:33–38

Horne J A, Percival J E & Traynor J R (1980)
Aspirin and human sleep
EEG and Clinical Neurophysiology, 49:409–413

Horne J A & Reid A J (1985a)
Night-time sleep EEG changes following body heating in a warm bath
EEG and Clinical Neurophysiology, 60:154–157

Hovland C I & Weiss W (1951)
The influence of source credibility on communication effectiveness
Public Opinion Quarterly, 15:635–650

Howell K & Martin A (1975)
An investigation of the effects of hearing protectors on vocal communication in noise
Journal of Sound and Vibration, 41:181–196

Howes D H (1957)
On the relation between the intelligibility and frequency of occurrence of English words
Journal of the Acoustical Society of America, 29:296

Hurst R (1985)
Verbal communication and hearing deficiency
The International Journal of Aviation Safety, 3, 1:62–65

IATA (1975)
Safety in flight operations
The 20th Technical Conference of IATA, Istanbul, 10–14 November 1975

Idzikowski C (1984)
Sleep and memory
British Journal of Psychology, 75, 4:439–449

International Civil Aviation Organization (1989)
Human Factors Digest No. 1
Circular 216-AN/131, Montreal

Jenkins J G & Dallenbach K M (1924)
Obliviscence during sleep and waking
American Journal of Psychology, 35:605–612

Johnson D A (1979)
An investigation of factors affecting aircraft passenger attention to safety information and presentations
Report No. IRC-79-1, Interaction Research Corporation, Stanton

Johnson L C (1982)
Sleep deprivation and performance
In: *Biological Rhythms, Sleep and Performance*, W B Webb (Ed), New York: John Wiley & Sons

Johnston A N (1992)
The Development and Use of a Generic Non-Normal Checklist with Applications in Ab Initio and Introductory AQP Programmes,
Dublin: Aviation Psychology Research Group, Trinity College

Johnston N (1985)
Occupational stress and the airline pilot. The role of the pilot advisory group (PAG)
Aviation, Space and Environmental Medicine, 56:633–637

Kalton G, Collins M & Brook L (1978)
Experiments in wording opinion questions
Journal of the Royal Statistical Society, C, 27, 2:149–161

Kalton G, Roberts J & Holt D (1980)
The effects of offering a middle option with opinion questions
The Statistician, 29, 1:65–78

Kalton G & Schuman H (1982)
The effect of the question of survey responses: a review
Journal of the Royal Statistical Society, A, 145, Part 1:42–73

Karacan I, Thirnby J I, Anch A M, Booth G H, Williams I L & Salis P J (1976)
Dose-related sleep disturbances induced by coffee and caffeine
Clinical Pharmacological Therapy, 20, 6:682–689

Khalil T M & Kurucz C N (1977)
The influence of 'Biorhythm' on accident occurrence and perfomance
Ergonomics, 20, 4:389–398

Khan R L, Zarit S H, Hilbert N M & Niederehe G (1975)
Memory complaint and impairment in the aged: the effect of depression and altered brain function
Archives of General Psychiatry, 32, 12:1569–1573

Klein K E, Wegmann H M & Hunt B I (1972)
Desynchronisation of body temperature and performance circadian rhythm as a result of outgoing and homegoing transmeridian flights
Aerospace Medicine, 43, 2:119–132

Klein K E & Wegmann H M (1980)
Significance of circadian rhythms in aerospace operations
NATO AGARD-AG-247

Klein K E, Wegmann H M, Athenassenas G, Hohlweck H & Kuklinski P (1976)
Air operations and circadian performance rhythms
Aviation, Space and Environmental Medicine, 47, 3:221–230

Klema *et al.* (1987)
Unpublished report
Oak Ridge, National Laboratories, Harvard and Taft Universities

Klemmer E T (1971)
Keyboard entry
Applied Ergonomics, 2, 1:2–6

Koffler Group (1985)
Glare
Office Systems Ergonomics Report, 4, 1

Koffler Group (1986)
The ergonomics of office keyboards
Office Systems Ergonomics Report, 5, 1

Kragt H (1978)
Human reliability engineering
IEEE Transactions on Reliability, R27, 3

Krueger W C F (1929)
The effect of overlearning on retention
Journal of Experimental Psychology, 12:71–78

Kryter K D (1972)
Speech communication
In: *Human Engineering Guide to Equipment Design*, H P Van Cott & R G Kinkade (Eds), Wash.: US Government Printing Office

Lader M H (1983)
Insomnia and short–acting benzodiazepine hypnotics
Journal of Clinical Psychiatry, 44:47–53

Lamers G L (1968)
A comparative study of circular and vertical tape jet engine instruments indicating five parameters of four engines
NLR Technical Report, TR 68 047, Amsterdam

Lauber J K & Foushee H C (1983)
Line-Oriented Flight Training
In: *The Flight Operations Symposium*, Irish Air Line Pilots Association/Aer Lingus, Dublin, 19–20 October 1983

Lautman L G & Gallimore P L (1987)
Control of the crew caused accident
Boeing Airliner, April/June 1987

Lavie P (1984)
The function of dreaming
Neurology, 34, 9:1271

Lederer J F & Enders J H (1987)
Aviation safety – the global conditions and prospects
In: *Transportation Deregulation and Safety Conference*, Northwestern University, Evanston, Ill, 23–25 June (1987)

Loftus E (1979)
Eye Witness Testimony
Cambridge, Mass: Harvard University Press

Lubin A, Hord D J, Tracy M L & Johnson L C (1976)
Effect of exercise, bedrest and napping on performance during 40 hours
Psychophysiology, 13:334–339

Luckiesh M & Moss F K (1927)
The new science of seeing
In: *Interpreting the Science of Seeing into Lighting Practice*, Vol. 1, General Electric Company, Cleveland

Lyman E G & Orlady H W (1981)
Fatigue and associated performance decrements in air transport operations
NASA CR 166167, NASA-Ames Research Center, Moffett Field

MacKenzie-Orr M H (1988)
Aviation security in an age of terrorism
FSF, Flight Safety Digest, 7, 12, December 1988

Mackworth N H (1950)
Researches on the measurement of human performance
MRC Special Report Series 268, HMSO, London

Mandelbaum J (1960)
An accommodation phenomenon
Archives of Opthalmology, 63:923–926

Martinez E & DiNunno G (1990)
Accident aftermath
Air Line Pilot, February, 1990

Maslow H (1943)
A theory of human motivation
Psychological Review, 50:370–396

Mason C (1984)
The emotionally disturbed passenger
In: *The First Annual International Aircraft Cabin Safety Symposium*, University of Southern California, 7–9 February 1984

McClelland D C (1965)
Achievement motivation can be developed
Harvard Business Review, 43, 6–14, 20–23, 178

McCormick E J & Sanders M S (1983)
Human Factors in Engineering and Design, (5th Edition)
Japan: McGraw-Hill

McFarland R A (1953)
Human Factors in Air Transportation
New York: McGraw-Hill

McGuire W J (1969)
The nature of attitudes and attitude change
In: *The Handbook of Social Psychology*, G Lindzey & E Aronson (Eds), Vol. 3, Reading, Mass.: Addison–Wesley

McSweeny T E (1985)
Status of FAA cabin safety efforts
In: Proceedings of *Cabin Safety Conference and Workshop*, Arlington, 11–14 December 1984, DOT/FAA/ASF100-85/01

Meier Muller H (1940)
Flugwehr Und Technik, 1:412–414 and 2:40–42

Meister D (1971)
Human Factors: theory and practice
New York: John Wiley & Sons

Meister D (1982)
The role of Human Factors in system development
Applied Ergonomics, 13, 2:119–124

Melton C E (1980)
Effects of long–term exposure to low levels of ozone: a review
FAA-AM-80-16

Mertens H W & Collins W E (1985)
The effects of age, sleep deprivation and altitude on complex performance
FAA-AM-85-3

Metz B & Marcoux F (1960)
Alcoolisation et accidents du travail
Revue de L'Alcoolisme, 6, 3

Miller R B (1954)
Psychological considerations in the design of training equipment
Report 54-563, Wright Air Development Center, USAF

Mills J N, Minors D S & Waterhouse J M (1978)
Adaptation to abrupt time-shifts of the oscillator(s) controlling human circadian rhythms
Journal of Physiology, (London), 285:455–470

Mohler S R (1969)
Crash protection in survivable accidents
Memorandum, Staff Study, FAA Office of Aviation Medicine, Washington DC, 4 February 1969

Mohler S R (1976)
Physiological index as an aid in developing airline pilot scheduling patterns
Aviation, Space and Environmental Medicine, 47, 3:238–247

Mohler S R (1978)
Modern concepts in pilot aging
FSF, Human Factors Bulletin, November/December 1978

Mohler S R (1980)
Pilots and alcohol: mix with caution
FSF, Human Factors Bulletin, September/October 1980

Mohler S R (1982)
Back problems and the pilot
FSF, Human Factors Bulletin, 28, 4

Mohler S R (1983)
Medicines and the pilot
FSF, Human Factors Bulletin, 29, 3

Mohler S R (1984)
Aircrew physical status and career longevity
FSF, Human Factors Bulletin, 31, 1

Monash (1985)
Resetting the body's clock
Monash Review, 2:1–2

Moore–Ede M C, Czeisler C A & Richardson G S (1983)
Circadian timekeeping in health and disease. Part 1. Basic properties of circadian pacemakers
The New England Journal of Medicine, 309, 8:469–476

Morgan W P (1984)
Physical activity and mental health
In: *Exercise and Health*, H M Eckert & H Montoye (Eds),
Champaigne, Ill.: Human Kinetics Publishers

Moser H M & Bell G E (1955)
Joint USA/UK report AFCRC-TN-55-56-1955, Air Force Cambridge Research Center, Cambridge, Mass.

Murray H A (1938)
Explorations in Personality
New York: Oxford Book

Murrell H (1976)
Motivation at Work
London: Methuen

Murrell H (1978)
Work stress and mental strain
WRU Occasional Paper No. 6, Work Research Unit, Department of Employment, London

Myles W S & Chin A K (1974)
Physical fitness and tolerance to environmental stresses: a review of human research on tolerance to and work capacity in hot, cold and high altitude environments
DCIEM Report No. 74-R-1008, Defence and Civil Institute of Environmental Medicine, Downsview, Ontario

Myrstan A L & Andersson K (1978)
Effects of cigarette smoking on human performance
In: *Smoking Behaviour. Physiological and Psychological Influences*, R E Thornton (Ed), Edinburgh & New York: Churchill Livingstone

Naitoh P (1981)
Circadian cycles and restorative power of naps
In: *Biological Rhythms, Sleep and Shift Work*, L C Johnson, D I Tepas, W P Colquhoun & M J Colligan (Eds), New York: Spectrum

NASA (1978)
Anthropometry source book, 3 Vols
NASA/RP/1024

NASA (1981)
NASA Aviation Safety Reporting System: Quarterly Report No. 13
NASA Technical Memorandum 81274, NASA-Ames Research Center, Moffett Field

National Design Council (1975)
Design for People
Cat. No. Id41-23/1975, Information Canada, Ottawa

Newbold E M (1964)
A contribution to the study of the human factor in the causation of accidents (1926)
In: *Accident Research*, W Haddon, E A Suchman & D Klein (Eds),
New York: Harper & Row

Newsham D B (1969)
The challenge of change to the adult trainee
Training Information, Paper No. 3, HMSO, London

Nicassio P & Bootzin R R (1974)
A comparison of Progressive Relaxation and Autogenic Training as a treatment for insomnia
Journal of Abnormal Psychology, 8:253–260

Nicholson A N, Pascoe P A, Roehrs T, Roth T, Spencer M B, Stone B M & Zorick F (1985)
Sustained performance with short evening and morning sleeps
Aviation, Space and Environmental Medicine, 56, 2:105–114

Nicholson A N & Stone B M (1982)
Sleep and wakefulness handbook for flight medical officers
NATO AGARD-AG-270(E)

Nordlund C L (1983)
Fear of flying in Sweden
Scandinavian Journal of Behaviour Therapy, 12:150–168

Norman S D and Orlady H W (1988)
Flight deck automation: promises and realities
Final report of a NASA/FAA/Industry workshop held at Carmel Valley, California
NASA-Ames Research Center, Moffett Field, California

NTSB (1977)
Human Factors specialist's factual report of investigation. Accident to Capitol Airways DC8, Baltimore, 16 January 1977
NTSB-DCA 77-A-A008

NTSB (1981)
Cabin safety in large transport aircraft: special study
NTSB-AAS-81-2

NTSB (1983)
National Transportation Safety Board: Investigation Manual Aircraft Accidents and Incidents (Board Order 6200.1A)

Nunn H T (1981)
Line-oriented flight training
In: *Guidelines for Line-Oriented Flight Training*, Vol. II, J K Lauber and H C Foushee (Eds), Proceedings of a NASA/Industry Workshop
NASA-Ames Research Center, Moffett Field, California

Oborne D J & Clarke M J (1975)
Questionnaire surveys of passenger comfort
Applied Ergonomics, 6, 2:97–103

Observer & Maxwell M A (1959)
Study of absenteeism, accidents and sickness payments in problem drinkers in industry
Quarterly Journal of Studies on Alcoholism, 20:302–307

O'Connor M & Woodford F P (1977)
Writing Scientific Papers in English
Tunbridge Wells: Pitman Medical Publishing

O'Hanlon J F, Royal J W & Beatty J (1977)
Theta regulation and radar vigilance performance
In: *Biofeedback and Behaviour*, J Beatty & H Legewie (Eds), New York: Plenum Press

Open University (1975a)
Attitudes and Beliefs
Course D101, Block 7, Units 21–24, Open University, Milton Keynes

Open University (1975b)
Communication
Course D101, Block 3, Units 7–10, Open University, Milton Keynes

Oppenheim A N (1966)
Questionnaire Design and Attitude Measurement
Aldershot: Gower

Orlady H W (1975)
The operational aspects of pilot incapacitation
In: *The IATA 20th Technical Conference*, Istanbul, 10–14 November 1975, WP/19

Orlady H W (1993)
Airline pilot training for today and tomorrow
In: *Cockpit Resource Management*, Wiener E L, Kanki B G & Helmreich R L (Eds).

Orlady H W, Hennessy R T, Obermayer R W, Vruels D, & Murphy M (1988)
Full-Mission Simulation in Human Factors Research for Air Transport Operations
(NASA TM 88330)
NASA-Ames Research Center, Mountain View, California

Orlady L M (1992)
C/L/R – how do we know that it works?
Safetyliner, III, 2, Chicago: United Airlines

References

Orr D B, Friedman H L & Williams J C C (1965)
Trainability listening comprehension of speeded discourse
Journal of Educational Psychology, 56:148–156

Ostberg O & Mills-Orring R (1980)
Cabin attendants working environment –
a questionnaire study
Technical Report No. 1980:74 T. Department of Human Work Sciences, University of Lulea, Sweden

Owens D A (1984)
The resting state of the eyes
American Scientist, 72:378–387

Parker E S & Noble E P (1977)
Alcohol consumption and cognitive functioning in social drinkers
Journal of Studies on Alcohol, 38:1224–1232

Payne J, Blake E & Fitzgerald C (1979)
The ordeal at Three-Mile Island
In: *Nuclear News Special Report*, American Nuclear Society, La Grange Park, Ill.

Petersen L (1979)
Non-flying public in The Netherlands
Scandinavian Airlines System marketing survey, (unpublished), SAS, Stockholm

Peterson A P G & Gross E E (1978)
Handbook of Noise Measurement (8th Edition)
New Concord, Mass.: General Radio Company

Peterson L R & Peterson M J (1959)
Short-term retention of individual items
Journal of Experimental Psychology, 58:193–198

Pheasant S T (1983)
Sex differences in strength – some observations on their variability
Applied Ergonomics, 14, 3:205–211

Phillips R J (1979)
Why lower case is better
Applied Ergonomics, 10, 4:211–214

Popplow J R (1984)
After the fire-ball
Aviation, Space, and Environmental Medicine, April 1984

Poulton E C (1965)
Letter differentiation and the rate of comprehension in reading
Journal of Applied Psychology, 49:358–362

Poulton E C (1970)
Environmental and Human Efficiency
Springfield, Ill.: Thomas

Poulton E C (1971)
Skilled performance and stress
In: *Psychology at Work*, P B Warr (Ed), Harmondsworth: Penguin Books

Poulton E C (1985)
Geometric illusions in reading graphs
Perception and Psychophysics, 37:543–548

Presidential Commission (1986)
Report of the Presidential Commission on the space shuttle Challenger accident
Hearings of the Presidential Commission on the Space Shuttle Challenger Accident: 6 February–2 May 1986, Vol. 1, Appendix G – Human Factors Analysis

Presser S & Schuman H (1980)
The measurement of a middle position in attitude surveys
Public Opinion Quarterly, 44:70–85

Preston F S (1972)
An investigation into the workload and working conditions of cabin crew in BOAC
Report of the Air Corporations Joint Medical Service, BOAC

Preston F S (1975)
Transport flying and circadian rhythms
In: *The IATA 20th Technical Conference*, Istanbul, 10–14 November 1975, WP/60

Preston F S (1979)
Aircrew Stress
In: *The Symposium on Human Factors in Civil Aviation*, The Hague, The VNV Dutch Airline Pilots Association, 3–7 September 1979

Randel H (1971)
Aerospace Medicine
Baltimore: Williams & Wilkins

Rechtschaffen A, Gilliland M A, Bergmann B M & Winter J B (1982)
Physiological correlates of prolonged sleep deprivation
Science, 221:182–184

Rechtschaffen A & Kales A (1968)
A manual of standardised terminology; techniques and scoring system for sleep stages of human subjects
National Institute of Health Publication No. 204, Public Health Service, USA

Reichmann W J (1961)
Use and Abuse of Statistics
Harmondsworth: Penguin Books

Reinberg A, Andlauer P, Guillet P, Nicolai A, Vieux N & Laporte A (1980)
Oral temperature, ciradian rhythm amplitude, aging and tolerance to shift work
Ergonomics, 23, 1:55–64

Reynard W D, Billings C E, Cheaney E S & Hardy R (1986)
The Development of the NASA Aviation Safety Reporting System (NASA Reference Publication 1114). NASA-Ames Research Center, Moffett Field, California.

Robinson J O (1972)
The Psychology of Visual Illusion
London: Hutchinson

Robson B M, Huddleston H F & Adams A H (1974)
Some effects of disturbed sleep on a simulated flying task
Technical report 74057, RAE Farnborough

Roethlisberger F J & Dickson W J (1939)
Management and the worker – an account of a research programme conducted by the Western Electric Company, Hawthorne Works, Chicago
Cambridge, Mass.: Harvard University Press

Rolfe J M (1972)
Ergonomics and air safety
Applied Ergonomics, 3, 2:75–81

Rolfe J M & Bekerian D A (1984)
Witnesses
In: *ISASI Forum*, Report of ISASI Conference,
London Airport, 1984
Rood G (1985)
Aircraft cockpit and flight deck noise
The International Journal of Aviation Safety,
3, 1:52–61
Ross H (1975)
Mist, murk and visual perception
New Scientist, 66, 954:658–670
Royal College of Physicians (1983)
Smoking or Health
London: Pitman Medical
Royal College of Psychiatrists (1979)
Alcohol and Alcoholism
London: Tavistock Publications
Rubin D H, Krasilnikoff P A, Leventhal J M,
Weile B & Berget A (1986)
Effect of passive smoking on birth-weight
The Lancet, 2, 8504:415–417
Ruffell Smith H P (1979)
A simulator study of the interaction of pilot
workload with errors, vigilance and decision
making
NASA Technical Memorandum 78482, NASA-
Ames Research Center, Moffett Field
Rundus D & Atkinson R C (1970)
Rehearsal processes in free recall: a procedure
for direct observation
*Journal of Verbal Learning and Verbal
Behaviour*, 9:99–105
Ryback R S & Dowd P J (1970)
After effects of various alcoholic beverages on
positional nystagmus and coriolis acceleration
Aviation Medicine, 41, 4:429–435
Sandelin J, Bennet S & Case D (Eds) (1984)
*Video Display Terminals; usability issues and
health concerns*
New Jersey: Prentice-Hall
Schachter S (1973)
Nesbitt's Paradox
In: *Smoking Behaviour: Motives and Incentives*,
W L Dunn (Ed),
Washington: V H Winston & Sons
Schwartz F R & Kidera G J (1978)
Method of rehabilitation of the alcohol-addicted
pilot in a commercial airline
Aviation, Space and Environmental Medicine,
49, 5:729–731
Seminara J L & Shavelson R J (1969)
Effectiveness of space crew performance
subsequent to sudden sleep arousal
Aerospace Medicine, 40, 7:723–727
Sexton G A (1983)
Pilot's desk flight station
In: Proceedings of *The 2nd Symposium on
Aviation Psychology*,
R S Jensen (Ed),
Columbus: Ohio State University Press
Shephard R J (1982)
The Risks of Passive Smoking
London: Croom Helm

Sinclair M A (1975)
Questionnaire design
Applied Ergonomics, 6, 2:73–80
Singleton W T (1974)
Man-Machine Systems
Harmondsworth: Penguin Books
Skegg D C G, Richards S M & Doll R (1979)
Minor tranquillisers and road accidents
British Medical Journal, 1:917–919
Sleight R B (1948)
The effect of instrument dial shape on legibility
Journal of Applied Psychology, 32:170–188
Sloane S J & Cooper C L (1984)
Health-related lifestyle habits in commercial
airline pilots
British Journal of Aviation Medicine, 2:32–41
Smith M L, Glass G V & Miller T I (1980)
The Benefits of Psychotherapy
Baltimore: Johns Hopkins University Press
Smith R L & Lucaccini L F (1969)
Vigilance research: its application to industrial
problems
Human Factors, 11, 2:149–156
Smith S L (1979)
Letter size and legibility
Human Factors, 21, 6:661–670
Snow C C, Carrol J J & Allgood M A (1970)
Survival in emergency escape from passenger
aircraft
FAA-AM-70-16
Snyder R G (1976)
Advanced techniques in crash protection and
emergency egress from air transport aircraft
NATO AGARD-AG-221
Snyder R G (1982)
Impact protection in air transport passenger seat
design
SAE Technical Paper No. 821391
Steen D (1979)
Canadian Pilot's Fitness Manual
Don Mills, Ontario: Fitzhenry & Whiteside
Stone R B & Babcock G L (1976)
Pilot error: human failure or system failure
In: *The 29th International Air Safety Seminar
of the Flight Safety Foundation*, Anaheim,
25–29 October 1976
Strentz T (1985)
Preparing victims of hostage situations
In: Proceedings of *The 2nd Annual
International Aircraft Cabin Safety Symposium*
Strentz T (1986)
Hostage psychological survival guide
In: Proceedings of *The 3rd Annual
International Cabin Safety Symposium*
Sturtevant F M & Sturtevant R P (1979)
Chronopharmacokinetics of ethanol
In: *Biochemistry and Pharmacology of Ethanol*,
Vol. 1, E Majchroviez & E P Noble (Eds),
New York: Plenum Press
Sumby W & Pollack I (1954)
Visual contribution to speech intelligibility in
noise
Journal of the Acoustical Society of America,
26:212–215

References

Swain A D (1974)
  Design techniques for improving human
  performance in production
  Industrial and Commercial Techniques Ltd,
  Fleet Street, London
Swissair (1979)
  Comparison of success between ab initio and
  other candidates and initial upgrading from co-
  pilot to captain
  In: *The IATA 5th General Flight Crew Training
  Meeting*, Miami, 15–19 October 1979, W/P 1.3
  and 6.9
Szlichcinski K P (1979)
  Telling people how things work
  Applied Ergonomics, 10, 1:2–8
Talland G A (1968)
  Age and span of immediate recall
  In: *Human Ageing and Behaviour*, G A Talland
  (Ed), New York: Academic Press
Taub J M & Berger R J (1973)
  Performance and mood following variations in
  the length of time asleep
  Psychophysiology, 10, 6:559–570
Taylor F W (1911)
  The Principles of Scientific Management
  New York: Harper & Row
Taylor M R (1984)
  DVI and its role in future avionic systems
  In: *The First International Conference on
  Speech Technology*, Brighton,
  23–25 October 1984
Taylor R M (1985)
  Colour design in aviation cartography
  Displays, 6, 4:187–201
Ternham K (1978)
  The automatic complacency
  Aerospace Safety, 35, 4:12–13
Thomson L J (1987)
  Who should treat medical emergencies on civil
  airlines?
  Aviation Medicine Quarterly, 1:125–129
Tinker M A (1963)
  The Legibility of Print
  Ames, Iowa: Iowa State University Press
Torsvall L, Akerstedt T & Lindbeck G (1984)
  Effects on sleep stages and EEG power density
  of different degrees of exercise in fit subjects
  EEG and Clinical Neurophysiology, 57:347–
  353
Troland L T (1934)
  Vision 1. Visual phenomena and their stimulus
  correlations
  In: *A Handbook of General Experimental
  Psychology*, Worcester: Clark University Press
Turner J W & Huntley M S Jr (1991)
  The use and design of flightcrew checklists and
  manuals
  DOT/FAA/AM-917, Washington, DC
United Airlines (1975)
  Approach plate questionnaire
  In: *The IATA 20th Technical Conference*,
  Istanbul, 10–14 November 1975, W/P 35

Ursano R J (1980)
  Stress and adaptation: the interaction of the
  pilot personality and disease
  Aviation, Space and Environmental Medicine,
  51, 11:1245–1249
US Army (1980)
  Aircraft crash survival guide
  USARTL-TR-79-22, Applied Technology
  Laboratory, US Army Research and
  Technology Laboratories (AVRADCOM),
  Fort Eustis, Va
Van Cott H P & Kinkade R G (1972)
  *Human Engineering Guide to Equipment
  Design*
  Washington DC 20402: Superintendant of
  Documents, US Government Printing Office
Vernon T (1980)
  Gobbledegook
  London: National Consumer Council
Veroff J (1957)
  Development and validation of a projective
  measure of power motivation
  Journal of Abnormal and Social Psychology,
  54:1–8
Vette G (1983)
  Impact Erebus
  Auckland: Hodder & Stoughton
Visser C J (1988)
  Civil aviation remains vulnerable to terrorism
  FSF, Flight Safety Digest, 7, 4, April 1988
de Vries H A & Adams G M (1972)
  Electromyographic comparison of single doses
  of exercise and meprobamate as to effects on
  muscular relaxation
  American Journal of Physical Medicine,
  51:130–141
Vroom V H (1964)
  Work and Motivation
  New York: John Wiley & Sons
Vroom V H & Deci E L (1970)
  Management and Motivation
  Harmondsworth: Penguin Books
Warr P B (Ed) (1971)
  Psychology at Work
  Harmondsworth: Penguin Books
Webb W (1971)
  Sleep behaviour as a biorhythm
  In: *Biological Rhythms and Human
  Performance*, W P Colquhoun (Ed),
  London: Academic Press
Wegmann H M, Hansenclever S, Michel C &
  Trumbach S (1985)
  Models to predict operational loads of flight
  schedules
  Aviation, Space and Environmental Medicine,
  56, 1:27–32
Wegmann H M, Klein K E, Conrad B & Esser P
  (1983)
  A model for prediction of resynchronisation
  after time-zone flights
  Aviation, Space and Environmental Medicine,
  54, 6:524–527

Welford A T (1966)
The ergonomic approach to social behaviour
Ergonomics, 9:357–369

Wheale J (1983)
Crew coordination on the flight deck of
commercial transport aircraft
In: *The Flight Operations Symposium*, Irish Air
Line Pilots Association/Aer Lingus, Dublin,
19–20 October 1983

Whiteside T C D (1952)
Accommodation of the human eye in a bright
and empty field
Journal of Physiology, 118:65

Wickens C D (1992)
*Engineering Psychology and Human
Performance*
New York: Harper Collins .

Wiener E L (1977)
Controlled flight into terrain accidents:
system-induced errors
Human Factors, 19, 2:171–181

Wiener E L (1983)
The role of the human in the age of the
microprocessor
In: *Proceedings of Princeton University
Symposium on Air Traffic Control*, Princeton,
New Jersey, September 1983

Wiener E L (1985)
Cockpit automation: in need of a philosophy
In: Proceedings of the *Fourth Behavioral-
Technology Conference*, Long Beach,
California. Warrendale, Pa: Society of
Automotive Engineers
SAE Technical Series Paper 851956

Wilkins A J (1985)
Discomfort and visual displays
Displays, 6, 2:101–103

Wilkinson R T (1969)
Sleep deprivation: Performance tests for partial
and selective sleep deprivation
In: *Progress in Clinical Psychology*, L E Abt &
B F Reiss (Eds),
New York: Grune & Stratton

Williams A R T & Blackler F H M (1971)
Motives and behaviour at work
In: *Psychology at Work*, P B Warr (Ed),
Harmondsworth: Penguin Books

Williams H L, Hammack J T, Daly R L, Dement
W C & Lubin A (1964)
Responses to auditory stimulation, sleep loss
and the EEG stages of sleep
EEG Clinical Neurophysiology, 16:269–279

Wilson J W & Bateman L F (1979)
Human Factors and the advanced flight deck
In: *The 32nd International Air Safety Seminar*,
London, 8–11 October 1979

Wilson J W & Hillman R E (1979)
The advanced flight deck
In: *Aerospace Electronics in the Next Two
Decades*, The Royal Aeronautical Society's
Spring Convention, London, 16–17 May 1979

Winkel J (1983)
On the manual handling of wide-body carts
used by cabin attendants in civil aircraft
Applied Ergonomics, 14, 3:162–168

Winkel J & Jorgensen K (1986)
Evaluation of foot swelling and lower limb
temperatures in relation to leg activity during
long term seated office work
Ergonomics, 29, 2:313–328

Wise L M (1980)
Residual effect of alcohol on aircrew
performance
SAFE Journal, 10, 2:28–31

Wolcott J H, McMeekin R R, Burgin R E &
Yanowitch R E (1977)
Correlation of general aviation accidents with
the Biorhythm Theory
Human Factors, 19, 3:283–293

Wright P (1977)
Presenting technical information: a survey of
research findings
Instructional Science, 6:93–134

Wright P (1981)
Five skills technical writers need
*IEEE Transactions on Professional
Communication*, PC-24, 1:10–16

Wright P & Barnard P (1975)
Just fill in this form: a review for designers
Applied Ergonomics, 6, 4:213–220

Wright P & Lickorish A (1983)
Proof reading texts on screen and paper
Behaviour and Information Technology,
2:227–235

Yerkes R M & Dodson J D (1908)
The relation of strength of stimulus to rapidity
of habit formation
*Journal of Comparative Neurological
Psychology*, 18:459–482

Index